Genetic Recombination

METHODS IN MOLECULAR BIOLOGY™

John M. Walker, Series Editor

282. **Apoptosis Methods and Protocols,** edited by *Hugh J. M. Brady, 2004*
281. **Checkpoint Controls and Cancer, Volume 2:** *Activation and Regulation Protocols,* edited by *Axel H. Schönthal, 2004*
280. **Checkpoint Controls and Cancer, Volume 1:** *Reviews and Model Systems,* edited by *Axel H. Schönthal, 2004*
279. **Nitric Oxide Protocols,** *Second Edition,* edited by *Aviv Hassid, 2004*
278. **Protein NMR Techniques,** *Second Edition,* edited by *A. Kristina Downing, 2004*
277. **Trinucleotide Repeat Protocols,** edited by *Yoshinori Kohwi, 2004*
276. **Capillary Electrophoresis of Proteins and Peptides,** edited by *Mark A. Strege and Avinash L. Lagu, 2004*
275. **Chemoinformatics,** edited by *Jürgen Bajorath, 2004*
274. **Photosynthesis Research Protocols,** edited by *Robert Carpentier, 2004*
273. **Platelets and Megakaryocytes, Volume 2:** *Perspectives and Techniques,* edited by *Jonathan M. Gibbns and Martyn P. Mahaut-Smith, 2004*
272. **Platelets and Megakaryocytes, Volume 1:** *Functional Assays,* edited by *Jonathan M. Gibbns and Martyn P. Mahaut-Smith, 2004*
271. **B Cell Protocols,** edited by *Hua Gu and Klaus Rajewsky, 2004*
270. **Parasite Genomics Protocols,** edited by *Stuart N. Isaacs, 2004*
269. **Vaccina Virus and Poxvirology:** *Methods and Protocols,*edited by *Stuart N. Isaacs, 2004*
268. **Public Health Microbiology:** *Methods and Protocols,* edited by *John F. T. Spencer and Alicia L. Ragout de Spencer, 2004*
267. **Recombinant Gene Expression:** *Reviews and Protocols, Second Edition,* edited by *Paulina Balbas and Argelia Johnson, 2004*
266. **Genomics, Proteomics, and Clinical Bacteriology:** *Methods and Reviews,* edited by *Neil Woodford and Alan Johnson, 2004*
265. **RNA Interference, Editing, and Modification:** *Methods and Protocols,* edited by *Jonatha M. Gott, 2004*
264. **Protein Arrays:** *Methods and Protocols,* edited by *Eric Fung, 2004*
263. **Flow Cytometry,** *Second Edition,* edited by *Teresa S. Hawley and Robert G. Hawley, 2004*
262. **Genetic Recombination Protocols,** edited by *Alan S. Waldman, 2004*
261. **Protein–Protein Interactions:** *Methods and Applications,* edited by *Haian Fu, 2004*
260. **Mobile Genetic Elements:** *Protocols and Genomic Applications,* edited by *Wolfgang J. Miller and Pierre Capy, 2004*
259. **Receptor Signal Transduction Protocols,** *Second Edition,* edited by *Gary B. Willars and R. A. John Challiss, 2004*
258. **Gene Expression Profiling:** *Methods and Protocols,* edited by *Richard A. Shimkets, 2004*
257. **mRNA Processing and Metabolism:** *Methods and Protocols,* edited by *Daniel R. Schoenberg, 2004*
256. **Bacterial Artifical Chromosomes, Volume 2:** *Functional Studies,* edited by *Shaying Zhao and Marvin Stodolsky, 2004*
255. **Bacterial Artifical Chromosomes, Volume 1:** *Library Construction, Physical Mapping, and Sequencing,* edited by *Shaying Zhao and Marvin Stodolsky, 2004*
254. **Germ Cell Protocols, Volume 2:** *Molecular Embryo Analysis, Live Imaging, Transgenesis, and Cloning,* edited by *Heide Schatten, 2004*
253. **Germ Cell Protocols, Volume 1:** *Sperm and Oocyte Analysis,* edited by *Heide Schatten, 2004*
252. **Ribozymes and siRNA Protocols,** *Second Edition,* edited by *Mouldy Sioud, 2004*
251. **HPLC of Peptides and Proteins:** *Methods and Protocols,* edited by *Marie-Isabel Aguilar, 2004*
250. **MAP Kinase Signaling Protocols,** edited by *Rony Seger, 2004*
249. **Cytokine Protocols,** edited by *Marc De Ley, 2004*
248. **Antibody Engineering:** *Methods and Protocols,* edited by *Benny K. C. Lo, 2004*
247. **Drosophila Cytogenetics Protocols,** edited by *Daryl S. Henderson, 2004*
246. **Gene Delivery to Mammalian Cells: Volume 2:** *Viral Gene Transfer Techniques,* edited by *William C. Heiser, 2004*
245. **Gene Delivery to Mammalian Cells: Volume 1:** *Nonviral Gene Transfer Techniques,* edited by *William C. Heiser, 2004*
244. **Protein Purification Protocols,** *Second Edition,* edited by *Paul Cutler, 2004*
243. **Chiral Separations:** *Methods and Protocols,* edited by *Gerald Gübitz and Martin G. Schmid 2004*

METHODS IN MOLECULAR BIOLOGY™

Genetic Recombination

Reviews and Protocols

Edited by

Alan S. Waldman

Department of Biological Sciences
University of South Carolina, Coumbia, SC

HUMANA PRESS ✱ TOTOWA, NEW JERSEY

999 Riverview Drive, Suite 208
Totowa, New Jersey 07512

www.humanapress.com

This publication is printed on acid-free paper. ∞
ANSI Z39.48-1984 (American Standards Institute) Permanence of Paper for Printed Library Materials

Cover design by Patricia F. Cleary.
Cover illustration: Detection of recombination events after histochemical staining. The *Arabidopsis* leaf on the left was treated with a DNA damage-inducing chemical, whereas the one on the right is from a control plant. Each dark spot is indicative of a recombination event. (Fig. 2, Chapter 3; *see* discussion on pp. 26–28.)

For additional copies, pricing for bulk purchases, and/or information about other Humana titles, contact Humana at the above address or at any of the following numbers: Tel.: 973-256-1699; Fax: 973-256-8341; E-mail: humana@humanapr.com; or visit our Website: www.humanapress.com

Photocopy Authorization Policy:

Printed in the United States of America. 10 9 8 7 6 5 4 3 2 1

ISSN 1064-3745

E-ISBN 1-59259-761-0

Library of Congress Cataloging in Publication Data

Genetic recombination : reviews and protocols / edited by Alan S. Waldman.
p. ; cm. — (Methods in molecular biology, ISSN 1064-3745 ; v. 262)
Includes bibliographical references and index.
ISBN 1-58829-236-3 (alk. paper)
1. Genetic recombination. I. Waldman, Alan S. II. Series: Methods in molecular biology (Totowa, N.J.) ; v. 262.
[DNLM: 1. Recombination, Genetic. 2. Gene Expression. 3. Gene Targeting. 4. Genetic Engineering. W1 ME9616J v.262 2004 / QH 443 G3284 2004]
QH443.G458 2004
572.8'77--dc22

2003023391

Preface

Genetic recombination, in the broadest sense, can be defined as any process in which DNA sequences interact and undergo a transfer of information, producing new "recombinant" sequences that contain information from each of the original molecules. All organisms have the ability to carry out recombination, and this striking universality speaks to the essential role recombination plays in a variety of biological processes fundamentally important to the maintenance of life. Such processes include DNA repair, regulation of gene expression, disease etiology, meiotic chromosome segregation, and evolution.

One important aspect of recombination is that it typically occurs only between sequences that display a high degree of sequence identity. The stringent requirement for homology helps to ensure that, under normal circumstances, a cell is protected from deleterious rearrangements since a swap of genetic information between two nearly identical sequences is not expected to dramatically alter a genome. Recombination between dissimilar sequences, which does happen on occasion, may have such harmful consequences as chromosomal translocations, deletions, or inversions. For many organisms, it is also important that recombination rates are not too high lest the genome become destabilized. Curiously, certain organisms, such as the trypanosome parasite, actually use a high rate of recombination at a particular locus in order to switch antigen expression continually and evade the host immune system effectively. Whether or not a particular recombination event is a "good thing" or a "bad thing," recombination is indeed a double-edged sword with a prominent position in life as we know it. There is thus ample reason to gain a better understanding of this essential process and how it is regulated.

To gain such an understanding requires the appropriate tools. Part I of *Genetic Recombination: Reviews and Protocols* outlines approaches and model systems for studying several aspects of recombination in a variety of eukaryotic organisms and in a mammalian parasite. Since a good collection of mutants is often the starting point for dissecting pathways, a chapter on the isolation of recombination mutants in mice is included.

Just as a properly regulated recombination pathway can help ensure genetic stability, so can an alteration in recombination serve as a sign of genomic instability. Part II describes approaches for using recombination as a reporter of genomic instability in lower and higher eukaryotes. These experimental systems can be put to use to identify chemical or environmental agents or genetic mutations that may increase the risk of cancer or other diseases associated with genomic instability.

Recombination normally involves interactions between sequences displaying a high degree of sequence identity. The homology recognition machinery that is such a central part of normal recombination also provides the potential for putting recombination to practical use in order to bring about highly specific genomic changes. Over the past several years, investigators have developed methodologies for using recombination, or recombination-related processes, as a tool for producing targeted genetic modification. New methods continue to be developed. Part III discusses various methods and approaches for targeted genomic manipulation in higher and lower eukaryotes.

To understand fully any cellular process in terms of basic biology or for the rational design of potential applications, it is ultimately important to study the process on a biochemical level. Part IV presents two types of biochemical analyses useful for furthering the understanding of recombination mechanisms.

In summary, *Genetic Recombination: Reviews and Protocols* should be of significant value to both the novice and the established researcher in the field of recombination. Although this is indeed a collection of protocols and methods, contained within the various chapters is a wealth of more general information about recombination. Our hope is that this book will inspire further advances in this vital field.

Alan S. Waldman

Contents

Contributors

ERIC ALANI • *Department of Molecular Biology and Genetics, Cornell University, Ithaca, NY*
TERRY ASHLEY • *Department of Genetics, Yale University School of Medicine, New Haven, CT*
JIRI AUBRECHT • *Central Research Division, Pfizer Inc, Groton, CT*
ADAM M. BAILIS • *Division of Molecular Biology, Beckman Research Institute, City of Hope National Medical Center, Duarte, CA*
MARK D. BAKER • *Department of Molecular Biology and Genetics, University of Guelph, Guelph, Ontario, Canada*
J. DAVID BARRY • *Welcome Centre for Molecular Parasitology, University of Glasgow, Glasgow, United Kingdom*
ALEXANDER J. R. BISHOP • *Department of Genetics and Howard Hughes Medical Institute, Harvard Medical School, Boston, Massachusetts*
MICHAEL BOSHART • *Department of Biology I, University of Munich, Munich, Germany*
RICHARD J. BRENNAN • *Cornerstone Pharmaceuticals, Stony Brook, NY*
MARK A. BRENNEMAN • *Department of Genetics, Rutgers University, Piscataway, NJ*
PETER BURTON • *Welcome Centre for Molecular Parasitology, University of Glasgow, Glasgow, United Kingdom*
DANA CARROLL • *Department of Biochemistry, University of Utah School of Medicine, Salt Lake City, UT*
ANGELOS CONSTANTINOU • *Institute of Biochemistry, University of Lausanne, Epalinges, Switzerland*
PETER M. GLAZER • *Departments of Therapeutic Radiology, and Genetics, Yale University School of Medicine, New Haven, CT*
TAMARA GOLDFARB • *Department of Molecular Biology and Genetics, Cornell University, Ithaca, NY*
SUE JINKS-ROBERTSON • *Department of Biology, Emory University, Atlanta, GA*
ERIC B. KMIEC • *Delaware Biotechnology Institute, Department of Biological Sciences, University of Delaware, Newark, DE*
JEAN Y. KUAN • *Departments of Microbiology, Therapeutic Radiology, and Genetics, Yale University School of Medicine, New Haven, CT*

Anzhela Kyryk • *Institut für Pflanzengenetik und Kulturpflanzenforschung, Gatersleben, Germany*
Katie K. Maguire • *Department of Biological Sciences, University of Delaware, Delaware Biotechnology Institute, Newark, Delaware*
Glenn M. Manthey • *Division of Molecular Biology, Beckman Research Institute, City of Hope National Medical Center, Duarte, CA*
Richard McCulloch • *Welcome Centre for Molecular Parasitology, University of Glasgow, Glasgow, United Kingdom*
Michelle S. Navarro • *Division of Molecular Biology, Beckman Research Institute, City of Hope National Medical Center, Duarte, CA*
Jac A. Nickoloff • *Department of Molecular Genetics and Microbiology, University of New Mexico School of Medicine, Albuquerque, NM*
Nadiya Orel • *Institut für Pflanzengenetik und Kulturpflanzenforschung, Gatersleben, Germany*
Holger Puchta • *Institut für Pflanzengenetik und Kulturpflanzenforschung, Gatersleben; and Universitat Karlsruhe, Karlsruhe, Germany*
Laura Reinholdt • *The Jackson Laboratory, Bar Harbor, ME*
Ramune Reliene • *Departments of Pathology, Environmental Health, and Radiation Oncology, UCLA School of Medicine and School of Public Health, Los Angeles, CA*
Robert H. Schiestl • *Departments of Pathology, Environmental Health, and Radiation Oncology, UCLA Medical School and School of Public Health, Los Angeles, CA*
John Schimenti • *The Jackson Laboratory, Bar Harbor, ME*
Waltraud Schmidt-Puchta • *Institut für Pflanzengenetik und Kulturpflanzenforschung, Gatersleben, Germany*
Naoko Shima • *The Jackson Laboratory, Bar Harbor, ME*
Jason A. Smith • *Department of Biological Sciences, University of South Carolina, Columbia, SC*
Rachelle Miller Spell • *Department of Biology, Emory University, Atlanta, GA*
Erik Vassella • *Institute of Cell Biology, University of Bern, Bern, Switzerland*
Alan S. Waldman • *Department of Biological Sciences, University of South Carolina, Columbia, SC*
Stephen C. West • *Cancer Research UK , London Research Institute, Clare Hall Laboratories, Herts, United Kingdom*

I

Studying Recombination Events in Eukaryotes

1

Determination of Mitotic Recombination Rates by Fluctuation Analysis in *Saccharomyces cerevisiae*

Rachelle Miller Spell and Sue Jinks-Robertson

Summary

The study of recombination in *Saccharomyces cerevisiae* benefits from the availability of assay systems that select for recombinants, allowing the study of spontaneous events that represent natural assaults on the genome. However, the rarity of such spontaneous recombination requires selection of events that occur over many generations in a cell culture, and the number of recombinants increases exponentially following a recombination event. To avoid inflation of the average number of recombinants by jackpots arising from an event early in a culture, the distribution of the number of recombinants in independent cultures (fluctuation analysis) must be used to estimate the mean number of recombination events. Here we describe two statistical analyses (method of the median and the method of p_0) to estimate the true mean of the number of events to be used to calculate the recombination rate. The use of confidence intervals to depict the error in such experiments is also discussed. The application of these methods is illustrated using the intron-based inverted repeat recombination reporter system developed in our lab to study the regulation of homeologous recombination.

Key Words: fluctuation analysis, method of the median, confidence intervals, spontaneous recombination, mutation rate, inverted repeats, intron-based recombination assay, homeologous recombination

1. Introduction

The study of DNA damage and subsequent repair by recombination utilizes systems that examine both spontaneous and induced damage. Although studies of induced damage (by exogenous DNA-damaging agents or endogenous expression of endonucleases) have the benefit of inflicting specific types of damage, sometimes at known sites in the genome, spontaneous damage represents normal assaults on DNA integrity. The rarity of spontaneous damage and its repair demands different methods for experimental detection and analysis

From: *Methods in Molecular Biology, vol. 262, Genetic Recombination: Reviews and Protocols*
Edited by: A. S. Waldman © Humana Press Inc., Totowa, NJ

than does the study of induced damage. Such methods examine the number of events in several cultures to reveal how the number of events fluctuates from culture to culture (hence, a fluctuation analysis). This chapter details the protocol on how to conduct and interpret a fluctuation analysis to determine the rate of occurrence of rare events, such as recombination or mutation. Data from our study of the effect of sequence nonidentity on recombination rate in the budding yeast, *Saccharomyces cerevisiae*, will be used to illustrate this type of analysis. However, the notes on the practical use of this analysis are useful for the study of any rare events in a population of cells.

Spontaneous events can be infrequent (e.g., one event per billion cells) and thus difficult to quantitate without looking at large numbers of cells. In addition, because a mutation or recombination event could occur at any point in the growth of a population of cells, the final number of mutants/recombinants in a culture does not necessarily reflect the number of initial events. For example, a cell that experiences a recombination event early in the growth of a culture would undergo clonal expansion, causing a jackpot that would inflate the calculated frequency (number of recombinants per total cells). Therefore, many independent, parallel cultures are used in a fluctuation analysis to calculate the occurrence of events per generation using statistical methods to estimate the mean number of recombination/mutation events from the distribution of the number of recombinants/mutants. By taking into account the number of cell doublings that occur during the growth of a culture from a single cell, the calculation reveals the rate (events/cell/generation) that would yield the observed number of events after the prescribed number of generations.

The statistical methods described here calculate the rate using either the method of the median or the method based on the proportion of cultures with zero events (p_0) ***(1)***. The latter method is based on the Poisson distribution and was famously used by Luria and Delbrück to show that mutations arise spontaneously and not by "adaptative mutation" in response to a selective agent ***(2)***. If most cultures produce no events, then the mutation/recombination rate should be calculated using a method based on the fraction of cultures with zero mutants/recombinants (p_0) and the total number of cells. If most cultures produce mutants/recombinants, then the method of the median should be used. Importantly, use of the median avoids the extremes of the numbers of events in the different cultures and thus helps to remove jackpots from consideration.

The significance of a rate value obtained by the method of the median is indicated by a confidence interval, which defines the boundaries within which the true rate would be expected to fall with a certain level of confidence. Because the confidence interval is not a standard error, the interval may be distributed asymmetrically around the median. For example, the recombination rate could be 4×10^{-6} events/cell/generation, with a 95% confidence inter-

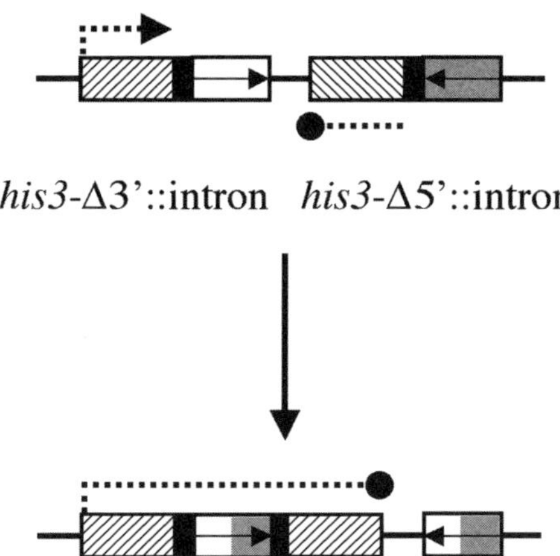

Fig. 1. Schematic of intron-based inverted repeat recombination substrates. Inverted repeats (open boxes with arrows) were fused to intron splice sites (black boxes) and placed next to the 5' and 3' halves of the coding sequence for *HIS3* (striped boxes). The direction of transcription of *HIS3* is indicated by a dashed line. Recombination between the repeats that leads to the reorientation of the sequence between the repeats allows expression of the full-length *HIS3* gene and selection on plates lacking histidine. The repeats can be engineered to have different levels of homology, different types of heterology, or different lengths. Comparison of the level of recombination between repeats that are similar but not identical (homeologous) and between identical (homologous) repeats reveals the relative suppression of homeologous recombination.

val of 1×10^{-6}–5×10^{-6} events/cell/generation. Two rates are considered statistically different if their confidence intervals do not overlap or if the distribution of the ranked, individual rates of each culture is nonrandom by χ^2 analysis.

Saccharomyces cerevisiae is especially useful for the study of spontaneous recombination because of the availability of selective systems that detect rare recombination events among billions of cells. Color assays or prototrophy selection can identify cells in which DNA damage has been repaired by specific recombination mechanisms and can be used to examine factors that affect those mechanisms, as reviewed by Symington *(3)*. The intron-based inverted repeat recombination assay system we use to study the effect of sequence identity on recombination is illustrated in **Fig. 1**. In this assay, repeats are placed within introns fused to the two halves of the coding sequence for a selectable marker (*HIS3*). The repeats can be manipulated to have different levels of sequence identity (e.g. 100 or 91% identity, called homologous and homeologous, respectively), different types of mismatches (base-base mismatches or insertion/deletion loops) or different lengths. Because the homeology is limited to

Name :rad51				
Strain : SJR1552				
	Dilution Factor:	1000000		1
	Fraction Plated:	0.2		0.5
10/19	**Culture #**	**C.F.U.**	**Rank**	**His⁺**
isolate a	1	223	5	223
	2	220	2	193
	3	238	3	206
	4	298	1	178
	5	272	7	256
	6	259	9	279
isolate b	7	207	4	214
	8	257	11	325
	9	282	6	243
	10	272	12	390
	11	247	10	291
	12	276	8	262
	Average =	254	**Median =**	250
	Corrected Average =	1.27E+9	**Corrected Median =**	499
			FREQUENCY =	3.93E-7
			RATE =	**4.76E-8**
			Rank	
	95% Confidence Interval=	low	3	4.05E-8
		high	10	5.43E-8

Fig. 2. Data spreadsheet from fluctuation analysis of homeologous recombination in a yeast strain lacking *RAD51*. Information pertinent to each experiment is entered in the light gray boxes of the spreadsheet. The data from the different isolates of each strain are differentiated by dark gray boxes. Appropriate dilutions of 12 independent cultures were plated on selective (SDGGE-His) and rich (YEPD) media to determine the number of His^+ recombinants (His^+) and the number of colony-forming units (C.F.U.), respectively. The median number of recombinants corrected for the dilution factor and fraction plated (Corrected Median) and the average number of colony-forming units corrected for the dilution factor and fraction plated (Corrected Average) were used to determine the recombination rate (events/cell/generation). The numbers of recombinants in the different cultures were ranked (Rank), and the numbers of recombinants ranked 3rd and 10th were used to calculate the rate values that define the lower and upper limits of the 95% confidence interval *(5)*.

the noncoding sequence, no types of recombinants are excluded by the requirement for prototrophy, rather, all recombination events can be detected that lead to the reorientation of the intervening sequence such that the selectable marker is transcribed. Such reorientation can occur by intramolecular interactions or by recombination between sister chromatids. The data presented here were generated to examine the effect of mutations in the recombination pathway on the normal suppression of homeologous recombination (**Fig. 2**).

2. Materials

The materials needed for fluctuation analysis differ according to the types of events that are measured and the method of selection/identification. In general, cells are grown nonselectively in liquid culture to allow recombinants/mutants to accumulate and then are plated on selective and nonselective media. The assay system described here uses cells grown nonselectively in YEP medium (1% yeast extract, 2% Bacto-peptone, 250 mg/L adenine; 2% agar for plates) supplemented with either 2% dextrose (YEPD) or 2% glycerol and 2% ethanol (YEPGE). Selective growth was done on synthetic complete (SC) media (0.17 % yeast nitrogen base, 0.5% ammonium sulfate, 2% agar) supplemented with 2% galactose, 2% glycerol, 2% ethanol, and 0.14% amino acid mix lacking histidine (SCGGE-His), as described in ref. ***4***. No special equipment or materials are required for dilution and plating of the cells, but analysis of the data is simplified by the use of a computer spreadsheet with a rate formula add-in, available upon request.

3. Methods

The methods described in the following subheadings outline (1) the growth of cultures and the preparation of dilutions to determine the number of recombinants and of total cells in a culture; and (2) the analysis of numbers generated from multiple cultures to determine the rate and confidence intervals.

3.1. Growth and Plating of Cultures

Independent cultures are started with a colony that grew from a single cell. The more cultures, the more significant the rate calculation will be. A pilot experiment with a few cultures may be necessary to determine the optimal dilutions needed for plating on rich and selective plates. Depending on the number of cultures to be tested, it is advisable to set up collection tubes, dilution tubes, and plates ahead of the day when dilutions and plating will occur.

1. Use a sterile toothpick to streak two isolates of each strain for single colonies on YEPD plates. Grow for 2 d at 30°C (*see* **Notes 1** and **2**).
2. For each culture, inoculate 5 mL YEPGE with an entire colony from the YEPD plate using a sterile toothpick (*see* **Notes 3–6**).

3. Grow cultures for 2–4 d on a roller drum at 30°C (*see* **Note 7**).
4. Transfer each culture to a sterile 15-mL conical tube and spin in a clinical centrifuge at room temperature. Remove supernatant, resuspend cells in 5 mL of sterile water, and spin again. Remove supernatant and resuspend cells in 1 mL of sterile water (*see* **Note 8**).
5. Make the appropriate serial dilutions of the washed cells into sterile water in sterile Eppendorf tubes such that plating 100 µL will give rise to 50–150 colonies per plate (*see* **Note 9**).
6. Plate 100 µL of the appropriate dilution of each culture on two plates each of YEPD and on two or more selective plates (*see* **Note 10**).
7. Incubate the plates at 30°C. The length of incubation will differ according to the type of media (*see* **Note 11**).
8. For each culture, count and total the number of colonies on the two YEPD plates for determining total cell number and on the two or more selective plates for determining the number of recombinants in each culture.

3.2. Rate Determination and Statistical Analysis

Analysis of the data from the fluctuation analysis is best done on a spreadsheet like Excel. The spreadsheet analysis assumes that all cultures of a strain were diluted and plated identically. A specific example of such a spreadsheet analysis is shown in **Fig. 2** and described in the following steps. Calculation of the rate by the spreadsheet requires a Mutation Rate Add-in, which is available upon request. Alternatively, the rate can be calculated manually.

1. For each strain, calculate the average of the number of cells that grew on the YEPD plates (in colony-forming units [CFU]). To determine the average number of cells in the total cultures, multiply the average from the YEPD counts by the dilution factor used and divide by the fraction of the dilution plated. See **Fig. 2** for an example of the YEPD counts from two plates with 100 µL each of the 10^{-6} dilution of 12 cultures. The corrected average total number of cells per culture is $254 \times 10^6/0.2 = 1.27 \times 10^9$ (*see* **Notes 12** and **13**).
2. For each strain, determine the median number of recombinants that grew on the selective plates. Multiply that median by the dilution factor used and divide by the fraction of the dilution plated to determine the median number of recombinants per total culture. See **Fig. 2** for example of the sum of the number of recombinants from five plates with 100 µL each of the 10^0 dilution of 12 cultures. The median is the average of the His$^+$ colonies counted in the cultures ranked sixth and seventh (Rank column). The corrected median number of recombinants per whole culture is $249 \times 10^0/0.5 = 498$.
3. If most cultures produce recombinants, one can estimate the mean using the median (*see* **Note 14**). We have set up an Excel spreadsheet using a Mutation Rate Add-in to calculate the mean and the rate based on formulas given in Lea and Coulson *(1)*, (*see* **Table 1** and **Notes 15** and **16**). The Mutation Rate Add-in makes possible reiterative calculations to achieve the best-fit median value.

Alternatively, one can manually determine the approximate mean (m) using Table 3 from Lea and Coulson ***(1)***. Given your experimentally determined median (r_0) and the corresponding r_0/m value from the table, determine the mean using the formula: $m = (r_0)/(r_0/m)$. Use this mean and the average number of cells per culture to calculate the rate, using the formula: rate = m (ln 2)/(average number of cells per culture). For example, for a median of 498, the approximate r_0/m from Table 3 is 5.7. Therefore, $m = 498/5.7 = 87.3$. With an average cell number of 1.27×10^9, the rate = 87.3 (0.693)/$1.27 \times 10^9 = 4.76 \times 10^{-8}$ events/cell/generation.

4. When the median is zero because most cultures produce no events, the rate must be calculated using the fraction of cultures with no events (p_0) to estimate the mean number of events, as described by Luria and Delbrück ***(2)***. This calculation requires plating of the entire culture. To calculate the rate in this case, use the formula: rate = [–ln (fraction of cultures with no recombinants)]/(average total number of cells). For example, for a strain for which 19 out of a total of 24 cultures had no recombinants and an average cell number of 6.67×10^8, the rate =[–ln (19/24)]/$6.67 \times 10^8 = 3.5 \times 10^{-10}$ events/cell/generation.
5. To determine whether the differences between two rates determined using the method of the median are statistically significant, calculate the confidence intervals. If the confidence intervals do not overlap, rates are statistically different.
   a. To determine the confidence intervals, sort the numbers of mutants in the cultures in ascending order. If using Excel:
      i. Highlight the column of data.
      ii. Click on Data.
      iii. Click on Sort (do not expand the current selection).
   b. Find the rankings to use for the interval calculation based on the number of cultures using Table B11 from *Practical Statistics for Medical Research* ***(5)***. For example, if 12 cultures were tested, the number of recombinants in the cultures ranked as 3rd and 10th should be used to calculate the 95% confidence intervals (as in **Fig. 2**).
   c. Substitute the number of mutants in the culture of the appropriate ranking for the median in the rate calculation in **step 3**. For example, in **Fig. 2**, the rate calculated using the third ranked number of recombinants ($206 \times 10^0/0.5$) with the average cell number (1.27×10^9) for all of the cultures defines the lower limit (4.05×10^{-8}) of the 95% confidence interval for the rate.
6. Another method using χ^2 analysis can be used to determine whether two rates (derived from two strains or a single strain grown under different conditions) are statistically different ***(6)***. For this method, calculate an individual rate for each culture by substituting the number of recombinants from that culture for the median and the total number of cells in that culture for the average cell number in the rate calculation described in **step 3**. Combine the individual rates from the two datasets and rank them as one dataset. If one strain has significantly more cultures in the top half of the rate values than the other strain, then the distribution of the rates from the two strains is nonrandom. Comparison of the expected

vs the observed distribution will indicate the χ^2 value and the probability that this distribution occurred by chance (*see* http://faculty.vassar.edu/lowry/VassarStats.html for templates for the goodness of fit test). Thus, this method indicates, like confidence intervals, whether the range of values included in rate calculations for two strains or two conditions overlaps.

4. Notes

1. When studying recombination or mutagenesis, it is important to have at least two isolates of each strain to be tested, especially when testing mutant backgrounds that may increase genome instability. If the recombination substrates are unstable in one of the isolates or if some other background difference between the two isolates affects the rate, the difference will become obvious in side-by-side comparison of the data from two different isolates.
2. Streak on YEPGE plates if petite formation (loss of mitochondrial function) is common in your strain. However, we generally find that the slow growth of a petite colony on YEPD is enough to prevent it from being used to inoculate a culture.
3. The upper and lower extreme of the numbers of recombinants in the dataset will be excluded from the 95% confidence intervals with a minimum of nine cultures. We routinely grow 14 cultures (7 of each isolate of each strain) and then proceed with the dilutions of 6 cultures of each isolate (*see* **Fig. 2**). This sample size allows the exclusion of the two lowest and two highest values from the determination of the confidence intervals ***(5)***.
4. Each culture is assumed to be the product of a single cell. Different techniques can optimize the chance that each culture starts with a single cell and that all the cells that grow from the initial cell are transferred to a liquid culture. One way to achieve this is to dilute a culture such that the number of cells per inoculation volume is less than 1. One can then assume that any culture that grows was derived from a single cell. Another approach is to inoculate each culture with a colony on a plug of agar cut from a plate to ensure that all the cells were transferred. Although these methods are not problematic, we find that such measures are unnecessary.
5. The volume of the cultures can be adjusted: smaller culture volumes for measurement of more frequent events or larger volumes for less dense cultures. If measuring very frequent events, the cells from an entire colony can be resuspended in water and plated directly. We routinely use 5-mL cultures grown to a density of approx 2×10^8 cells/mL because we often need one billion cells to measure recombination rates.
6. YEPGE liquid medium is used for the cultures to prevent the growth of petites, which could affect the rate of growth and of recombination and, therefore, skew the results. We have found that, for wild-type backgrounds, use of different media and different duration of growth affects the maximum level of growth but not the rate (R.M Spell, unpublished data). For example, cultures grown in YEPD or YEPGal reach higher cell density but have the same rate of recombination as

cultures grown in YEPGE. However, it is important to maintain the same conditions for all the cultures in one experiment and to reach the total expected cell concentration to be able to predict the correct dilutions.

7. We routinely grow cultures for 3 d, or 4 d if the culture grows slowly. Cultures grown for less time may still be in logarithmic growth and therefore may be at different cell densities (*see* **Note 12**). Growth to stationary phase ensures a somewhat consistent cell density from culture to culture. Shorter growth times can be used only if the final cell density is the same for all the cultures of a strain.
8. If your strain background has agglutination problems, brief sonication before diluting and plating may be necessary to separate clumped cells.
9. Fewer than 20 colonies per plate can increase variability, and counting more than 200 colonies per plate is difficult. In our experience, after growth in 5 mL YEPGE for 3 d and resuspension in 1 mL (approx 10^9 cells/mL), plating 100 µL of a 10^{-6} dilution on YEPD produces good colony counts for determining the number of cells in a culture. We have used 10^{-4}–10^0 (i.e., undiluted) dilutions for plating on selective plates. For example, we often make dilutions of 10^{-1} (100 µL washed cells + 900 µL sterile water), 10^{-2} (10 µL washed cells + 990 µL sterile water), 10^{-4} (10 µL of the 10^{-2} dilution + 990 µL sterile water), and 10^{-6} (10 µL of the 10^{-4} dilution + 990 µL sterile water). Transferring less than 10 µL when making dilutions produces variable results. Be sure to train new bench workers to change the pipette tip before every transfer.
10. For some events with very low rates, we plate more than two plates per culture. For example, plating the entire culture on 10 plates may be necessary. However, we find that plating more than 10^8 cells on one plate (i.e., more than 100 µL of 10^0 dilution) can inhibit the growth of selected cells.
11. We routinely incubate for only 2 d after colonies first become visible, to avoid counting events that occurred after the culture was plated *(7)*.
12. The total number of cells in the different parallel cultures of a strain must be similar. Otherwise, the median number of events will not represent a true median. For example, strains that experience significant cell death may give misleading numbers, making fluctuation analysis impossible. A clue that this is happening would be extreme variability in the cell densities in the cultures of a strain. Exclude data from cultures whose YEPD counts differ from the average number of cells by more than 2 standard deviations. The data from different isolates or from experiments done on different days can be pooled only if the YEPD counts (i.e., the number of cell divisions) are similar.
13. Because we resuspend the whole culture in 1 mL, the fraction of the total culture plated (20%, or 0.2) is the same as the volume plated (0.2 mL).
14. The spreadsheet add-in program does not work for low median numbers (less than 2). You have two options in that case: (1) for very low rates, the frequency (total events/per total cells) approximates the rate; or (2) you can calculate the rate manually.
15. Because of the number of data entry points, the number of different strains tested, and the number of experiments, transcription errors from the original data to the

spreadsheet, improper links in the spreadsheet, and mistakes in data management are unfortunately very common. Be cautious, review data entries, and use a standard, well-checked spreadsheet for each experiment.

16. We distinguish the data from different isolates and different experiments on the spreadsheet, so that any skew in the data (from a bad isolate, error in dilution, and so on) is easily detectable when the data are sorted.

Acknowledgments

The authors thank David Steele for generation of the rate program and the SJR lab for critical reading of this manuscript. This work was supported by NIH-NRSA grants GM20753 (to R.M. Spell), and GM38464 and GM064769 (to S. Jinks-Robertson).

References

1. Lea, D. E. and Coulson, C. A. (1949) The distribution of the numbers of mutants in bacterial populations. *J. Genet.* **49,** 264–285.
2. Luria, S. E. and Delbrück, M. (1943) Mutations of bacteria from virus sensitivity to virus resistance. *Genetics* **28,** 491–511.
3. Symington, L. S. (2002) Role of *RAD52* epistasis group genes in homologous recombination and double-strand break repair. *Microbiol. Mol. Biol. Rev.* **66,** 630–670.
4. Welz-Voegele, C., Stone, J. E., Tran, P. T., et al. (2002) Alleles of the yeast *PMS1* mismatch-repair gene that differentially affect recombination- and replication-related processes. *Genetics* **162,** 1131–1145.
5. Altman, D. G. (1990) *Practical Statistics for Medical Research.* Chapman & Hall/CRC, www.crcpress.com.
6. Wierdl, M., Greene, C. N., Datta, A., Jinks-Robertson, S., and Petes, T. D. (1996) Destabilization of simple repetitive DNA sequences by transcription in yeast. *Genetics* **143,** 713–721.
7. Steele, D. F. and Jinks-Robertson, S. (1992) An examination of adaptive reversion in *Saccharomyces cerevisiae*. *Genetics* **132,** 9–21.

2

Determination of Intrachromosomal Recombination Rates in Cultured Mammalian Cells

Jason A. Smith and Alan S. Waldman

Summary

Recombination is involved in many important biological processes including DNA repair, gene expression, and generation of genetic diversity. Recombination must be carefully regulated so as to prevent the deleterious consequences that may result from rearrangements between dissimilar sequences in a genome. It is of considerable interest to study the mechanisms by which genetic rearrangements in mammalian chromosomes are regulated in order to understand better how genomic integrity is normally maintained and to gain insight into the types of genetic mutations that may destabilize the genome. To explore such issues in mammalian chromosomes, a suitable experimental system must be developed. In this chapter, we describe a model system for studying intrachromosomal recombination in cultured mammalian cells. We discuss two model recombination substrates, a method for stably introducing the substrates into cultured Chinese hamster ovary cells, and a method for determining rates of intrachromosomal recombination between sequences contained within the integrated substrates. The general approach described here should be applicable to the study of a variety of aspects of recombination in virtually any cultured mammalian cell line.

Key Words: homologous recombination, fluctuation analysis, cell culture, DNA transfection

1. Introduction

Homologous recombination is defined as an exchange of genetic information between nearly identical DNA sequences. Homologous recombination can serve as a mechanism to repair double-strand breaks and other forms of DNA damage in mammalian cells. Recombination also plays roles in gene expression and genome evolution. One important aspect of recombination is that it typically occurs only between sequences that display a high degree of sequence identity. In this way, the cell usually manages to avoid the potentially harmful consequences of recombination between dissimilar sequences (*homeologous*

From: *Methods in Molecular Biology, vol. 262, Genetic Recombination: Reviews and Protocols*
Edited by: A. S. Waldman © Humana Press Inc., Totowa, NJ

recombination). Consequences of homeologous recombination may include chromosomal translocations, deletions, or inversions. The same proteins that catalyze homologous recombination may function to suppress homeologous recombination.

Homologous recombination in mammalian cells is indeed strongly dependent on sequence identity; as heterology increases, rates of homologous recombination and conversion tract length decrease ***(1–4)***. Waldman and Liskay ***(2)*** have shown that intrachromosomal recombination between two linked sequences sharing 81% homology was reduced 1000-fold compared with recombination between sequences displaying near-perfect homology. Lukacsovich and Waldman ***(5)*** reported that a single nucleotide heterology is sufficient to reduce recombination by about 2.5-fold, and a pair of nucleotide heterologies can act to suppress recombination from 7-fold to as much as 175-fold. It has been learned that mismatch repair (MMR) systems in bacteria, yeast, and mouse embryonic stem cells suppress homeologous recombination, and if any MMR components are lacking, rates of homeologous recombination increase ***(6–10)***. Gaining a more complete understanding of how cells normally regulate recombination and prevent unwanted homeologous exchanges is of fundamental importance to an understanding of how genome stability is maintained.

To explore spontaneous homologous and homeologous recombination in mammalian chromosomes, our lab developed a model system utilizing a gain-of-function assay. The system described in this chapter involves a pair of isogenic Chinese hamster ovary (CHO) cell lines designated MT+ and Clone B (generously provided by Margherita Bignami). The Clone B cell line is defective for an MMR protein named Msh2. MT^+ cells are wild-type for Msh2. To study spontaneous intrachromosomal homologous or homeologous recombination, plasmids pLB4 and pBR3 (**Fig. 1**) were constructed to serve as recombination substrates, and MT^+ as well as Clone B cells were stably transfected with these plasmid substrates. In this chapter we describe the isolation of stably transfected cell lines containing recombination substrates and fluctuation analysis to calculate intrachromosomal recombination rates. Although we describe work done with a specific set of CHO cell lines, the substrates used and the general approach discussed should be applicable to the study of a variety of issues relevant to intrachromosomal homologous and homeologous recombination in virtually any cultured mammalian cell line.

2. Materials

1. Plasmids pLB4 and pBR3 (**Fig. 1**) serving as substrates for homologous and homeologous recombination, respectively.
2. TE buffer: 10 m*M* Tris-HCl, 1 m*M* EDTA, pH 8.0.
3. Bio-Rad Gene Pulser (or other electroporator).
4. 40-cm Gap cuvets for electroporator.

pLB4:

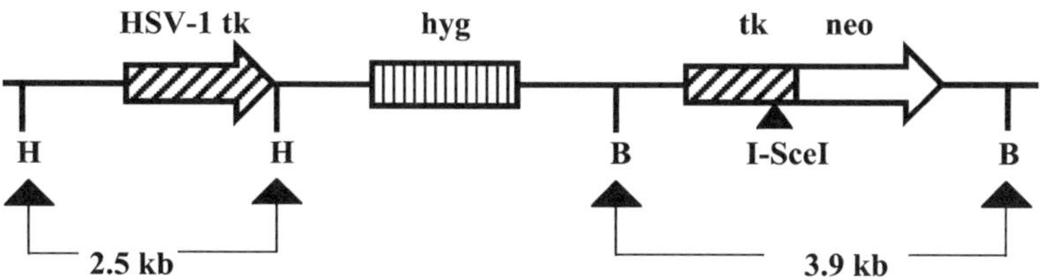

pBR3:

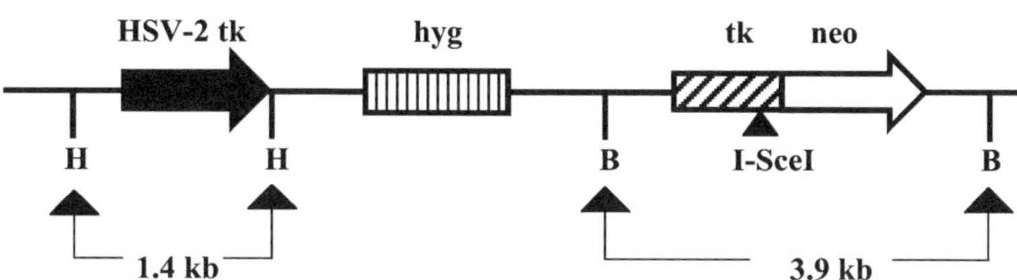

Fig. 1. Recombination substrates pLB4 and pBR3. Substrates pLB4 and pBR3 are suitable for the study of intrachromosomal homologous and homeologous recombination, respectively. Both substrates contain a tk-neo fusion gene that is disrupted by the insertion of an I-*Sce*I recognition site in the tk portion of the fusion gene. Each substrate also contains an additional "donor" tk sequence. The tk portion of the tk-neo fusion gene is from herpes simplex virus type 1 (HSV-1). The donor tk sequence on pLB4 is from HSV-1, and the donor tk sequence on pBR3 is from herpes simplex virus type 2 (HSV-2). Both substrates contain a hygromycin resistance gene (hyg), which allows for the isolation of stable transfectants. For cells containing either substrate, recombination between the tk donor and the disrupted tk-neo fusion gene can eliminate the I-*Sce*I site, restore function to the fusion gene, and produce a G418^r phenotype. Also shown in the figure are the locations of *Bam*HI (B) and *Hin*dIII (H) restriction sites.

5. G418 and hygromycin.
6. CHO cell lines MT$^+$ and Clone B.
7. Alpha-modified minimal essential medium (αMEM), supplemented with 10% fetal bovine serum (heat-inactivated).
8. Trypsin-EDTA solution (GIBCO, cat. no. 15400-054).
9. Phosphate-buffered saline (PBS): 137 m*M* NaCl, 2.7 m*M* KCl, 10 m*M* Na_2HPO_4, 1.5 m*M* KH_2PO_4.
10. Cell culture flasks (25 cm^2, 75 cm^2, and 150 cm^2 surface area).

11. 24-Well tissue culture plates.
12. Hemacytometer.
13. Sterile cotton swabs.
14. Fixative/stain solution: 0.04% methylene blue in 20% ethanol.
15. Dimethyl sulfoxide (DMSO).
16. Cryovials, 2 mL (for freezing cell lines).
17. Multiprime labeling kit (Amersham Biosciences cat. no. RPN 1601Z).
18. Restriction enzymes *Bam*HI and *Hin*dIII
19. Endonuclease I-*Sce*I
20. Agarose

3. Methods

3.1. Plasmid Constructs for Studying Intrachromosomal Recombination

To evaluate spontaneous homologous and homeologous recombination in mammalian cells, the plasmids pLB4 and pBR3 were constructed (**Fig. 1**) (*see* **Note 1**). A hygromycin resistance gene is included on each plasmid for stably installing the plasmid into mammalian cells. Plasmids pLB4 and pBR3 both contain a herpes thymidine kinase (tk) sequence (flanked by *Hin*dII sites) that serves as a potential "donor" sequence for recombination. Each construct also contains a nonfunctional tk/neomycin-resistance fusion gene (flanked by *Bam*HI sites). The "tk-neo" fusion gene is nonfunctional because a 22-bp insertion containing the 18 bp I-*Sce* I endonuclease recognition site has been incorporated into the tk portion of the fusion gene (*see* **Note 2**). The nonfunctional tk-neo gene in either substrate can be corrected (that is, the I-*Sce*I site can be eliminated) via recombination with the donor tk sequence, and recombinants can be recovered as $G418^r$ segregants. In pLB4, the donor tk sequence shares greater than 97% sequence homology with the tk portion of the tk-neo fusion gene sequence, and this construct is used to study homologous recombination. (The donor and the tk-neo gene on pLB4 do not share *perfect* homology, but the very limited number of scattered nucleotide differences allows for unambiguous identification of conversion tracts upon DNA sequencing.) In pBR3, the donor shares only about 80% sequence homology with the tk-neo gene and this construct is used to study homeologous recombination. As described in **Subheading 3.2.**, cells are stably transfected with pLB4 or pBR3 to study intrachromosomal recombination.

3.2. Establishing Cell Lines for Studying Homologous Recombination and Homeologous Recombination by Stable Transfection With Recombination Substrates

1. Grow MT^+ and Clone B CHO cells in αMEM to near confluence in 75-cm^2 flasks, trypsinize, and count cells using a hemacytometer.
2. Resuspend MT^+ or Clone B CHO cells (5×10^6 cells) in 800 µL of PBS, mix with 3 µg of plasmid pLB4 or pBR3 (DNA should be added in a minimal volume of

water, PBS or, TE buffer, not to exceed 50 µL), and electroporate in a 40-cm gap cuvet using a Bio-Rad Gene Pulser set at 1000 V, 25 µF (*see* **Note 3**).

3. Following electroporation, plate cells into a 150-cm^2 flask and allow cells to grow for 2 d without selection to permit recovery from electroporation.
4. After 2 d, plate 1×10^6 cells per 75-cm^2 flask containing αMEM supplemented with either 500 µg/mL of hygromycin (Clone B cell lines) or 400 µg/mL of hygromycin (MT$^+$ cell lines) (*see* **Note 4**).
5. Allow cells to grow until colonies are visible. Typically, CHO colonies are clearly visible after 8–10 d.
6. Draw circles around colonies on the outside of the flasks using a marking pen. Pick hygromycin-resistant colonies with sterile cotton swabs dipped in trypsin-EDTA solution. A swab is aimed at the center of a circle drawn around a colony; by gently brushing the cells in a colony with the cotton swab, the cells are released from the flask and adhere to the cotton. Transfer cells from each colony to a different, single well in a 24-well plate by dipping the cotton swab containing the cells into a well containing 1.5 mL of medium and gently rubbing the swab against the bottom of the well.
7. Incubate the wells at 37°C. When a well becomes full of cells, transfer cells from that well to a 25-cm^2 flask; when that flask is full, transfer cells to one 25-cm^2 and one 75-cm^2 flask.
8. When the 25-cm^2 and 75-cm^2 flasks of cells for a particular colony become full, prepare genomic DNA from the cells in the 75-cm^2 flask and freeze down the cells in the 25-cm^2 flask. Freeze cells at –80°C in 1 mL of αMEM supplemented with 10% DMSO in a 2-mL cryovial.
9. Analyze genomic DNA by Southern blot analysis to determine which cell lines contain a single unrearranged copy of the plasmid construct with the correct restriction fragment sizes (*see* **Note 5**). Using a tk-specific probe (labeled to greater than 1×10^9 per µg using a Multiprime labeling kit) and a DNA digestion with *Hin*dIII plus *Bam*HI, cell lines containing pLB4 should display a 3.9-kb and a 2.5-kb band and cell lines containing pBR3 should display a 3.9-kb and a 1.4-kb band (*see* **Fig. 1**).
10. After identifying one or more suitable cell lines, remove the vial(s) containing the desired frozen culture(s) from the –80°C freezer and thaw the cells. Propagate the cells and conduct fluctuation analysis as described below.

3.3. Recovery of G418^r Colonies From a Fluctuation Test

To determine spontaneous intrachromosomal homologous and homeologous recombination rates, fluctuation tests are performed. Single-copy cell lines containing pLB4 or pBR3 previously identified by Southern blotting are used. Each cell line is initially sensitive to G418; recombinants from a cell line are recovered as G418^r segregants in a fluctuation test as follows:

1. Separate a given cell line into 10 subclones containing 100 cells per subclone, and plate each subclone into a separate well of a 24-well plate (*see* **Note 6**).

2. Grow each subclone to confluence in a well, and then transfer to a 25-cm^2 flask. When the 25-cm^2 flask is full, transfer cells to a 75-cm^2 flask. Continue to culture cells until a sufficient number of cells is obtained per subclone. For the experiments described here, 4 million cells per subclone are required (*see* **Note 7**).
3. For each subclone, plate 1×10^6 per 150-cm^2 flask in αMEM supplemented with 1000 µg/mL G418 to select for G418^r segregants arising from homologous or homeologous recombination. We routinely use between four and eight 150-cm^2 flasks per subclone in our work.
4. Incubate cells for about 10 d, until colonies are visible.
5. As described in **Subheading 3.2., step 6**, pick several colonies per flask using sterile cotton swabs dipped in trypsin/EDTA solution and transfer cells from each individual colony into a separate well of a 24-well plate. Propagate cells and extract genomic DNA from a full 75-cm^2 flask of cells.
6. Fix and stain any colonies that were not picked with swabs by adding 10 mL of fixative/stain solution per flask and incubating at room temperature for 10 min. Wash out the solution with tap water. Colonies should be stained blue and should be easily visible.
7. Count all stained colonies, and be sure to add to your count the number of colonies that had been picked.

3.4. Calculation of Recombination Rate

Table 1 displays rates of recombination calculated by the "method of the median" for four different cell lines containing pLB4 (*see* **Notes 8** and **9**). The reader is referred to Lea and Coulson ***(11)*** for further details and the mathematical theory behind the rate calculation. Here we present a "cookbook" approach to calculating rate:

1. Calculate the median number of colonies per subclone. For example, cell line MT$^+$pLB4-22 (**Table 1**) had a median subclone colony number of 7.5. This value is referred to as r_0.
2. Next, using the value of r_0, an estimated value of r_0/m is interpolated from Table 3 in Lea and Coulson ***(11)***. The value of m is the average number of recombination events per subclone. In our example for cell line MT$^+$pLB4-22, where $r_0 = 7.5$, the estimated value of r_0/m was found to be 2.38.
3. Using the values of r_0 and r_0/m, calculate the value of m as $[(r_0) \div (r_0/m)]$. In our example, $m = 7.5 \div 2.38 = 3.15$.
4. Calculate the (estimated) rate of recombination by dividing m by the number of cells plated per subclone. For our example using cell line MT$^+$pLB4-22, rate = $(3.15) \div (4 \times 10^6) = 7.87 \times 10^{-7}$ recombination events/cell/generation. It is customary to divide this number by the number of copies of integrated recombination substrate to yield recombination rate in terms of recombination events/cell/generation/locus.

Table 1
Intrachromosomal Homologous Recombination Rates

Cell line	Cells plated, total ($\times 10^{-6}$)[a]	Median no. G418^r colonies per subclone	Colonies analyzed by *Alu*I digestion	Recombinant colonies[b]	Recombination rate[c]
MT^+pLB4-3	40	4	19	17	5.12 (4.58)
MT^+pLB4-22	40	7.5	20	15	7.87 (5.90)
CBpLB4-9	36	7	18	18	7.5
CBpLB4-20	40	4	16	16	5.12

[a]Independent subclones of 4×10^6 cells each were plated into G418 selection.

[b]The number of G418^r colonies analyzed that displayed the recombinant *Alu*I digestion pattern (*see* **Note 10**).

[c]Calculated by method of the median ***(11)***. Presented in parentheses are rates that were corrected by multiplying the initially calculated rate by the percentage of clones determined to actually have arisen by recombination, based on the *Alu*I digestion pattern.

3.5. Analysis of Recombinants

It is important to ascertain that G418^r colonies recovered from a fluctuation test were indeed produced by recombination events rather than by some unexpected event that fortuitously produced a G418^r phenotype. This can be accomplished in a number of ways by analyzing samples of genomic DNA isolated from G418^r colonies picked from the fluctuation test. Polymerase chain reaction (PCR) amplification of a portion of the tk-neo fusion gene spanning the I-*Sce*I site is one approach we have used. The first level of analysis should be to confirm that the I-*Sce*I site has indeed been eliminated. This is easily accomplished by digesting PCR products with I-*Sce*I and displaying the products on an agarose gel.

A second level of analysis takes advantage of two *Alu*I restriction sites that immediately flank the I-*Sce*I site in pLB4 and pBR3. Both of these *Alu*I sites will be eliminated by either homologous recombination (in the case of pLB4) or homeologous recombination (in the case of pBR3). Digestion of PCR products generated from G418^r segregants with *Alu*I therefore provides an expedient and reliable screen for recombinants (*see* **Note 10**).

The ultimate analysis comes from DNA sequence determination, which can be performed directly on PCR products. The donor tk sequences on pLB4 as well as on pBR3 display nucleotide differences when compared with the tk portion of the tk-neo gene. (There are *many* more differences between the donor and the tk-neo sequence on pBR3 than on pLB4 and, hence, pBR3 is useful for studying homeologous recombination). Recombination events can result in the transfer of some of the nucleotide differences from the donor tk sequence to the tk-neo gene, which will be detectable upon DNA sequencing. Detection of the transfer of sequence information between the donor sequence and the tk-neo gene can provide unambiguous confirmation of recombination (*see* **Note 10**).

It should be noted that there at least two different types of recombination events that can be recovered from cells containing pLB4 or pBR3. One type of event is a nonreciprocal exchange, also known as a gene conversion. In this case, information is transferred from the donor to the tk-neo gene with no other change. The other type of event is a crossover or "pop-out" in which the donor is essentially "spliced" to the tk-neo gene and the genetic information between the donor and the tk-neo gene (including the hygromycin resistance gene) is "popped-out" or deleted. The two different types of events produce very different restriction patterns on a Southern blot, and it is a good idea to perform Southern blotting to distinguish between these two types of events before attempting to interpret DNA sequence information (*see* **Note 10**). In our experience, gene conversions comprise at least 80% of events recovered from CHO cells. (Chapter 4 in this volume provides an excellent discussion and further consideration of a variety of types of recombination events that may occur among mammalian chromosomal sequences.)

After analyzing recombination events, it is useful to correct recombination rate by multiplying the calculated rate by the percentage of recovered clones that were determined to have actually arisen via recombination. This correction has been made to the data presented in **Table 1**. It should be noted that the data in **Table 1** suggest that there is no significant difference between the homologous recombination rate in MT^+ cells (Msh2 wild-type) vs Clone B (CB) cells (Msh2-deficient).

4. Notes

1. Plasmids pLB4 and pBR3 are available from the authors on request.
2. In this chapter, we describe the use of plasmids pLB4 and pBR3 for the study of spontaneous recombination. The presence of the recognition site for endonuclease I-*Sce*I also makes these plasmids useful for the study of double-strand break-induced recombination. This would be accomplished by an experimental design that includes the introduction of I-*Sce*I into cells to induce a break in a recombination substrate. The reader is referred to Chaps. 4 and 12 in this volume for additional information about strategies for studying break-induced recombination.
3. We routinely use electroporation to transfect mammalian cells with DNA. The optimal conditions for transfection via electroporation will vary depending on cell type and must be determined empirically. However, in our hands, the conditions we describe work reasonably well for a variety of cell types. Other transfection methods, such as liposome-mediated transfection, may be used but, in our experience, electroporation is the method of choice when low-copy-number integrations are desired. We also find that linearization of DNA prior to electroporation somewhat enhances transfection efficiency, but linearization is not necessary.
4. The proper level of hygromycin (or any selective agent) to use will vary by cell type and must be empirically determined.
5. It is not trivial to determine the number of copies of integrated substrate. If a cell line known to contain a single integrated copy of pLB4 or pBR3 is available, restriction digestions of DNA from that cell line can be displayed on a blot along with the DNA samples to be analyzed. Comparison of the hybridization intensity of an experimental sample with that of the established single-copy cell line will allow an estimate of copy number for the sample. Additionally, restriction digestions can be done that are predicted to produce a single junction fragment per integrated copy of substrate. (A junction fragment is a restriction fragment having one terminus within the integrated construct and the other terminus within adjacent genomic DNA). Samples that display only a single junction fragment on a Southern blot would be candidate single-copy lines. Single-copy cell lines sometimes occur at a relatively low frequency among stable transfectants. It is therefore advisable to analyze many (more than 20) transfectants. It is possible to

use cell lines that contain two or three integrated copies of the recombination substrate, but analysis of recombination events is somewhat confounded by the presence of multiple copies of substrate. One can actually only estimate copy number on a Southern blot. Ultimately, copy number is ascertained after recombinants are recovered. In a true single-copy line, the single integrated copy of substrate will be altered by recombination. In a multi-copy line, a single copy of the substrate will be altered by recombination whereas the remaining copies present in a recombinant will remain unaltered. Such a situation is readily revealed during analysis of recombinants.

6. Ideally, one should start with a single cell per subclone to initiate a fluctuation test. Practically speaking, all that is important is that the initial number of cells per subclone is small enough to effectively preclude the presence of a recombinant in the starting population. Starting with 100 cells per subclone satisfies this criterion and helps to expedite the experiment.
7. The appropriate number of cells needed per subclone depends on an approximation of recombination rate, which may not be known in advance. Small pilot experiments involving a couple of subclones may be used to estimate roughly the frequency of occurrence of recombinants. Essentially, for the method of rate determination presented here, one should plate enough cells per subclone to try to ensure the recovery of recombinants in all subclones.
8. There are several other methods for calculating recombination rates other than the method of the median. (*See* Chap. 1 in this volume for a second rate calculation method.) We find the method of the median to be very easy. Additionally, by virtue of using the *median* number of colonies per subclone, this method avoids potential complications introduced by calculation methods that average in data from "jackpot" subclones, that is, subclones that produce inordinately high numbers of recombinants because of recombination relatively early in the growth of the subclone.
9. Since the recombination substrates are randomly integrated, it is certain that the site of integration is different in each cell line. It is therefore advisable to determine the recombination rate for a given parental cell line using at least two or three stably transfected cell lines for each substrate in order to see if there is any significant position effect on recombination.
10. Detailed sequence information for plasmids pLB4 and pBR3 and further information helpful for analyzing recombinants by *Alu*I digestion or other approaches are available from the authors upon request.

Acknowledgments

We are grateful to Margherita Bignami for providing the CHO cell lines, to Laura Bannister and Brady Roth for constructing pLB4 and pBR3, and to Raju Kucherlapati for providing the original tk-neo fusion gene.

This work was supported by Public Health Service grant GM47110 from the National Institute of General Medical Sciences to A.S.W.

References

1. Rubnitz, J. and Subramani, S. (1984) The minimum amount of homology required for homologous recombination in mammalian cells. *Mol. Cell Biol.* **4,** 2253–2258.
2. Waldman, A. S. and Liskay, R. M. (1987) Differential effects of base-pair mismatch on intrachromosomal versus extrachromosomal recombination in mouse cells. *Proc. Natl. Acad. Sci. USA* **84,** 5340–5344.
3. Waldman, A. S. and Liskay, R. M. (1988) Dependence of intrachromosomal recombination in mammalian cells on uninterrupted homology. *Mol. Cell Biol.* **8,** 5350–5357.
4. Yang, D. and Waldman, A. S. (1997) Fine-resolution analysis of products of intrachromosomal homeologous recombination in mammalian cells. *Mol. Cell. Biol.* **17,** 3614–3628.
5. Lukacsovich, T. and Waldman, A. S. (1998) Suppression of intrachromosomal gene conversion in mammalian cells by small degrees of sequence divergence. *Genetics* **151,** 1559–1568.
6. Worth, L., Clark, S., Radman, M., and Modrich, P. (1994) Mismatch repair proteins MutS and MutL inhibit RecA catalyzed strand transfer between diverged DNAs. *Proc. Natl. Acad. Sci. USA* **91,** 3238–3241.
7. Chambers, S. R., Hunter, N., Louis, E. J., and Borts, R. H. (1996) The mismatch repair system reduces meiotic homeologous recombination and stimulates recombination and stimulates recombination-dependent chromosome loss. *Mol. Cell Biol.* **16,** 6110–6120.
8. Nicholson, A., Hendrix, M., Jinks-Robertson, S., and Crouse, G. F. (2000) Regulation of mitotic homeologous recombination in yeast. Functions of mismatch repair and nucleotide excision repair genes. *Genetics* **154,** 133–146.
9. Rayssiguier, C., Thaler, D. S., and Radman, M. (1989) The barrier to recombination between *Escherichia coli* and *Salmonella typhimurium* is disrupted in mismatch repair mutants. *Nature* **342,** 396–401.
10. Elliot, B. and Jasin, M. (2001) Repair of double strand breaks by homologous recombination in mismatch repair defective mammalian cells. *Mol. Cell. Biol.* **8,** 2671–2682.
11. Lea, D. E. and Coulson, C. A. (1949) The distribution of the number of mutants in bacterial populations. *J. Genet.* **49,** 264–285.

3

Intrachromosomal Homologous Recombination in *Arabidopsis thaliana*

Waltraud Schmidt-Puchta, Nadiya Orel, Anzhela Kyryk, and Holger Puchta

Summary

Because of the availability of the complete sequence of the genome of the model plant *Arabidopsis* and of insertion mutants for most genes in public mutant collections, the elucidation of the particular role of different factors involved in DNA recombination and repair processes, an important task for plant biology, is becoming feasible. An assay system based on transgenes harboring homologous overlaps of the β-glucuronidase (*uid*A) gene is available to determine recombination behavior in various mutant backgrounds. Restoration of the marker gene by homologous recombination can be detected by histochemical staining *in planta*. Inclusion of a site of the rare cutting restriction enzyme I-*Sce*I in the transgene construct enables the determination of recombination frequencies after induction of double-strand breaks. In this chapter we describe how the respective transgene is transferred by transformation or crossing into the mutant background, how recombination frequencies are determined, and, if necessary, how cells carrying a restored *uid*A gene can be isolated and propagated for molecular analysis of the particular recombination event.

Key Words: plants, homologous recombination, double-strand break repair, I-*Sce*I, β-glucuronidase, transformation

1. Introduction

Many plant species contain genomes with large amounts of repetitive DNA. Therefore, the frequency of somatic homologous recombination has to be tightly regulated to obtain genome stability. To characterize basic aspects of somatic homologous recombination, different marker(s) (genes) have been developed (for review, *see* ref. ***1***). In most experiments the recombination of nonfunctional overlapping parts of a marker into a functional unit is used to detect recombination events. Some years ago we had set up a nonselective

From: *Methods in Molecular Biology, vol. 262, Genetic Recombination: Reviews and Protocols*
Edited by: A. S. Waldman © Humana Press Inc., Totowa, NJ

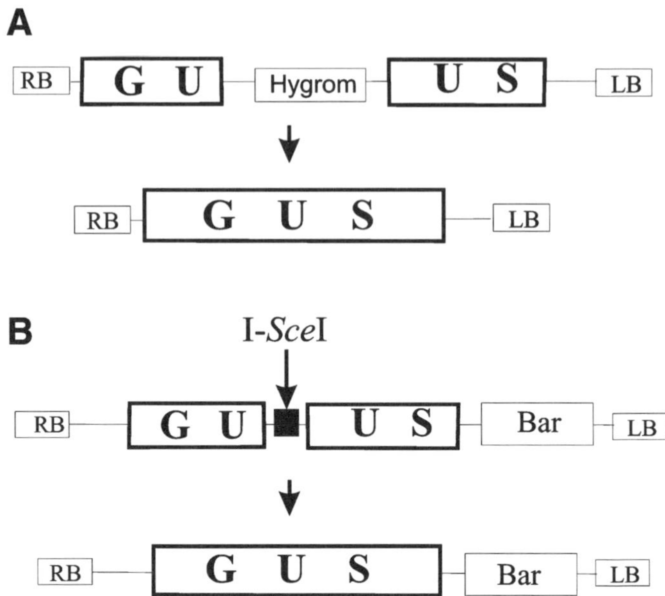

Fig. 1. Schematic representation of the recombination substrates pGU.US (**A**) and pDGU.US (**B**) used to monitor intrachromosomal homologous recombination in *Arabidopsis thaliana*. GUS, β-glucuronidase gene; Hygrom, hygromycin resistance gene; Bar, phosphinotricin resistance gene; LB, left border; RB, right border.

assay system that enabled us to visualize intrachromosomal homologous recombination events throughout the whole life cycle and in all organs of the plants *Arabidopsis thaliana* and tobacco *(2,3)*. The assay system employs a disrupted chimeric β-glucuronidase (*uid*A) gene as a genomic recombination substrate (**Fig. 1**). In cells with a restored *uid*A gene, recombination events have occurred. This could be demonstrated by polymerase chain reaction (PCR) and Southern blot experiments. Cells expressing β-glucuronidase, and their progeny, could be precisely localized upon histochemical staining of the whole plant, thereby enabling the quantification of recombination frequencies. As every stained sector represents an independent recombination event, the total number of recombination events per plant can be determined. Recombination can be detected in all examined organs, from seeds to flowers *(2)*. Recombination frequencies of around 10^{-6} events per cell division have been found. Small deviations in recombination frequencies may be caused, for example, by the genomic locus, different copy numbers of the transgenic units, the exact configuration of the recombination substrate used, or the plant species analyzed.

Using this system it could be demonstrated that the frequency of intrachromosomal homologous recombination can be enhanced by the application

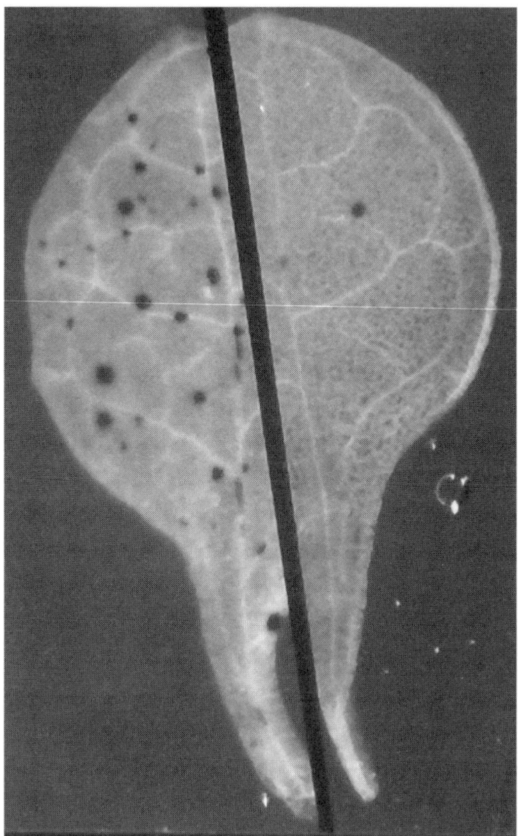

Fig. 2. Detection of recombination events after histochemical staining. The *Arabidopsis* leaf on the left was treated with a DNA damage-inducing chemical, whereas the one on the right is from a control plant grown under similar conditions. Each blue (dark spots) sector is indicative of a recombination event.

of physical and chemical agents that damage DNA, such as X-rays, methyl methansulfate (MMS), and ultraviolet (UV) irradiation ***(3,4)*** (**Fig. 2**). Indeed, the system was proved to be sensitive enough to detect genome instabilities induced by environmental factors, e.g., pollution by radioactive waste ***(4)***, higher doses of UV-B radiation resulting from the depletion of the ozone layer ***(5)***, or pathogen attack ***(6)***.

Especially important for practical application is the use of the assay in connection with the elucidation of gene functions in model plants. Enhanced recombination frequencies could be detected after the use of enzyme inhibitors (e.g., against poly(ADP)ribose polymerase ***[3]***), by the overexpression of genes involved in recombination (e.g., the Mim gene of *Arabidopsis* ***[7]***), and in mutant backgrounds (e.g., a Rad50 insertion mutant of *Arabidopsis* ***[8]***). A major prospect for future applications of this assay will be to study mutants. Therefore, recombination substrates will be routinely introduced in mutant backgrounds of *Arabidopsis thaliana,* as will be described in detail, either by transformation or by crossing with already existing reporter lines. Since a feasible gene targeting procedure is not yet available for plants, many approaches are based on the expectation that plants with higher intrachromosomal recombination behavior might in the long run be useful to establish a directed gene knockout technique (for a recent review, *see* ref. ***9***).

Rare cutting restriction enzymes (like I-*Sce*I or HO nuclease from yeast) or transposable elements have been used to induce double-strand breaks (DSBs) at specific loci in the plant genome and to study their repair (for review, *see* ref. ***10***). Using a *uid*A construct disrupted by an internal HO site, it was reported that DSB induction after HO expression enhances intrachromosomal recombination by one order of magnitude ***(11)***. By integrating and activating a transposable element between the overlapping halves of a *uid*A construct, Xiao and Peterson ***(12)*** demonstrated that recombination could be induced by up to 3 orders of magnitude.

Recently, we could demonstrate with a similar construct using the rare cutting endonuclease I-*Sce*I that, depending on the recombination substrate, in up to one-third of cases homologous sequences can be used for DSB repair ***(13)***. The basic principle of DSB-induced recombination studies is that in a controlled manner at a given time point an open reading frame of the restriction enzyme is expressed in a transgenic plant line containing a recognition site for the enzyme integrated in the genome. Expression of the enzyme can be achieved either by the use of a specific promoter (e.g., inducible or organ specific) within a stably integrated transgene or via transient expression using a constitutive promoter. Transient expression of I-*Sce*I in plants has already been described in another volume of this series ***(14)***. Crossing *Arabidopsis* plants containing a recognition site with plants expressing a restriction endonuclease (or in case of a nonfunctional transposon in the marker construct with plants expressing a functional transposase) will result in progeny in which DSBs are induced continuously during growth until the respective sites are destroyed either by end-joining or by homologous recombination. Thus, DSB-induced repair events can be monitored.

Owing to the availability of the genomic sequence and insertion mutants for almost all open reading frames, the model plant *Arabidopsis thaliana* has become the major research object for most plant biologists, including those working on genome stability (for a recent review, *see* ref. ***15***). Laboratory manuals for the performance of genetic experiments with *Arabidopsis* are at hand ***(16,17)***, and the reader is advised to consult them for a more in-depth description on how to handle this plant. The protocols described in **Subheading 3.** focus exclusively on the steps directly relevant for assaying intrachromosomal recombination.

In **Fig. 1** two different recombination substrates are shown. The substrate pGU.US ***(18)*** contains a directly oriented overlap of 557 bp of the β-glucuronidase gene interrupted by a hygromycin gene. In case of pDGU.US ***(19)***, an I-*Sce*I site is inserted between the overlaps and as a selection marker the phosphinotrycin-resistance gene is positioned next to the 3' end of the GUS gene. pDGU.US can be used for measuring intrachromosomal recombination with and without DSB induction. The recombination substrate has to be brought into the respective *Arabidopsis* background either by transformation or by crossing. The first part of the protocol (**Subheading 3.1.**) describes transformation of *Arabidopsis* plants with *Agrobacterium* using an *in planta* protocol. The second part describes the performance of a crossing experiment (**Subheading 3.2.**). At the time point when the recombination frequencies have to be determined, the plants are harvested and histochemically stained. The description of this procedure can be found in **Subheading 3.3.** Often the molecular nature of the recombination events that lead to the restoration of the reporter is of interest. In principle, it is also possible to propagate plant material that has been identified by a short staining procedure that does not kill all cells. How such clonal material can be amplified by tissue culture is described in **Subheading 3.4.** Finally, some specific hints are given in the Notes.

2. Materials

1. *Arabidopsis thaliana c. Columbia* seeds.
2. *Agrobacterium* strains carrying as binary vectors pGU.US and pDGU.US.
3. YEB medium: 5 g/L Bacto-beef extract, 1 g/L Bacto yeast extract, 5 g/L peptone, 5 g/L sucrose, 0.493 g/L $MgSO_4 \times 7\ H_2O$ sterilized by autoclaving.
4. Infiltration medium: ½ MS salts (MS salts: 1650.00 mg/L ammonium nitrate, 332.02 mg/L calcium chloride anhydrous, 180.70 mg/L magnesium sulfate anhydrous, 1900.00 mg/L potassium nitrate, 170.00 mg/L potassium phosphate monobasic, 6.20 mg/L boric acid, 0.025 mg/L cobalt chloride·$6H_2O$, 0.025 mg/L cupric sulfate·$5H_2O$, 37.26 mg/L Na^{2-} EDTA, 16.90 mg/L manganese sulfate-H_2O, 0.250 mg/L molybdic acid sodium salt, 0.83 mg/L potassium iodide, 27.80 mg/L ferrous sulfate·$7H_2O$, 8.60 mg/L zinc sulfate·$7H_2O$), vitamin B_5 ***(19)***; sucrose 5%, 6-benzylaminopurine (BAP) 0.0187 μM, Acetosyringone (AS, 3,5-dimethoxy-4-hydroxy-acetophenone, Fluka, Buchs, Switzerland) stock solution (100 mg/mL dimethyl sulfoxide [DMSO], diluted for use at 1:1000), Silwet L-77 (Lehle Seeds) 0.05%; pH 5.7.
5. Selection medium (SM) plates: ½ MS salts, vitamin B_5, 0.8% agar, pH 5.7, supplied with hygromycin 25 mg/L or phosphinotricin (Riedl de Haen, Seelze, Germany) 16 mg/L.

6. Germination medium (GM) plates plus ticarcillin: ½ MS salts, vitamin B_5, Fe-EDTA ***(20)***, sucrose 1%, 2-(*N*-morpholino)ethanesulfonic acid (MES) 500 mg/L, agar 0.8%, ticarcillin 250 mg/L, pH 5.7.
7. Drying pearls blue (Roth).
8. 0.1 *M* Phosphate buffer, pH 7.0.
9. 1% X-Glu (5-bromo-4-chloro-3-indolyl-β-D-glucuronide), stock solution in dimethylformamide (DMF); keep at –20°C.
10. 5% Sodium azide stock solution in H_2O; keep at –20°C. ***Caution: extremely toxic.***
11. Callus-inducing medium (CIM) plates: ½ MS salts, vitamin B_5, Fe-EDTA ***(20)***, 1% sucrose, 500 mg/L MES 8 g/L agar (Difco); adjust to pH 5.7, autoclave, and add 2,4-D (2,4-dichlorophenoxyacetic acid) 1 mg/L stock solution in a dilution of 1:1000 and kinetin (6-furfurylamino purine) 0.2 mg/L stock solution in a dilution of 1:1000.
12. Shoot-inducing medium (SIM) plates: like CIM plates but instead of 2,4-D and kinetin add IAA (indole-3-acetic acid) 0.15 mg/L stock solution in a dilution of 1:10,000 and 2iP (6-dimethylallylamino-purine) 5 mg/L stock solution in a dilution 1:2000, and the appropriate antibiotic. Antibiotics and plant-specific compounds were obtained from Duchefa, Haarlem, The Netherlands, if not stated differently.

3. Methods

3.1. In planta *Transformation of* Arabidopsis *With* A. tumefaciens

1. Grow *Arabidopsis* plants with pots in soil under long day condition (16-h day length) until they are flowering.
2. Prepare a liquid culture of the *Agrobacterium tumefaciens* strain carrying the gene of interest. Grow the culture in YEB medium supplied with the corresponding antibiotics. Inoculate at 28°C and with vigorous agitation. The culture should be in the mid-log phase or recently stationary.
3. Spin down *Agrobacterium* and resuspend in infiltration medium. Add Silwet L-77 0.05% just before use (*see* **Note 1**). We normally used 400 mL of an overnight culture and resuspended the cells in 800 mL infiltration medium.
4. Cut off already existing siliques of the plants.
5. Dip the plants up to the soil in the *Agrobacterium* suspension for 2–3 s with gentle agitation.
6. Position pots horizontally and keep plants under a dome for 24 h to provide high humidity.
7. Remove the dome and after a few days put aracons over the plants for collecting seeds.
8. Grow plants until seeds ripen. Stop watering and collect dried seeds.
9. For selection take about 50 mg of seeds per 15-cm plate. Sterilize them for 1 min with 70% ethanol, for not more than 10 min with 4% sodium hypochlorite. Wash four times with sterile water and resuspend in 0.1% agarose solution. Spread them over selection plates containing SM medium with 25 mg/L hygromycin or 16 mg/L phosphinotricin (*see* **Note 2**).
10. Synchronize germination at 4°C for 48 h. Transfer the plates to a growth chamber (light: 6000 lx intensity 16 h/22°C, dark: 8 h/18°C).

11. Under these conditions all seeds will germinate and develop cotelydons. After 2 wk, resistant plantlets develop leaves and long roots; the susceptible plants turn yellow.
12. Transfer resistant plantlets to GM plates plus ticarcillin.
13. Transfer to soil when they are sufficiently developed (*see* **Note 3**).
14. Collect leaf material to isolate DNA to verify by polymerase chain reaction (PCR) that transformants contain the construct of interest, and finally harvest seeds for further analyses.

3.2. Setting Up Crosses With Arabidopsis *Mutants*

1. Identify the most suitable flowers for crossing. On the female parent, the flowers have to be closed, with the first tips of the petals just visible. Healthy plants have about three flowers on the main shoot suitable for crossing at a certain time point. Flowers from the male parent should be widely open and visibly shedding pollen (*see* **Note 4**).
2. Cut off siliques and flowers below the selected ones on the female parent with scissors and also remove all the flowers above with a forceps.
3. Remove sepals, petals, and anthers with a sharp forceps and leave carpels intact. It is helpful to use a magnifying device.
4. Remove an open flower from the male parent and squeeze it near the base with a forceps. This will spread out the anthers.
5. Brush the surface of the anthers against the stigmas of the femal parents.
6. Label the crosses with a piece of sewing thread.
7. After 3 d you can see if the crosses were successful. In this case the siliques will have elongated.
8. When the siliques turn yellow they are ready to harvest. Dry seeds for 2 wk at room temperature with, e.g., drying pearls blue, before planting.
9. Sow seeds on plates harboring both antibiotics for selection of the recombination construct as well as of the inserted transfer DNA (leading to the mutant phenotype). Bring resistant seedlings to the soil and let them grow until seed are set. Harvest seeds for each plant individually.
10. Bring out seeds from the individual plants on plates harboring either one or the other antibiotic. Depending on the number of alleles for the repair construct as well as for the mutant insertion, the seedlings are either resistant, segregated at a 3:1 ratio for resistance, or sensitive to the respective antibiotic. For the analysis, select seedlings of plants that contain the recombination substrate and either the mutant or the wild-type background in a homozygous state. The seedlings obtained can be stained to determine the recombination frequencies in wild-type and mutant backgrounds (*see* **Subheading 3.3.**).

3.3. Histochemical Staining and Determination of Recombination Frequencies

1. The staining reactions are normally carried out in Falcon tubes.
2. Pipet in each tube 4.65 mL 0.1 *M* phosphate buffer, pH 7.0, 250 µL 1% X-Glu stock solution, and 100 µL 5% Na-azide stock solution.

3. Add plant material to the tubes.
4. Apply a vacuum for 5 min.
5. Incubate tubes overnight at 37°C.
6. Remove staining buffer and destain plant material with 70% ethanol.
7. Incubate for 20 min at 60°C. Repeat this step 3–4 times until the chlorophyll is removed.
8. Count blue spots under a binocular (*see* **Note 5**).

3.4. Regeneration of Plant Tissue With the Recombined Reporter Gene

1. Carry out a very short GUS staining of about half an hour (*see* **Note 6**).
2. Cut out the "blue sectors" and transfer the material to CIM plates for a pretreatment with phytohormones. A close contact between agar and plant material is important.
3. To control regeneration efficiency, include some untreated plant material in the experiment.
4. After 3 d in a growth chamber (16 h of light), plant material is transferred to SIM plates.
5. Transfer the material to new SIM plates every second week.
6. After 2 to 3 wk, the first calli should be seen.
7. Take part of the callus to prepare DNA *(21)*. Grind the callus in liquid nitrogen. Add 3 mL isolation buffer, vortex, and incubate the mixture at 65°C for 1 h. Add 3 mL chloroform isoamylalcohol 24:1. Mix well and spin at 10 min at 4°C. Take off the upper (water) phase and transfer to a new tube. Add 20 µL RNase A (10 mg/mL) and incubate for 15–30 min at room temperature. Add 3 mL cold isopropanol and centrifuge for 5 min at 4°C. Wash with 70% ethanol, spin down, and let the DNA pellet dry. Resuspend the pellet in 200–500 µL TE buffer (overnight). The DNA is now ready for further analysis by PCR or other methods.

4. Notes

1. If after the transformation procedure the plants do not look healthy, reduce the Silwet concentration to 0.025%.
2. Resuspending the seeds in 0.1% agarose during the selection step guarantees that seeds are equally distributed on the plates. This reduces the number of false positives.
3. It is helpful to transfer resistant plantlets from selection plates to GM plates prior to soil. Plantlets become stronger and roots longer. The survival rate is much higher than without this step.
4. Do not take the first flowers on the first inflorescence for crossing. They might be infertile, but do not wait too long because flowers become smaller and the failure rate higher.
5. An important point in the determination of recombination frequencies is the analysis on the population level as well as on the level of the individual organism. Without DSB induction, on average one to two recombination events per seedling can be detected. For a statistical analysis 30–100 seedlings should be

analyzed. A surplus of information can be gained by the determination and statistical evaluation of the number of recombination events per plant. Under normal growth conditions a stochastic distribution of the events per plant occurs. This is represented by a Poisson distribution if the number of events per plant is plotted against the respective fraction of the plant population showing this number (at least for most transgenic reporter lines; *see* ref. ***21***). However, specific mutant backgrounds or specific environmental stimuli can result in a nonstochastic distribution of the recombination events. This indicates that different plants (or parts of them) might be in different states of "competence," possibly because of epigenetic phenomena. Alternatively, exposure to DNA damaging agents like UV irradiation might yield a nonstochastic distribution. Bar diagrams are especially useful for the presentation and evaluation of such distributions on the population level (*see*, e.g., refs. ***8*** and ***21***).

6. In the long run the histochemical staining procedure is lethal for the plant cells. Therefore a compromise has to be found between the efficiency of detection of recombined sectors and the preservation of viability of the stained material. The longer the plants are kept in the staining solution, the more recombination events can be identified. However, after longer exposure, fewer cells survive the treatment. In our previous experiments, 30 min of incubation was used for staining. At that time a reasonable number of recombination events could already be identified. The efficiency of regeneration was about 1% of the identified events ***(2)***.

References

1. Puchta, H. and Hohn, B. (1996) From centiMorgans to basepairs: homologous recombination in plants. *Trends Plant Sci.* **1,** 340–348.
2. Swoboda, P., Gal, S., Hohn, B., and Puchta, H. (1994) Intrachromosomal homologous recombination in whole plants. *EMBO J.* **13,** 484–489.
3. Puchta, H., Swoboda, P., and Hohn, B. (1995) Induction of intrachromosomal homologous recombination in whole plants. *Plant J.* **7,** 203–210.
4. Kovalchuk, I., Kovalchuk, O., Arkhipov, A., and Hohn, B. (1998) Transgenic plants are sensitive bioindicators of nuclear pollution caused by the Chernobyl accident. *Nat. Biotechnol.* **11,** 1054–1059.
5. Ries, G., Heller, W., Puchta, H., Sandermann, H. J., Seidlitz, H. K., and Hohn, B. (2000) Elevated UV-B radiation reduces genome stability in plants. *Nature* **406,** 98–101.
6. Lucht, J. M., Mauch-Mani, B., Steiner, H. Y., Metraux, J. P., Ryals, J., and Hohn, B. (2002) Pathogen stress increases somatic recombination frequency in *Arabidopsis*. *Nat. Genet.* **30,** 311–314.
7. Hanin, M., Mengiste, T., Bogucki, A., and Paszkowski, J. (2000) Elevated levels of intrachromosomal homologous recombination in *Arabidopsis* overexpressing the MIM gene. *Plant J.* **24,** 183–189.
8. Gherbi, H., Gallego, M. E., Jalut, N., Lucht, J. M., Hohn, B., and White, C. I. (2001) Homologous recombination in planta is stimulated in the absence of Rad50. *EMBO Rep.* **2,** 287–291.

9. Puchta, H. (2002) Gene replacement by homologous recombination in plants. *Plant Mol. Biol.* **48,** 173–182.
10. Ray, A. and Langer, M. (2002) Homologous recombination: ends as the means. *Trends Plant Sci.* **7,** 435–440.
11. Chiurazzi, M., Ray, A., Viret, J.-F., et al. (1996) Enhancement of somatic intrachromosomal homologous recombination in *Arabidopsis* by HO-endonuclease. *Plant Cell* **8,** 2057–2066.
12. Xiao, Y. L. and Peterson, T. (2000) Intrachromosomal homologous recombination in *Arabidopsis* induced by a maize transposon. *Mol Gen Genet.* **263,** 22–29.
13. Siebert, R. and Puchta, H. (2002) Efficient repair of genomic double-strand breaks via homologous recombination between directly repeated sequences in the plant genome. *Plant Cell* **14,** 1121–1131.
14. Puchta, H. (1999) Use of I-*Sce*I to induce double-strand breaks in *Nicotiana*. *Methods Mol. Biol.* **113,** 447–451.
15. Hays, J. B. (2002) *Arabidopsis thaliana*, a versatile model system for study of eukaryotic genome-maintenance functions. *DNA Repair* **1,** 579–600.
16. Martínez-Zapater, J. and Salinas, J., eds. (1998) *Arabidopsis* protocols. In: *Methods in Molecular Biology*, vol. 82, Humana, Totowa, NJ.
17. Weigel, D. and Glazebrook, J. (2002) Arabidopsis. *A laboratory manual.* Cold Spring Harbor Press, Cold Spring Harbor, NY.
18. Tinland, B., Hohn, B., and Puchta, H. (1994) *Agrobacterium tumefaciens* transfers single stranded T-DNA into the plant cell nucleus. *Proc. Natl. Acad. Sci. USA* **91,** 8000–8004.
19. Orel, N., Kirik, A., and Puchta, H. (2003) Different pathways of homologous recombination are used for the repair of double-strand breaks within tandemly arranged sequences in the plant genome. *Plant J.* **35,** 604–612.
20. Murashige, T. and Skoog, F. (1962) A revised medium for rapid growth and bioassays with tobacco tissue culture. *Physiol. Plant.* **15,** 473–497.
21. Fulton, T. M., Chunwongse, J., and Tanksley, S. D. (1995) Microprep protocol for extraction of DNA from tomato and other herbaceous plants. *Plant Mol. Biol. Rep.* **13,** 207–209.
22. Puchta, H., Swoboda, P., Gal, S., Blot, M., and Hohn, B. (1995) Intrachromosomal homologous recombination events in populations of plant siblings. *Plant Mol. Biol.* **28,** 281–292.

4

Analysis of Recombinational Repair of DNA Double-Strand Breaks in Mammalian Cells With I-*Sce*I Nuclease

Jac A. Nickoloff and Mark A. Brenneman

Summary

Eukaryotes repair DNA double-strand breaks (DSBs) by homologous recombination (HR) or by nonhomologous end-joining (NHEJ). DSBs are a natural consequence of DNA metabolism, occurring, for example, during DNA replication and meiosis. DSBs are also induced by chemicals and radiation. I-*Sce*I endonuclease recognizes an 18-bp sequence with little degeneracy; therefore I-*Sce*I is highly specific, and its recognition sequence is predicted to occur by chance less than once in even the largest known genomes. As such, I-*Sce*I can be used to introduce a DSB into a defined (engineered) site in a mammalian chromosome, and this facilitates detailed studies of DSB repair. DSBs induced in repeated regions can be repaired by several different HR processes, including gene conversion with or without associated crossovers, or single-strand annealing. The specific types of HR events that can be scored depend on the configuration of the repeated regions and whether selection for recombinants is imposed. Nonselective assays detect both HR and NHEJ events. This chapter focuses on the systems for delivering I-*Sce*I nuclease to mammalian cells and the strategies for detecting various outcomes of DSB repair.

Key Words: DNA repair, homologous recombination, gene conversion, nonhomologous end-joining, DNA double-strand breaks

1. Introduction

1.1. Development of I-SceI for Analysis of Homologous Recombination in Mammalian Cells

Eukaryotes repair DNA double-strand breaks (DSBs) by homologous recombination (HR) or by nonhomologous end-joining (NHEJ). DSBs are a natural consequence of DNA metabolism, occurring, for example, during DNA

From: *Methods in Molecular Biology, vol. 262, Genetic Recombination: Reviews and Protocols*
Edited by: A. S. Waldman © Humana Press Inc., Totowa, NJ

replication and meiosis. In the early 1980s it became clear that DSBs could be repaired by a recombinational mechanism ***(1)***. DSBs are also induced directly by ionizing radiation and indirectly by chemotherapeutic DNA crosslinking agents, such as cisplatin and mitomycin C. It is difficult to study the repair of chemical- or radiation-induced DSBs because the lesions occur randomly within a cell population, and because these agents can produce several other forms of DNA damage including base damage and single-strand breaks ***(2,3)***.

Nucleases produce DSBs without any collateral damage and are therefore well suited for studies of DSB repair at defined sites. Yeast mating-type switching involves recombinational repair of DSBs introduced into the *MAT* locus by HO nuclease ***(4)***. The HO nuclease recognition site in *MAT* is quite long, on the order of 18–24 bp depending on assay conditions ***(5)***, and this suggested its use to study defined DSBs at engineered loci in mitotic and meiotic yeast ***(5,6)***. The power of this general approach has been documented in a wide variety of yeast studies of HR mechanisms, DSB repair by NHEJ, and cellular responses to DSBs including checkpoint activation and adaptation. During the late 1980s and 1990s, several labs attempted to adapt the HO system to mammalian cells, but there has only been one such report in which viral DNA was cleaved by HO nuclease expressed from an adenovirus vector in human cells ***(7)***.

The Dujon lab developed I-*Sce*I endonuclease as an alternative to the HO system. I-*Sce*I endonuclease is encoded by the group I intron of the mitochondrial LSU gene of *Saccharomyces cerevisiae*, which belongs to a group of mobile introns that propagate by inserting themselves into DSBs created by encoded endonucleases. The HO and I-*Sce*I nucleases share several features including some similarity at the amino acid sequence level ***(8)***. Like HO, I-*Sce*I recognizes a long (18-bp) sequence, but the HO recognition sequence is quite degenerate ***(9)***, whereas the I-*Sce*I recognition sequence shows minimal degeneracy ***(10)***. Because the I-*Sce*I recognition site is so long, it is expected to occur in random DNA less than once in even the largest genomes. The I-*Sce*I gene is encoded in mitochondrial DNA, so it has nonuniversal codons, and these were modified to allow nuclear expression ***(11)***. I-*Sce*I was then used to create DSBs in bacteriophage, yeast, and plant chromosomes ***(12,13)***. The first reports describing I-*Sce*I cleavage of mammalian chromosomes were in 1994–1995 ***(14,15)***, and as with HO in yeast, I-*Sce*I has since been used in a wide variety of HR and DSB repair studies in mammalian cells. Here we review the systems for delivering I-*Sce*I nuclease to mammalian cells and the strategies for detecting various outcomes of DSB repair, focusing primarily on HR. For additional information about the use of I-*Sce*I in mammalian cells, see the review by Jasin ***(16)***.

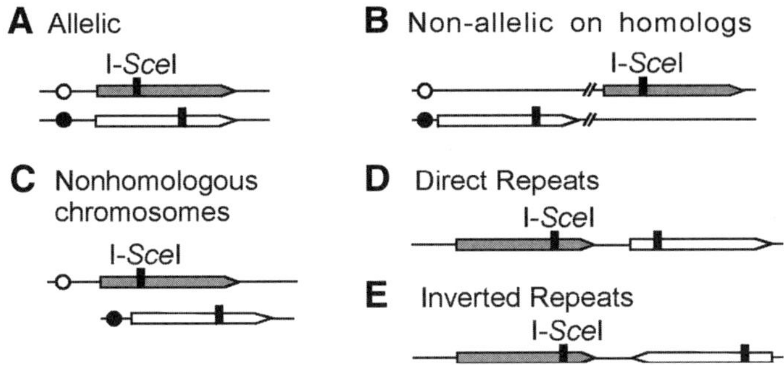

Fig. 1. Different arrangements of homologous sequences in recombination substrates. Genes are indicated by open or shaded boxes, inactivating mutations by black bars, and centromeres by circles. **(A)** Allelic substrates score interactions between genes at allelic positions on homologous chromosomes. In this arrangement, homology extends the entire length of the chromosomes. In the arrangements shown in **(B–E)**, homology is limited to the length of the repeated region, typically the size of the gene under study. **(B)** Interacting regions present on homologs at nonallelic loci. These are shown on the same chromosomal arm but could be present on different arms. **(C)** Interacting regions on nonhomologous chromosomes. In this arrangement, crossovers produce translocations. **(D,E)** Direct and inverted repeats present on a single chromosome.

1.2. Features of HR Substrates

HR occurs between DNA sequences sharing significant lengths of homology, and the interacting regions can be arranged in many configurations (**Fig. 1**). Interactions between allelic regions on homologous chromosomes is termed allelic HR. All other interactions are ectopic, including those between nonallelic repeats on homologs, repeats on nonhomologous chromosomes, linked repeats arranged in direct or inverted orientation, and between a chromosomal locus and an exogenous, plasmid-borne copy (gene targeting). Modifications to endogenous loci for analysis of allelic HR, or interactions between loci on heterologous chromosomes, have been reported in mouse embryonic stem cells ***(17,18)***, as gene targeting is fairly efficient in this cell type. In other cell types, HR substrates often consist of duplicated, selectable genes (like *neo*) on plasmids that are integrated into a random position in a mammalian genome, or they can be targeted to a specific locus by using flp or cre recombinase systems ***(19,20)***. Each of the duplicated genes is inactivated by specific mutations (i.e., frameshift insertions or deletions of coding or promoter sequences), and HR between these genes creates a functional gene that can be selected. Linked repeat substrates typically include a wild-type copy of a different selectable gene that is used to identify transfectants carrying the substrate. It is important to confirm that the integrated HR substrate is structurally intact, as HR between the repeats can occur prior to, or after integration, or the plasmid may integrate in such a way that one or both repeats are lost, truncated, or rearranged. In most cases it is desirable to identify transfectants with a single copy of the HR substrate as this greatly simplifies subsequent analysis of HR products. Transfectants carrying intact, single copies of HR substrates are identified by using polymerase chain reaction (PCR) and Southern hybridization assays. Genetic tests are used to confirm that the substrate can recombine spontaneously to produce selectable products at low frequency (typically approx 10^{-5}), and that expression of I-*Sce*I stimulates HR by two to three orders of magnitude above the spontaneous frequency. Additional information can be gained when one repeat carries phenotypically silent mutations, although the extra heterologies may reduce spontaneous HR below the limit of detection *(21)*.

HR includes reciprocal (crossover) and nonreciprocal exchange (gene conversion). Gene conversion involves precise information transfer from a donor locus to a recipient and results in localized loss of heterozygosity; the donor locus is unchanged. Gene conversion is strongly enhanced by DSBs, and broken alleles almost always act as recipients. Gene conversion alone does not cause gross alteration of chromosomal structures, but it is frequently associated with crossing over, which can lead to gross chromosomal rearrangements. For example, crossovers between nonhomologous chromosomes can produce balanced or unbalanced translocations, the latter yielding dicentric and acentric fragments. Crossovers between linked repeats result in repeat deletion or addition, or inversion of sequences between repeats, as discussed later, this subheading. Gene conversion and crossovers are conservative events, but HR via single-strand annealing (SSA) is nonconservative. SSA between linked direct repeats results in repeat deletion, and SSA between genes on different chromosomes can produce translocations. *See* Nickoloff *(22)* for additional information about these HR mechanisms and their outcomes.

The configuration of the interacting regions determines the type of DSB repair events that can be detected. Because I-*Sce*I produces ligatable, cohesive four-base overhangs, these DSBs can be repaired by precise NHEJ that is usually not detected; however, precise NHEJ can be detected when two closely linked I-*Sce*I sites were cleaved *(23)*. Allelic substrates score HR between homologs; HR between sister chromatids is not detected as it restores the parental structure. In addition, allelic gene conversion can be scored when conversion tracts are short, but tracts that extend to an inactivating mutation in the donor allele do not produce a functional allele (**Fig. 2A**) and will not be detected unless a non-selective assay is used (*see* **Subheading 1.4.**). Crossovers between allelic loci can be scored by monitoring markers flanking the crossover point (**Fig. 2B**).

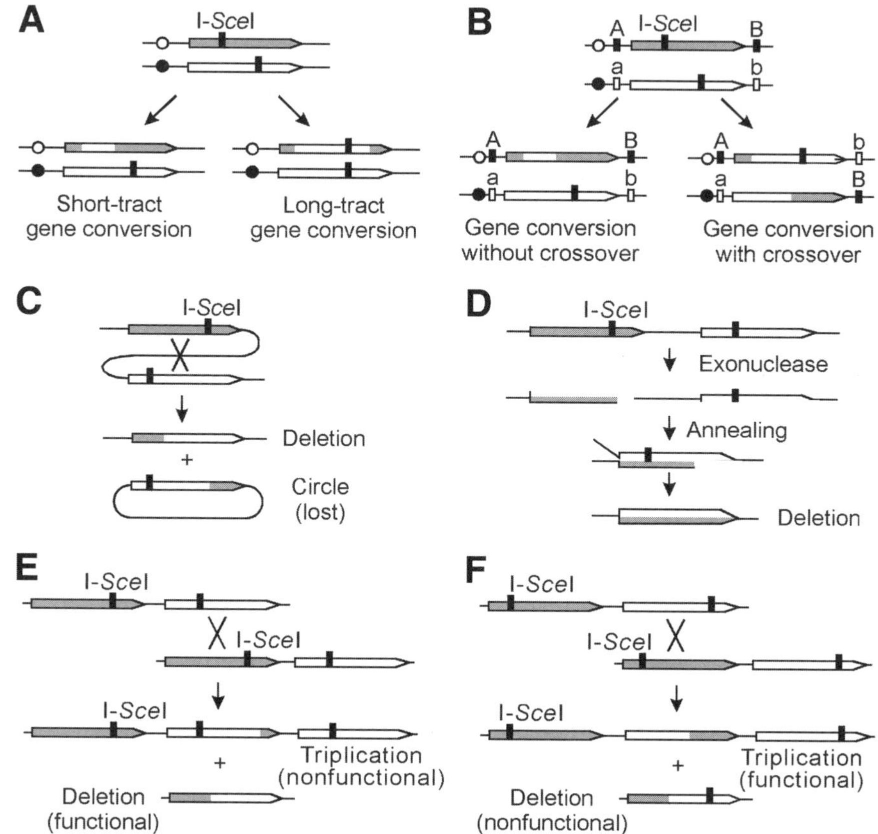

Fig. 2. Detectable HR events. **(A)** Short-tract gene conversion leads to loss of the I-*Sce*I site and produces a selectable product. Long-tract gene conversion also eliminates the I-*Sce*I site, but tracts that extend to the inactivating mutation in the donor allele do not produce a functional gene and go undetected in selective assays. **(B)** Detection of crossovers by monitoring flanking markers (A/a and B/b). Note that conversion tracts may include flanking markers and in this case, crossovers may go undetected. **(C)** Crossovers (shown by X) between direct repeats produce a selectable deletion product and circular product that is typically undetected. **(D)** SSA between direct repeats also produces deletions, but unlike crossovers, no circular product is formed. **(E,F)** Unequal exchange between direct repeats in sister chromatids produces deletion and triplication products that are selectable or not, depending on the arrangement of the inactivating mutations within the repeats.

Linked repeats can be arranged in direct or inverted orientation. Crossovers between inverted repeats invert the sequences between the repeats, but such inversions are not definitive for crossovers because sister chromatid conversion can produce the same outcome *(24)*. Crossovers between direct repeats delete one repeat, and the sequence between the repeats, with the deleted sequences found on a circular product (**Fig. 2C**). However, the circular product usually goes undetected because it is mitotically unstable unless it carries an origin of replication that functions in the host cell. With direct repeats, deletions can also arise by SSA or by unequal sister chromatid exchange (**Fig. 2D** and **E**), so these deletions are not definitive for crossovers. Deletions that arise by unequal sister chromatid exchange produce a functional (selectable) product when the mutation in the upstream gene is downstream (3') relative to the mutation in the second copy (**Fig. 2E**); the associated triplication product is nonfunctional unless an independent event converts one of three genes. On the other hand, if the mutation in the upstream gene is in the 5' position, the triplication product will carry a functional gene (but the deletion is nonfunctional). Therefore, this arrangement provides a definitive measure of crossovers between sister chromatids (**Fig. 2F**). Note that in **Fig. 2**, the I-*Sce*I recognition site is present in the upstream gene, but these same general features are apparent if the I-*Sce*I site is present in the downstream gene (not shown). Johnson et al. *(25)* used a system with the I-*Sce*I site in the downstream gene to detect triplications arising by crossovers between sister chromatids.

Another type of HR substrate has a donor with deletions at both 5' and 3' ends. In this case, gene conversions without crossovers produce a functional gene (**Fig. 3A**), but crossovers between sister chromatids (**Fig. 3B**) or intrachromosomal crossovers (not shown) produce only nonfunctional genes. Thus, this arrangement is well suited for studies of gene conversion without crossovers. By contrast, if the upstream gene has a 3' deletion and the downstream gene has a 5' deletion, only crossovers or SSA produce a functional (deletion) product; gene conversion without a crossover does not produce a functional product (**Fig. 3C**). Gene conversion of a chromosomal locus suffering a DSB can also be effected by an exogenous (transfected) copy of the gene; these events are analogous to gene conversion without a crossover and are a form of gene targeting (**Fig. 3D**) *(26)*. These gene targeting systems are relatively easy to construct and test, and they often lead to the same conclusions as studies of chromosomal HR events *(21,27–29)*. Nonetheless, some aspects of DSB repair are likely to be different in a chromosomal context vs a chromosome interacting with an exogenous, naked DNA fragment.

In addition to the constraints imposed by the specific arrangement of repeated loci, two other parameters may influence the choice of repair pathway: the lengths of the repeated regions and the distance separating the repeats. In yeast, 1.2-kbp direct repeats usually produce deletions *(30)*, whereas deletions were very rare with 6.5-kbp repeats *(31)*. Similarly, in mammalian cells, deletions were common with 0.7-kbp repeats *(32)* but very rare with 1.4-kbp

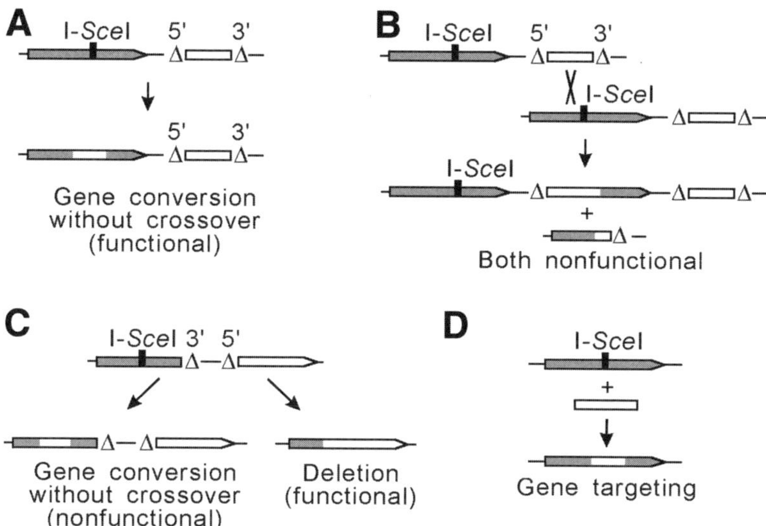

Fig. 3. Exclusive detection of crossover/noncrossover events. **(A)** Gene conversions without crossovers are detected when donor loci have 5' and 3' deletions. **(B)** Unequal exchange between sister chromatids produces nonfunctional deletion and triplication products when donor loci have 5' and 3' deletions. With this arrangement, intrachromosomal crossovers (as diagrammed in **Fig. 2C**) or SSA (as diagrammed in **Fig. 2D**) also produce nonfunctional deletions (not shown). **(C)** Direct repeats in which the upstream copy has a 3' deletion and the downstream copy has a 5' deletion does not detect gene conversion without a crossover. Deletions resulting from intrachromosomal crossovers, unequal sister chromatid exchange, or SSA are detected. **(D)** Gene targeting in which a chromosomal locus suffering a DSB is corrected by an exogenous (transfected) DNA fragment. The correcting DNA may be present in a circular plasmid.

repeats ***(21)***. The idea that deletion frequency depends on repeat length remains tentative, however, because a systematic study at a single chromosomal locus has not yet been reported. With regard to the distance between repeats, our preliminary analysis (using targeted substrates to avoid position effects) indicates that deletions increase as the distance between repeats decreases (E. Schildkraut and J.A.N., unpublished results).

1.3. Systems for Delivering I-SceI to Mammalian Cells

I-*Sce*I expression vectors typically employ the strong, constitutive cytomegalovirus (CMV) promoter, and expression is improved by the inclusion of one or more nuclear localization signals fused to the I-*Sce*I reading frame ***(33)***. A variety of methods can be used to transfect I-*Sce*I expression vectors into mammalian cells, including calcium phosphate co-precipitation ***(34)***, diethylaminoethyl (DEAE) dextran aggregation *(35)*, microinjection *(36,37)*, electroporation *(38)*, and lipofection *(39)*; to date, most studies of DSB repair with I-*Sce*I have employed electroporation or lipofection. Electroporation is effective with a wide range of cell types, but it is fairly inefficient. (For this reason, it is an excellent choice for constructing cell lines with single copies of HR substrates.) The efficiency of electroporation for delivery of I-*Sce*I expression vectors can be improved by using large quantities of vector (*see* **Subheading 1.4.**). In addition to vector transfection, electroporation can be used to transfect I-*Sce*I protein directly, or other restriction enzymes *(40)*. Lipofection is more limited in the types of cells that can be transfected, but with most permissive cell types, large fractions of cells are transfected (typically >70%). In **Subheadings 2.–4.** we present a detailed protocol for the analysis of I-*Sce*I-induced HR using lipofection-mediated transfection of I-*Sce*I expression vectors.

1.4. Selective vs Nonselective Assays

As discussed in **Subheading 1.2.**, appropriate substrates coupled with selective assays can be used to examine a wide or very narrow range of DSB repair mechanisms. To detect the widest possible range of DSB repair events, however, a nonselective assay must be used, and this requires high-efficiency cleavage by I-*Sce*I. The efficiency of I-*Sce*I cleavage depends on many factors, including transfection efficiency, the I-*Sce*I expression level (reflecting transcription, mRNA stability, translation, and nuclear import), and the accessibility of the enzyme to its target site (reflecting the chromatin environment and perhaps the local DNA sequence or topology). Thus, the efficiency of I-*Sce*I-induced HR can vary widely in cells carrying randomly integrated, but otherwise identical, HR substrates. These "position effects" are well documented (e.g., ref. ***21***), but the specific underlying causes have not been clearly defined.

Liang et al. *(33)* described a nonselective, PCR-based assay for mammalian DSB repair that provides a measure of the relative rates of HR and NHEJ. To achieve high-efficiency cleavage by I-*Sce*I nuclease, cells were transfected with 100 μg of I-*Sce*I expression vector by electroporation. A limitation of this approach is that events are only scored in populations, and certain events are not detected, such as NHEJ events that delete one or both sites complementary to the PCR primers, or very large sequence capture events that separate the PCR primer target sites by distances that are too large to amplify by conventional PCR. This procedure was efficient, with DSBs introduced into a chromosomal I-*Sce*I site in approx 50% of cells. However, the use of very large quantities of I-*Sce*I expression vector could result in artifactually high levels of vector sequence capture.

As mentioned in **Subheading 1.3.**, a larger fraction of cells can be transfected by lipofection than by electroporation. However, lipofection has strict limits on the amount of DNA that can be transfected (on the order of 2 μg). We recently developed methods for high-efficiency cleavage by I-*Sce*I nuclease and for efficient analysis of large numbers of individual DSB repair products. Our strategy was to transfect cells with a vector that coexpresses I-*Sce*I and green fluorescent protein (GFP) by lipofection, incubate cells for 24 h, and then sort those that express high levels of GFP using a MoFlo flow cytometer (Cytomation, Ft. Collins, CO). The MoFlo is a dual-laser, high-speed sorter capable of single-cell deposition into individual microplate wells. Control (non-GFP-transfected) cells are processed first to calibrate autofluorescence and set gates. Gates are set to eliminate cell debris and aggregated cells, ensuring efficient sorting of single cells. GFP-positive cells are then sorted such that a single cell is delivered to a single well of a 96-well plate in nonselective growth medium. Gates are set so that only the top 5–10% of GFP-expressing cells are recovered. Because some cells are killed by transfection and mechanical stress, this procedure yields an average of 8 viable cells per 96-well plate; this provides another level of assurance (by Poisson statistics) that nearly all wells receive at most one viable cell. After 7–10 d, wells with single colonies are identified (typically >95%), expanded, and genomic DNA is prepared and analyzed by PCR and Southern blotting. In our initial tests with this procedure, detectable DSB repair events were found in approx 40% of colonies, including HR, deletions via NHEJ ranging from a few bp to >10 kbp, and complex rearrangements ***(41)***.

2. Materials

1. Plasmid vector carrying HR substrate (*see* **Note 1**).
2. I-*Sce*I expression vector (*see* **Note 2**) and expression vectors for other transgenes of interest.
3. Colony fixation/staining solution: 1% (w/v) crystal violet in 100% methanol (or alternatively, 0.4% Coomassie blue in 50% methanol, 10% glacial acetic acid, 40% water).
4. Selective antibiotics, e.g., G418, hygromycin, puromycin, blasticidin (*see* **Note 3**).
5. Standard molecular biology and mammalian cell culture materials and equipment.

3. Methods

3.1. Installing an HR Substrate in Mammalian Cells

As discussed in the Introduction, recombination substrates may consist of recombining sequences positioned on two homologous (or nonhomologous) chromosomes, as direct or inverted repeats located on a single chromosome, or present as extrachromosomal elements. Installation of substrates at allelic positions on homologous chromosomes can be accomplished only through gene targeting, and readers interested in this approach should consult manuals on gene targeting methodology. Here we describe the installation of direct or inverted repeat substrates at a random position in a single host cell chromosome.

The cell line(s) in which the HR substrate is to be installed may be wild type or mutated for a particular gene of interest, or may be from a particular tissue or tumor type. In the case of a mutant cell line, a wild-type control can be problematic unless there is some assurance that the two lines are directly related (isogenic). Even for matched cell lines, it is necessary to measure HR in multiple substrate-bearing clones from both lines in order to confirm statistically that any differences seen between mutant and parental lines do not reflect position effects. For these reasons, a preferable approach is to install a recombination substrate only in the mutant cell line, and then measure recombination phenotypes with and without complementation for the mutated gene ***(42,43)***. Another way to avoid position effects is to install the HR substrate at the same chromosomal locus in parental and mutant cells by gene targeting, or by using cre or flp recombinases.

Electroporation is the method of choice for stably integrating a recombination substrate into mammalian cell lines. Under suitably chosen conditions, electroporation will often result in the chromosomal integration of a single, intact copy of the substrate construct. This is in contrast to chemical transfection methods, which tend to produce integration of multiple, concatamerized copies of any transfected DNA. In outline, the approach is as follows:

1. Linearize the plasmid vector containing the HR substrate (*see* **Note 4**).
2. Transfect host cells by electroporation, using limiting amounts of HR substrate DNA (*see* **Note 5**).
3. Select stable transfectants, using the drug-resistance "installation" marker (*see* **Note 6**).
4. Isolate a panel of resistant colonies by moving them into individual wells of a 24- or 48-well tray, using ring transfer or a similar colony transfer method. Expand each clone, and cryopreserve two aliquots of each (*see* **Note 7**).
5. Screen the panel of clones for the presence of a single, non-rearranged copy of the recombination substrate by Southern blotting (*see* **Note 8**).
6. Screen the panel of clones for ability to undergo I-*Sce*I-induced HR, using a simplified version of the HR frequency assay described below in **Subheading 3.2.** (*see* **Note 9**).

3.2. Measuring HR Frequencies

Once cell lines bearing a stably integrated HR substrate have been established, HR frequencies can be measured after transfecting for transient expression of I-*Sce*I. Expression vectors for additional transgenes may be cotransfected with the I-*Sce*I vector. For example, cells mutated for a gene of interest may be cotransfected with a complementing cDNA. Conversely, wild-type cells may be cotransfected with a dominant-negative allele of a gene of interest. To date, there have been no reports of gene knockdown using small inhibitory RNA (siRNA) in studies of I-*Sce*I-induced HR. Transient knockdown by cotransfection of an siRNA, or an siRNA expression vector, with an I-*Sce*I expression vector may not be effective because I-*Sce*I is likely to cleave its target before significant knockdown of the siRNA target protein occurs. It may be possible to use transient knockdown if the siRNA is transfected or expressed for a few days prior to transfection of the I-*Sce*I expression vector. A better alternative is to develop cell lines that stably express the siRNA.

The following protocol is for chemically mediated transfection methods (liposome, polycationic amine reagents, and so on). The protocol assumes an adherent cell line, so that recombination events can be scored as discrete colonies of antibiotic-resistant cells (*see* **Note 10**).

3.2.1. Transfection for Transient Expression of I-SceI With or Without Additional Transgenes

1. Prepare expression vector DNA(s). Plasmid DNA should be of high quality and largely free of bacterial protein and endotoxin; DNA prepared using commercial kits, such as those marketed by Qiagen (Valencia, CA), is usually adequate. As these vectors are intended for transient expression, not stable integration, it is not necessary or desirable to linearize the plasmid DNA prior to transfection. An accurate determination of DNA concentration is important. Remember to prepare appropriate control vectors, e.g., the expression vector(s) lacking I-*Sce*I or other transgene.
2. Plate cells in preparation for transfection the next day. Plate cells in six-well trays (the wells are equivalent to 35-mm cell culture dishes), at a density such that they will reach approx 75% confluence by the following morning, e.g., $3–5 \times 10^5$ cells per well (*see* **Note 11**). Plate three (or more) replica wells for each transfection group.
3. Prepare transfection mixtures, according to the guidelines furnished by the manufacturer of the chosen transfection reagent. The amount of I-*Sce*I vector used, the total amount of DNA used (I-*Sce*I plus other transgene or control), and the ratio of DNA to transfection reagent must be constant across all transfection groups. Wash cell monolayers, apply transfection mixtures in a minimal volume of medium (e.g., 1 mL), and return to the incubator for 3–6 h (*see* **Note 12**).
4. Remove transfection mixtures from cells, and refeed with complete, nonselective medium. Incubate for approx 24 h to allow recovery and expression of I-*Sce*I and/or other transgenes.

3.2.2. Select Recombinants and Measure Cloning Efficiency

1. After recovery, wash cells with PBS, trypsinize, and resuspend in 5 mL of growth medium with serum in a 15-mL conical tube.

2. Count cells with a hemocytometer or, preferably, a Coulter counter.
3. For each transfection, plate three (or more) 10-cm dishes, at a density suitable for antibiotic selection, e.g., 10^5 cells per dish or less (*see* **Note 13**). This plating should be done in nonselective medium and then supplemented with the selective drug the following day (*see* **Note 14**). These selection cultures may require one refeeding or more over the following growth period of 7–14 d, depending on the cell line and selective agent used (*see* **Note 15**). If performing molecular analysis of individual recombinants, it may be useful to include dishes plated at low density to facilitate the isolation of individual colonies.
4. For each transfection, plate three dishes at low density (approx 200 cells per dish) to measure cloning efficiency (CE) (*see* **Note 16**); this will require an intermediate cell dilution. These cells are maintained in nonselective medium and, because of the low initial density, they do not require refeeding.

3.2.3. Score Colonies and Calculate Recombination Frequency

1. When colonies are well developed (*see* **Note 17**), fix and stain them in one step: drain the dish, apply a few milliliters of crystal violet stain solution, allow to stand for about a minute, then rinse in a basin of tap water and leave to dry.
2. Count colonies in each CE dish (*see* **Note 18**). CE can be calculated for each transfection as the total number of colonies in CE dishes divided by the total number of cells initially plated. For example:

   CE = 234 colonies (in 3 dishes)/600 cells plated (in 3 dishes) = 0.39

3. Count colonies in selection dishes. HR frequency can then be calculated for each transfection as the total number of colonies in selection dishes, divided by the total number of cells initially plated, with a correction for CE:

   HR frequency = 637 colonies (in 3 dishes)/[3×10^5 cells plated (in 3 dishes) $\times$ 0.39 (CE)] = 5.44×10^{-2}

   In this case, the HR frequency is 5.44×10^{-2} recombinants per viable cell plated for a single transfection. The HR frequency for the transfection group can then be calculated as the mean of HR frequencies of the three (or more) replicate transfections that comprise the group, with an associated standard deviation.

4. Notes

1. HR substrates with I-*Sce*I recognition sites can be constructed using any selectable marker as a reporter gene, in accordance with the principles outlined in the Introduction. For the installation marker, be careful to choose a different selectable gene that will not interfere with downstream plans such as expression of another transgene. Alternatively, HR substrates previously reported in the literature can usually be obtained from the originating laboratory on request.
2. The most widely used expression vector for I-*Sce*I is pCMV(3xNLS)I-SceI, constructed by Greg Donoho in the laboratory of Paul Berg. It contains the I-*Sce*I open reading frame fused at its 5' end to a triplicated nuclear localization signal and equipped with the cytomegalovirus immediate early region promoter/enhancer,

and the bovine growth hormone polyadenylation signal region. The vector can be obtained by request from this lab.

3. Of the four aminoglycoside antibiotics listed, G418 (a neomycin analog effective against mammalian cells) has been most often used in recombination substrates, for historical reasons. The other three antibiotics are equally (or more) effective as selective drugs. Plasmid vectors carrying resistance markers for hygromycin (*hyg*), puromycin (*pac*), and blasticidin (*BSD*) are available commercially from Invitrogen (Carlsbad, CA) and Clonetech (Palo Alto, CA). ***All these drugs are highly toxic; handle them only with gloves, and observe all precautions recommended by the supplier.***
4. To increase the likelihood that a direct or inverted repeat HR substrate will integrate into a host chromosome with the donor and recipient alleles intact and in the intended orientation, linearize the plasmid at a site that does not lie within or between the donor and recipient alleles. The linear form of the plasmid should contain the entire repeat structure in a continuous sequence bounded on each end by buffer regions of non-essential sequence. Buffer regions of 900 bp are very effective ***(21)***, but buffers of 150 bp were not (D. Taghian and J.A.N., unpublished results).
5. Electroporation conditions can vary greatly among cell lines. Choose relatively mild electroporation conditions that do not produce much cell killing. The goal is not the highest possible frequency of stable transfection, but rather to integrate a single copy of the substrate. Perform at least three transfections, beginning with a very low amount of linearized plasmid, e.g., 40 ng, 200 ng, and 1.0 μg.
6. The installation marker is a selectable gene other than that used to score HR. Plate at moderate density, e.g., 100,000 cells per dish, in 5–10 dishes per transfection. Include control dishes of nonelectroporated cells to ensure that the drug selection is effective. Isolate colonies resulting from the lowest possible amount of transfecting DNA, e.g., from the 40-ng transfection in preference to the 200-ng or 1-μg transfections, if all yield resistant colonies. If no colonies result, increase the amount of DNA.
7. Typically, a small fraction of the candidate colonies will prove suitable; a panel of 24–48 colonies should yield several that are usable for analysis of HR.
8. After expansion of each colony under continuing drug selection, extract genomic DNA and digest with restriction enzymes that will yield one to three diagnostic fragments of the HR substrate. Include the plasmid vector as control, and hybridize with a probe prepared from the plasmid vector (e.g., the gene used for donor and recipient alleles).
9. Screen candidates for I-*Sce*I-induced HR using the protocol in **Subheading 3.2.**, but with the following modifications: (1) omit replicates; transfect only a single well of a six-well tray for each clone to be screened; (2) plate only for selection of HR; do not measure cloning efficiency; (3) plate cells from a nontransfected well of each candidate for selection to determine the baseline spontaneous (noninduced) HR frequency and to identify candidates that underwent HR during HR substrate installation. In some situations, it may be more efficient to screen first

for I-*Sce*I-induced HR and then confirm that the HR substrate is intact and single copy by Southern blotting.

10. The general experimental scheme could be adapted for use with nonadherent cell lines. This would require plating of cells for selection in microtiter trays at low densities, to reduce the likelihood of recovering more than one recombinant colony in a single microtiter well, but this is labor-intensive.
11. Cells should be plated far enough in advance that they have ample time to recover from the stress of trypsinization and be present as a dense but subconfluent and highly mitotic population at the time transfection begins. Too low a cell density may result in low transfection efficiency, and too high a density may result in arrested growth owing to contact inhibition or medium depletion.
12. Transfection efficiency varies between cell lines and may vary with different transfection reagents. For any particular reagent, give careful consideration to the manufacturer's guidelines. Variables such as the amount of DNA used, the ratio of DNA to transfection reagent, the length of time cells are exposed to DNA/transfection reagent mixtures, and the presence or absence of serum during transfection may be critical and require some optimization. However, it may be useful to diverge from the manufacturer's guidelines. For example, some transfection reagents are intended for use in the presence of serum but may actually perform better in the absence of serum if the cells tolerate serum deprivation well. Incubation times in excess of the recommended period may give better results in certain cell lines. To optimize transfection conditions, use a rapidly measurable surrogate endpoint such as expression of green fluorescent protein detected by flow cytometry.
13. The density chosen represents a compromise. It must be high enough to recover sufficient numbers of recombinants for statistical analysis. It must also be low enough to permit scoring of discrete colonies. When in doubt, plate at lower density. Plate all selection dishes for all transfection groups at a uniform density, as initial density can influence the dynamics of cell survival and colony formation under selection. If you anticipate a low HR frequency in a particular treatment group, plate more dishes for selection in that group, but maintain uniform density across groups.
14. Adding the selective drug 24 h after transfection allows time for expression of the selectable marker gene following HR. Depending on the selection marker used, this expression period may not be necessary, and it is less effort to plate directly into selective medium.
15. Whether and how often selection cultures will need refeeding will depend on the initial plating density, the growth rate, and the selection used. Some selective agents, such as puromycin, clear the background of nonresistant cells very quickly, thereby slowing depletion of the medium and reducing the need for refeeding. Others, such as the G418, clear nonresistant cells much more slowly. Generally, however, one to three refeedings over the course of colony formation will suffice. Do not handle cultures any more than absolutely necessary to minimize "satellite" colonies that arise when cells become dislodged from a colony and then reattach and grow elsewhere.

16. It is necessary to measure cloning efficiency for each transfection. This allows HR frequencies to be expressed in terms of events per viable cell plate in selective medium. HR frequencies can be compared between transfection groups in a single experiment, and between experiments. Cell viability after transfection cannot be assumed to be constant between groups of replicate transfections, or even within them. Differences in cloning efficiency can sharply alter the apparent HR frequencies obtained from raw counts of drug-resistant colonies.
17. The time needed for colonies to form will vary among cell lines, but is generally between 7 and 14 d after plating. There is usually some heterogeneity in the size of mammalian cell colonies, presumably reflecting clonal differences in growth rate. If dishes are fixed and stained too early, smaller colonies may be missed. Neither should colony growth be allowed to go on too long, as this may increase the incidence of satellite colonies, or allow closely spaced primary colonies to merge, making an accurate count difficult.
18. Colony counting can be done with the unaided eye. However, decide at the outset upon some standard for what constitutes a true, proliferating colony, as opposed to a nascent colony that has arrested, or few loosely clustered cells that might or might not be of clonal origin. It is customary to count colonies that have at least 50 cells. With practice, it will be easy to recognize colonies that meet your criteria with only occasional recourse to a microscope for confirmation.

References

1. Orr-Weaver, T. L., Szostak, J. W., and Rothstein, R. J. (1981) Yeast transformation: a model system for the study of recombination. *Proc. Natl. Acad. Sci. USA* **78,** 6354–6358.
2. Ward, J. (1998) The nature of lesions formed by ionizing radiation. In: *DNA Damage and Repair: DNA Repair in Higher Eukaryotes* (Nickoloff, J. A. and Hoekstra, M. F., eds.). Humana, Totowa, NJ, pp. 65–84.
3. Kartalou, M. and Essigmann, J. M. (2001) Recognition of cisplatin adducts by cellular proteins. *Mutat. Res.* **478,** 1–21.
4. Strathern, J. N., Klar, A. J. S., Hicks, J. B., et al. (1982) Homothallic switching of yeast mating type cassettes is initiated by a double-stranded cut in the *MAT* locus. *Cell* **31,** 183–192.
5. Nickoloff, J. A., Chen, E. Y. C., and Heffron, F. (1986) A 24-base-pair sequence from the *MAT* locus stimulates intergenic recombination in yeast. *Proc. Natl. Acad. Sci. USA* **83,** 7831–7835.
6. Kolodkin, A. L., Klar, A. J. S., and Stahl, F. W. (1986) Double-strand breaks can initiate meiotic recombination in *S. cerevisiae*. *Cell* **46,** 733–740.
7. Nicolas, A. L., Munz, P. L., Falck-Pedersen, E., and Young, C. S. H. (2000) Creation and repair of specific DNA double-strand breaks *in vivo* following infection with adenovirus vectors expressing *Saccharomyces cerevisiae* HO endonuclease. *Virology* **266,** 211–224.
8. Monteilhet, C., Perrin, A., Thiery, A., Colleaux, L., and Dujon, B. (1990) Purification and characterization of the *in vitro* activity of I-*Sce*I, a novel and highly

specific endonuclease encoded by a group I intron. *Nucleic Acids Res.* **18,** 1407–1413.
9. Nickoloff, J. A., Singer, J. D., and Heffron, F. (1990) *In vivo* analysis of the *Saccharomyces cerevisiae* HO nuclease recognition site by site-directed mutagenesis. *Mol. Cell. Biol.* **10,** 1174–1179.
10. Colleaux, L., d'Auriol, L., Galibert, F., and Dujon, B. (1988) Recognition and cleavage site of the intron-encoded omega transposase. *Proc. Natl. Acad. Sci. USA* **85,** 6022–6026.
11. Colleaux, L., d'Auriol, L., Betermier, M., et al. (1986) Universal code equivalent of a yeast mitochondrial intron reading frame is expressed in *E. coli* as a specific double strand endonuclease. *Cell* **44,** 521–533.
12. Thierry, A., Perrin, A., Boyer, J., et al. (1991) Cleavage of yeast and bacteriophage T7 genomes at a single site using the rare cutter endonuclease I-*Sce*I. *Nucleic Acids Res.* **19,** 189–190.
13. Puchta, H., Dujon, B., and Hohn, B. (1993) Homologous recombination in plant cells is enhanced by in vivo induction of double strand breaks into DNA by a site-specific endonuclease. *Nucleic Acids Res.* **21,** 5034–5040.
14. Rouet, P., Smih, F., and Jasin, M. (1994) Introduction of double-strand breaks into the genome of mouse cells by expression of a rare-cutting endonuclease. *Mol. Cell. Biol.* **14,** 8096–8106.
15. Choulika, A., Perrin, A., Dujon, B., and Nicolas, J.-F. (1995) Induction of homologous recombination in mammalian chromosomes by using the I-*Sce*I system of *Saccharomyces cerevisiae. Mol. Cell. Biol.* **15,** 1968–1973.
16. Jasin, M. (1996) Genetic manipulation of genomes with rare-cutting endonucleases. *Trends Genet.* **12,** 224–228.
17. Richardson, C. and Jasin, M. (2000) Frequent chromosomal translocations induced by DNA double-strand breaks. *Nature* **405,** 697–700.
18. Moynahan, M. E. and Jasin, M. (1997) Loss of heterozygosity induced by a chromosomal double-strand break. *Proc. Natl. Acad. Sci. USA* **94,** 8988–8993.
19. O'Gorman, S., Fox, D. T., and Wahl, G. M. (1991) Recombinase-mediated gene activation and site-specific integration in mammalian cells. *Science* **251,** 1351–1355.
20. Fukushige, S. and Sauer, B. (1992) Genomic targeting with a positive-selection *lox* integration vector allows highly reproducible gene expression in mammalian cells. *Proc. Natl. Acad. Sci. USA* **89,** 7905–7909.
21. Taghian, D. G. and Nickoloff, J. A. (1997) Chromosomal double-strand breaks induce gene conversion at high frequency in mammalian cells. *Mol. Cell. Biol.* **17,** 6386–6393.
22. Nickoloff, J. A. (2002) Recombination: mechanisms and roles in tumorigenesis. In: *Encyclopedia of Cancer*, 2nd ed. (Bertino, J. R., ed.). Elsevier Science, San Diego, pp. 49–59.
23. Lin, Y., Lukacsovich, T., and Waldman, A. S. (1999) Multiple pathways for repair of double-strand breaks in mammalian chromosomes. *Mol. Cell. Biol.* **19,** 8353–8360.

24. Chen, W. and Jinks-Robertson, S. (1998) Mismatch repair proteins regulate heteroduplex formation during mitotic recombination in yeast. *Mol. Cell. Biol.* **18,** 6525–6537.
25. Johnson, R. D. and Jasin, M. (2000) Sister chromatid gene conversion is a prominent double-strand break repair pathway in mammalian cells. *EMBO J.* **19,** 3398–3407.
26. Smih, F., Rouet, P., Romanienko, P. J., and Jasin, M. (1995) Double-strand breaks at the target locus stimulate gene targeting in embryonic stem cells. *Nucleic Acids Res.* **23,** 5012–5019.
27. Xia, F., Taghian, D. G., DeFrank, J. S., et al. (2001) Deficiency of human BRCA2 leads to impaired homologous recombination but maintains normal nonhomologous end joining. *Proc. Natl. Acad. Sci. USA* **98,** 8644–8649.
28. Elliott, B., Richardson, C., Winderbaum, J., Nickoloff, J. A., and Jasin, M. (1998) Gene conversion tracts in mammalian cells from double-strand break repair. *Mol. Cell. Biol.* **18,** 93–101.
29. Moynahan, M. E., Pierce, A. J., and Jasin, M. (2001) BRCA2 is required for homology-directed repair of chromosomal breaks. *Mol. Cell* **7,** 263–272.
30. Cho, J. W., Khalsa, G. J., and Nickoloff, J. A. (1998) Gene conversion tract directionality is influenced by the chromosome environment. *Curr. Genet.* **34,** 269–279.
31. Ray, A., Siddiqi, I., Kolodkin, A. L., and Stahl, F. W. (1988) Intrachromosomal gene conversion induced by a DNA double-strand break in *Saccharomyces cerevisiae*. *J. Mol. Biol.* **201,** 247–260.
32. Dronkert, M. L. G., Beverloo, H. B., Johnson, R. D., Hoeijmakers, J. H. J., Jasin, M., and Kanaar, R. (2000) Mouse *RAD54* affects DNA double-strand break repair and sister chromatid exchange. *Mol. Cell. Biol.* **20,** 3147–3156.
33. Liang, F., Han, M. G., Romanienko, P. J., and Jasin, M. (1998) Homology-directed repair is a major double-strand break repair pathway in mammalian cells. *Proc. Natl. Acad. Sci. USA* **95,** 5172–5177.
34. Graham, F. L. and van der Eb, A. J. (1973) Transformation of rat cells by DNA of human adenovirus 5. *Virology* **52,** 456–467.
35. McCutchen, J. H. and Pagano, J. S. (1968) Enhancement of the infectivity of SV40 deoxyribonucleic acid with diethylaminoethyl-dextran. *J. Natl. Cancer Inst.* **41,** 351–357.
36. Huberman, M., Berg, P. E., Curcio, M. J., Dipietro, J., Henderson, A. S., and Anderson, W. F. (1984) Fate and structure of DNA microinjected into mouse TK{+-} L cells. *Exp. Cell Res.* **153,** 347–362.
37. Folger, K. R., Wong, E. A., Wahl, G., and Capecchi, M. R. (1982) Patterns of integration of DNA microinjected into cultured mammalian cells: evidence for homologous recombination between injected plasmid DNA molecules. *Mol. Cell. Biol.* **2,** 1372–1387.
38. Nickoloff, J. A. (ed.) (1995) *Animal Cell Electroporation and Electrofusion Protocols*. Humana, Totowa, NJ.
39. Felgner, P. L., Gadek, T. R., Holm, M., et al. (1987) Lipofection: a highly efficient, lipid-mediated DNA-transfection procedure. *Proc. Natl. Acad. Sci. USA* **84,** 7413–7417.

40. Brenneman, M., Gimble, F. S., and Wilson, J. H. (1996) Stimulation of intra-chromosomal homologous recombination in human cells by electroporation with site-specific endonucleases. *Proc. Natl. Acad. Sci. USA* **93,** 3608–3612.
41. Allen, C., Miller, C. A., and Nickoloff, J. A. (2003) The mutagenic potential of a single DNA double-strand break is not influenced by transcription. *DNA Repair* **2,** 1147–1156.
42. Allen, C., Kurimasa, A., Brennemann, M. A., Chen, D. J., and Nickoloff, J. A. (2002) DNA-dependent protein kinase suppresses double-strand break-induced and spontaneous homologous recombination. *Proc. Natl. Acad. Sci. USA* **99,** 3758–3763.
43. Brenneman, M. A., Wagener, B. M., Miller, C. A., Allen, C., and Nickoloff, J. A. (2002) XRCC3 controls the fidelity of homologous recombination: roles for XRCC3 in late stages of recombination. *Mol. Cell* **10,** 387–395.

5

Transformation of Monomorphic and Pleomorphic *Trypanosoma brucei*

Richard McCulloch, Erik Vassella, Peter Burton, Michael Boshart, and J. David Barry

Summary

African trypanosomes, such as *Trypanosoma brucei*, are protozoan parasites of mammals that were first described over 100 hundred years ago. They have long been the subjects of biological investigation, which has yielded insights into a number of fundamental, as well as novel, cellular processes in all organisms. In the last decade or so, genetic manipulation of trypanosomes has become possible through DNA transformation, allowing yet more detailed analysis of the biology of the parasite. One facet of this is that DNA transformation has itself been used as an assay for recombination and will undoubtedly lead to further genetic approaches to examine this process. Here we describe protocols for DNA transformation of *Trypanosoma brucei*, including two different life cycle stages and two different strain types that are distinguished by morphological and developmental criteria. We consider the application of transformation to recombination, as well as the uses of transforming the different life cycle stages and strain types.

Key Words: trypanosome, *Trypanosoma brucei*, parasite, transformation, recombination, VSG, antigenic variation

1. Introduction

Genetic manipulation of African trypanosomes by DNA transformation was first described in 1990 and has revolutionized our ability to ask biological questions of this protozoan parasite that causes sleeping sickness in humans and Nagana in cattle. The uses of transformation in *Trypanosoma brucei* have included the creation and complementation of gene mutants ***(1,2)***, altering the level and timing of gene expression ***(3)***, the generation of gene knockdowns by induced double-stranded RNA interference ***(4)***, and induced or constitutive expression of foreign RNAs and proteins ***(5,6)***. Stable transformation of *T. brucei*

From: *Methods in Molecular Biology, vol. 262, Genetic Recombination: Reviews and Protocols*
Edited by: A. S. Waldman © Humana Press Inc., Totowa, NJ

apparently occurs exclusively by integration of DNA via homologous recombination. We now understand at least some of the cellular factors that control and catalyze this process, and their analysis has allowed DNA transformation to be used as an assay to dissect the controls and mechanisms of recombination in this ancient parasite.

T. brucei is transmitted between mammals by the biting insect the tsetse fly. In the bloodstream and tissue fluids of the mammal the parasite proliferates as a "long-slender" form and differentiates into a growth-arrested "short stumpy" form that is preadapted for survival in the fly. When bloodstream-stage parasites are ingested by a tsetse fly, the short stumpy form differentiates to generate procyclic-stage cells and continues its life cycle through a number of further developmental forms until differentiation in the fly's salivary glands to a metacyclic stage, which is infective for a new mammalian host (for a recent description of the life cycle, *see* ref. ***7***). An axenic culture system is available for both bloodstream- and procyclic-stage *T. brucei*, whereas the entire life cycle can be reproduced in vitro for *T. congolense*, a livestock pathogen ***(8,9)***.

Most genetic analyses have been conducted in *T. brucei* strains adapted to laboratory growth by serial syringe passages between rodents, with the result that these two life cycle stages can be cultured in vitro to relatively high densities. However, such strains often suffer from the drawback that they appear unable to differentiate to the stumpy form and to complete their life cycle in the tsetse fly; these are termed monomorphic strains. Much more difficult has been the genetic manipulation of what are termed pleomorphic *T. brucei* strains, which are fully competent to differentiate. Recently, transformation of pleomorphic parasites, again in both bloodstream and procyclic life cycle stages, has been achieved, as has transformation of *T. congolense* ***(10,11)***. Here, we describe the methodologies used for transformation of the different *T. brucei* strain types and consider their experimental uses. For *T. congolense*, which we will not discuss, similar approaches and considerations are likely to apply.

2. Materials

1. *T. brucei* strains (both bloodstream- and procyclic-stage cells).
2. *T. brucei* culture media (HMI-9 for bloodstream-stage parasites and SDM-79, DTM or Cunningham's medium for procyclic-stage parasites; *see* **Notes 1** and **2**), methylcellulose, and glycerol.
3. Incubators and plasticware for *T. brucei* culture.
4. DNA electroporation equipment (e.g., Bio-Rad gene pulser II or Genetronics BTX ECM630) and cuvets.
5. Electroporation buffers (*see* **Note 3**): Zimmerman's postfusion medium (132 m*M* NaCl, 8 m*M* Na_2HPO_4, 0.5 m*M* Mg.acetate, 0.09 m*M* Ca acetate, pH 7.0), or Cytomix (120 m*M* KCl, 15 m*M* $CaCl_2$, 10 m*M* K_2HPO_4, 25 m*M* HEPES, 2 m*M* EDTA, 5 m*M* $MgCl_2$, pH 7.6).

6. Plasmid DNA (*Escherichia coli* strains, such as DH5α or XL1-Blue, and growth media)
7. Polymerase chain reaction (PCR) equipment (thermostable DNA polymerase, oligonucleotides, buffers, thermocycler).
8. Restriction enzymes, T4 DNA ligase.
9. Agarose and gel electrophoresis apparatus.

3. Methods

3.1. Preparation of DNA for T. brucei *Transformation*

3.1.1. Transient Transformation

The approaches outlined in this review represent methodologies for stable transformation of *T. brucei*, which occurs by integration of foreign DNA into the parasite's genome. Transient, or episomal, transformation will not be discussed in detail but has previously been used successfully. In *T. brucei*, transient transformation has generally been used for the introduction of supercoiled bacterial plasmids containing *T. brucei* DNA and a reporter gene in order to assay, for instance, promoter or terminator activity ***(12–17)***. Since *T. brucei* replication functions have yet to be described, the transformed plasmids are inevitably lost in the absence of drug selection as the parasite cell divides, which contrasts with the stable inheritance of similar bacterial plasmids in the related kinetoplastid parasite *Leishmania* ***(17)***. Drug selection has been reported, however, to yield plasmid multimers and, occasionally, integrated plasmid DNA ***(18,19)***. Two groups have developed plasmids containing *T. brucei* sequences that are not obviously replication-related but nevertheless promote episome maintenance ***(20,21)***, and one has been used to demonstrate the essentiality of a gene via complementation ***(22)***. Artificial, linear extrachromosomal episomes have also been described for *T. brucei*, but these too are lost during cell divisions in the absence of drug selection ***(23,24)***. Episomal plasmid transformation remains the exception in *T. brucei* genetic manipulation and is currently of limited use as a route to assay recombination, as well as other processes, although perhaps the recent identification of naturally occurring, stably inherited circular DNA in the *T. brucei* nucleus will provide impetus in this area of research ***(25)***.

3.1.2. Plasmid DNA Integration

Two distinct types of plasmid construct, *ends-in* and *ends-out*, have been described that can stably integrate into the genome of *T. brucei* (**Figs. 1** and **2**). Integration of both relies on recombination of the transformed DNA into the genome via sequence homology. Many such integrations have been performed, and it appears that most of the *T. brucei* genome can be targeted, including the megabase (housekeeping) chromosomes ***(1,26)***, intermediate chromosomes

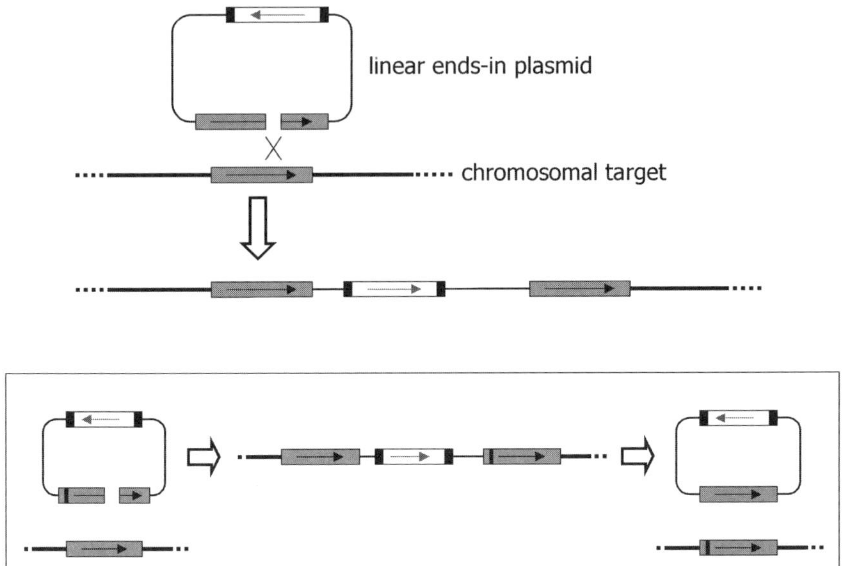

Fig. 1. Ends-in recombination to integrate a transformed plasmid into a homologous locus in the *T. brucei* genome. The region of homology between the *T. brucei* genome and a transformed plasmid (shown as a gray box containing an arrow) is duplicated upon plasmid integration (the cross represents recombination); the plasmid is linearized in the homologous sequence prior to transformation (shown by a gap in the box). The plasmid backbone (thin line) is also integrated in this process, which is selected for via a cassette containing a selectable marker (white box with arrow) and flanking processing signals (black boxes). The insert shows "hit and run mutagenesis," whereby selection of recombination events that precisely remove the integrated plasmid and recombination target (gray box) can leave a specific mutation (indicated by a vertical line) in the genomic copy of that sequence if it was present in the transformed plasmid.

(27,28), and high-copy-number minichromosomes ***(29)*** (*see* refs. ***30*** and ***31*** for a review of chromosome organization in *T. brucei*). Recently, the natural, episomal *T. brucei* nuclear DNA has also been targeted ***(25)***. Within the megabase chromosomes, both telomeric ***(32–34)*** and chromosomal-internal loci ***(35)*** have been targeted, as well as actively transcribed or silent regions (e.g., the rDNA "spacer" region ***[5,36]***). It appears, therefore, that most, if not all, of the *T. brucei* genome can be genetically manipulated.

The two types of targeting vector differ in their mechanisms of DNA integration. One of the integration mechanisms, shown in **Fig. 1**, occurs by ends-in recombination (reviewed in refs. ***37*** and ***38***), and this type of construct is occasionally referred to as being of the *single-crossover type* ***(39)***. A DNA double-

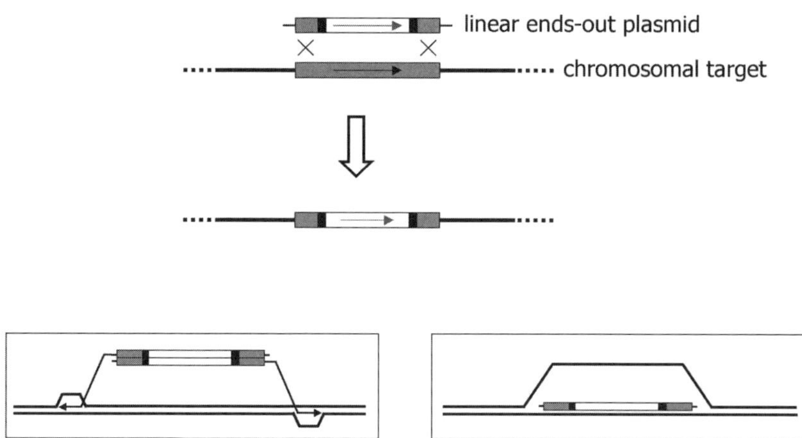

Fig. 2. Ends-out recombination. (All components are as drawn in **Fig. 1**.) Regions of sequence homology flanking a selectable marker cassette allow recombination (crosses) into a chromosomal locus, deleting any sequences between the flanks. The left insert depicts an integration event whereby the ends of the transformed DNA (shown double-stranded) initiate DNA synthesis (arrows) at regions of homology in the genome; the right insert depicts assimilation of a complete single strand of the transformed linear DNA molecule into genomic double-stranded DNA.

strand break (DSB) is created within a cloned *T. brucei* DNA sequence, typically by restriction digestion. After transformation of the linear plasmid into the parasite, the DSB is repaired by gene conversion from a homologous chromosomal locus. This repair could occur without crossing over to recreate the circular plasmid (not shown), but in the absence of replication functions this product cannot readily be recovered in *T. brucei* and has not been observed. Instead, integration of the plasmid into the homologous locus as a result of gene conversion with associated crossing over is selected for, normally via antibiotic resistance. Ends-in recombination strategies result in duplication of the target sequence. This is most commonly used in *T. brucei* for insertion of parasite-derived or foreign DNA sequences into novel locations, or to provide novel regulation of parasite gene expression. A prominent example is targeting of the silent intergenic region of the *rDNA* locus to provide controllable gene expression via bacteriophage T7 RNA polymerase ***(3,5,40)***.

In theory, however, the plasmid-borne gene could have been artificially mutated prior to transformation and, by selecting for a reverse recombination event that leaves the mutant gene copy in the genome and removes the intact gene and plasmid, defined mutations could be created (**Fig. 1**). Such a strategy of "hit and run" mutagenesis has been used in *Leishmania* (H. Denise and J. Mottram, personal communication) but not yet in *T. brucei* and relies on positive and then negative selection, markers for which are available (*see* third paragraph following).

Figure 2 shows gene targeting by ends-out recombination, often called gene replacement. Here, a linear DNA fragment is transformed that normally has two DNA ends (and therefore, effectively, 2 terminal DSBs) with sequence homology to a genomic locus. Recombination on the terminal *T. brucei* sequences inserts the linear DNA into the genome, deleting any intervening sequence. In **Fig. 2**, the terminal targeting sequences are shown separated by an antibiotic resistance marker, and this strategy is most commonly used to perform deletions or disruptions of a characterized locus. However, the same recombination pathway can also be used to create simple insertions, for example, to place genes in novel locations or alter the expression elements of genes, if the terminal targeting flanks are derived from a contiguous sequence. In addition, the interior of the targeting DNA fragment could contain other sequences, such as novel genes or promoter elements, to provide altered gene expression controls.

Ends-out recombination is often considered to occur via two co-ordinated crossovers between the terminal targeting flanks and the homologous chromosomal *T. brucei* genome sequence. In fact, two other mechanisms can also account for ends-out integration (**Fig. 2**) *(37,38)*. In one, a single strand of the linear DNA is incorporated into the homologous chromosome and mismatch repair then corrects in favor of the resident, unbroken DNA strand or the assimilated strand. The former outcome is the more likely *(41)*, but antibiotic pressure would select for the latter. In the second alternative ends-out mechanism, each DNA end invades a homologous sequence and initiates DNA synthesis. This could then give rise to crossovers that integrate the transformed DNA, or the DNA synthesis could proceed to the chromosome ends, creating a new chromosome copy. That this last possibility occurs in trypanosomatids is supported by evidence for chromosome trisomy following attempted disruption of essential genes *(42–44)*. The creation of a new chromosome copy illustrates the difference between ends-in and ends-out recombination: DNA synthesis during repair of the DSB in ends-in recombination proceeds from one broken end of the plasmid towards the other, whereas DNA synthesis and replication moves away from the two DNA fragment ends in ends-out recombination.

Ends-out or ends-in plasmid constructs for transformation are constructed by standard molecular biology cloning approaches. Currently, six antibiotic resistance genes have been shown to function as positive selection markers in *T. brucei*. These are metabolically distinct, allowing them to be used in combination, and encode resistance to puromycin, hygromycin, G418, phleomycin (or zeocin)/bleomycin, blasticidin, or nourseothricin. Sources of the resistance genes and antibiotics are given in **Table 1**. The amount of antibiotic needed for selection is highly variable, depending on the drug, life cycle stage, and target locus, and should therefore be empirically determined, but Prof. G.A.M. Cross has compiled some useful general guidelines for selection conditions (http://tryps.rockefeller.edu/). Viral thymidine kinase genes have been shown to be effective as a negative selective markers ***(45)*** and could therefore be used to select for reverse recombinants of ends-in transformants. One potential drawback, however, is that in some circumstances the rate of inactivating mutations in such genes may be as high as the rate of recombination. Marker genes that are not based on drug selection but that are suitable for screening and sorting by fluorescence-activated cell sorting (FACS), such as encoding autofluorescent proteins, may also be possible. Although the principal of using such markers for cell sorting has been demonstrated in *Leishmania* ***(46)***, no reports have yet been made showing they can be used to select for stable transformation in *T. brucei* or related parasites.

There is a complication in stable transformation approaches in trypanosomes in that all mature mRNAs in *T. brucei* and other kinetoplastids are generated from a polycistronic precursor RNA by upstream *trans*-splicing, to provide a 5'cap, and downstream polyadenylation. For this reason, antibiotic resistance genes must be expressed as cassettes comprising the gene's open reading frame (ORF) and upstream and downstream processing signal sequences. A convenient approach is to use intergenic regions from *T. brucei* multigenic arrays such as the *TUBULIN*, *ACTIN*, *PROCYCLIN*, and *CALMODULIN* loci. In the case of ends-out constructs, however, the endogenous processing signals from the gene or region being targeted could be employed.

Most of the *T. brucei* genome is arranged in multigenic transcription units, and if transformation constructs integrate in the same translational direction as the chromosomal genes, read-through transcription is normally sufficient to drive marker gene expression. If the constructs are to be integrated in reverse orientation to chromosomal transcription, or integrated into silent regions (for example, silent variant surface glycoprotein [VSG] gene arrays or minichromosomes), a promoter must be included in the construct. Currently only *T. brucei* RNA Polymerase I promoters have been used in this way, although bacterial promoters (such as from bacteriophage T7) can also function if the *T. brucei* strain has been engineered to express the cognate RNA polymerase ***(36)***. Recently, the first example of an RNA Polymerase II promoter has been described in *T. brucei*, so perhaps this could now be employed also ***(47,48)***.

In plasmid constructs, the sequences used for homologous targeting can be as large as the bacterial vector can maintain, or as there is available sequence from which to clone. (In theory, the imminent completion of the sequencing of

Table 1
Markers for Selection of Transformants in *T. brucei*[a]

Selectable marker	Mode of selection	Drug (source)	*T. brucei* ref.
Hygromycin phosphotransferase (HYG)	Positive	Hygromycin B (Roche)	***107***
Neomycin phosphotransferase (NEO)	Positive	G418 sulphate/geneticin (Sigma)	***18***
Bleomycin resistance gene (BLE)	Positive	Phleomycin D1/zeocin (Cayla/Invitrogen)	***92***
Puromycin acetyltransferase (PAC/PUR)	Positive	Puromycin HCl (Calbiochem)	***108***
Blasticidin deaminase (BSD/BSR)	Positive	Blasticidin S HCl (Calbiochem)	***109***
Streptothricin acetyltransferase (SAT)	Positive	Nourseothricin sulphate (clonNAT) (Werner Bioagents)	***110***
Thymidine kinase (TK)	Negative	5' Bromo-2'-deoxyuridine (Sigma)	***45***
Green fluorescent protein[b] (GFP)	FACS	—	—

[a]Commonly used abbreviations of the marker genes are given in parentheses. FACS, fluorescence-activated cell sorting.
[b]Autofluorescent proteins other than GFP could be used.

the *T. brucei* genome should remove this limitation; *see* **Note 4**.) No systematic work has been conducted in *T. brucei* to establish the relative efficiency of gene targeting over a large range of DNA substrate sizes. Nevertheless, recombination flanks as small as around 100 bp have been shown to function efficiently in generating transformants. Indeed, smaller targeting flanks appear to work also (*see* **Subheading 3.1.3.**). It remains to be seen whether larger recombination substrates would improve targeting efficiency, as seems to be the case in *Leishmania* ***(49)***, or whether the efficiency is related to the locus being targeted for integration. Decreasing amounts of DNA sequence identity between the transformed construct and chromosomal target clearly adversely affect transformation efficiency ***(50)***. Recently, this has been shown to be partly a consequence of the mismatch repair system of *T. brucei*, and the level of this constraint has been quantified (J. Bell and R. McCulloch, ms. submitted). A sequence divergence of 1% causes approx a threefold decrease in transformation, and this drops exponentially to around a 100-fold decrease at 11% divergence. This may have important implications when, for instance, the recombination flanks used are generated from a *T. brucei* strain that is not isogenic to the one being transformed, or when imperfectly conserved repeat sequences are used as recombination flanks.

3.1.3. PCR-Based Integration

As outlined in **Fig. 3**, PCR can also be employed to generate linear DNA molecules for transformation, without the need to construct a separate plasmid for each transformation experiment. This has been used both to create epitope tagged gene variants ***(51)*** and to create gene disruptions ***(52)***. In this approach an antibiotic resistance cassette, or simply the antibiotic gene's ORF, is PCR-amplified with oligonucleotide primers that contain 5' sequences derived from regions that flank the chromosomal sequence to be targeted. After amplification, the 5' oligonucleotide sequences create upstream and downstream flanks on the PCR product that act as substrates for ends-out recombination. The length of such recombination flanks is dictated by the amount of sequence incorporated into the primers, so this approach could, in theory at least, provide ends-out recombination substrates of similar size to those in plasmid constructs. In practice, PCR products with 90-, 75-, 50-, and 42-bp recombination flanks have been used to transform procyclic-stage *T. brucei*, and all appeared to act at similar efficiencies ***(51,52)*** that were comparable to those obtained with longer plasmid constructs. In the same experiments, reducing the recombination flanks to 30 bp reduced the transformation efficiency by around five-fold ***(51)***. In monomorphic bloodstream-stage cells, a minimum of 24 bp of recombination flank was shown to be capable of acting during ends-out inte-

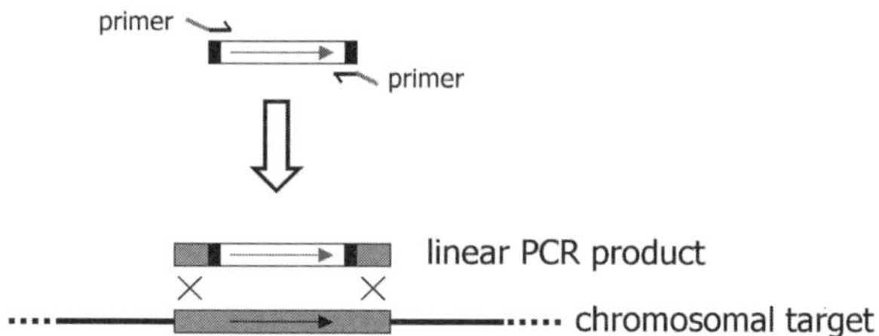

Fig. 3. Polymerase chain reaction (PCR)-based integration. Half-arrows represent primers used to amplify a PCR product that, upon transformation, can be integrated by ends-out recombination (the integrated DNA is not shown but would be equivalent to **Fig. 2**). In each primer, the 3' sequence (black lines) base pairs with the processing flanks of a selectable marker cassette (shown as in **Figs. 1** and **2**), whereas non-complementary 5' ends are derived from the genomic loci to be targeted (shaded gray); after PCR amplification, the 5' primer ends form the homologous targeting flanks in the linear product.

gration, and this was also around fivefold less efficient than recombination flanks of greater than 100 bp ***(53)***.

Three points should be stressed about this PCR approach. First, a systematic analysis of the efficiency of this type of targeting to different loci and using different lengths or base composition of flank has not been conducted in *T. brucei*. In the yeasts *Saccharomyces cerevisiae* and *Schizosaccharomyces pombe*, a direct relationship between targeting flank length and integration efficiency has been seen for at least silent, or lowly transcribed, regions ***(37,54)***. Second, although the integrations first described occurred by homologous recombination into the expected loci, there is evidence that short substrates can lead to DNA integration into aberrant locations ***(53)***, and we have found that variable numbers of attempted gene disruptions by this approach integrate into incorrect chromosomal loci (C. Proudfoot and R. McCulloch, unpublished data). Third, this approach has not yet been applied to transformation of pleomorphic bloodstream cells.

3.1.4. Preparation of DNA for Transformation

The initial work on *T. brucei* transformation demonstrated that supercoiled plasmids are poor substrates for integration following electroporation, in contrast to linear plasmid DNA ***(18,55,56)***. This is in keeping with the expectation that the DNA ends act as DSBs that initiate homologous recombination. For either ends-out or ends-in plasmid constructs, the simplest means of creating linear plasmid is by restriction digestion. Others means, for example, utilizing site-specific recombination enzymes and their cognate sites, have not been attempted until now. For ends-in constructs, a unique restriction site within the targeting sequence is preferable. For ends-out recombination constructs, restriction sites around each flank are generally used to generate terminal homology flanks at both ends of the linear molecule. It should be noted, however, that linearization at only one end (creating one terminal homology flank and one internal flank, and leaving one end of the construct as nonhomologous plasmid sequence) is sufficient to allow stable integration ***(19)***. For ends-out PCR products, no such restriction digestion is needed. All reports of PCR-based transformation have used *Taq* DNA polymerase for amplification, and it is not clear whether using alternative thermostable DNA polymerases with higher replication fidelities would improve transformation efficiency.

To prepare either plasmids or PCR products for transformation, all reported methods involve removing restriction enzymes or DNA polymerase by, for example, phenol/chloroform extraction, concentrating the DNA by ethanol precipitation, and finally resuspending it in sterile distilled water following 70% ethanol washing. We routinely use plasmid DNA prepared from *Escherichia coli* by alkaline denaturation, renaturation, and anion exchange chromatography (such as in kits purchased from Qiagen or other manufacturers). Plasmid DNA purified by CsCl gradients appears to be comparable in transformation efficiency. No reports have been made that the bacterial strain used to propagate plasmid constructs has any influence on transformation efficiency. Note that the liberated bacterial plasmid backbone in linearized ends-out plasmids need not be removed from the DNA to be transformed (for instance, by electrophoretic separation), as there is no evidence that this is capable of stable integration at significant levels.

The amount of DNA needed for *T. brucei* transformation has not been assessed systematically. Early reports used around 5 μg of plasmid DNA ***(19)***, and higher amounts gave rise to multiple integration events, at least when targeting a tandem array locus ***(55)***. Using less than 5 μg is clearly sufficient to generate transformants, however, as one report used only 1 μg of PCR product ***(53)***. It may be that transformation efficiency is not simply a consequence of the amount of DNA used but instead results from a combination of flank length and sequence, chromosome location and expression level of the targeting loci, and the condition of the cells being used for transformation.

3.2. Growth of T. brucei *for Transformation*

3.2.1. A Comparison of Monomorphic and Pleomorphic T. brucei

Most genetic analyses of bloodstream-stage *T. brucei* have been performed with monomorphic strains of this life cycle stage, most notably with the stock

Lister 427 *(57)* and its many genetically manipulated derivatives that have been created in different laboratories. These are well adapted to laboratory conditions, including growth in rodents and axenic culture, and are convenient for transformation (*see* second paragraph following), but they suffer from the drawback that they probably exist only as the replicative long slender form and are unable to differentiate to the nondividing short stumpy form that is the natural precursor to tsetse fly transmission. In contrast, pleomorphic bloodstream-stage trypanosomes, which are not so adapted to the laboratory, are fully competent to perform this differentiation and thus complete their life cycle. In addition, the rate of antigenic variation, a crucial immune evasion process in African trypanosomes that has functional similarities with processes in many pathogenic micro-organisms, is much higher in pleomorphic strains than in monomorphic strains *(7)*. Whether or not alterations to the phenotypes of antigenic variation and differentiation have a common cause is unclear. Depending on the biological process being studied, one or other of the two cell line types might be more suitable.

Despite their inability to differentiate to the stumpy form, monomorphic bloodstream-stage parasites have proved useful for studying the differentiation steps from the bloodstream to the procyclic stage ***(58–60)***. However, in determining the molecular mechanisms controlling bloodstream form differentiation and development of the parasite in the tsetse fly, pleomorphic lines are preferable ***(61–67)***. The high rate of antigenic variation in pleomorphic cells relative to monomorphic cells means that it is important to consider what might drive the process in the former ***(68)***. On the other hand, the high rate of variation complicates analysis of clonal antigen switching events, and monomorphic cells have been invaluable in detailing the reactions used ***(7,69–72)***. If we consider recombination as an example of a fundamental process of all organisms, then monomorphic cells are likely to be valuable in detailing the underlying processes ***(53)***, but we must consider that there may be alterations to any activities in pleomorphic cells, since recombination is an important driver of antigenic variation.

Pleomorphic bloodstream-stage cells can be harvested from the blood of infected mice or rats but until recently have proved difficult to culture axenically. However, the demonstration that high-molecular-mass agarose ***(73)*** or methylcelluose ***(74)*** supports growth of pleomorphic trypanosomes made these parasites amenable to in vitro culture. In the presence of either matrix, pleomorphic cells are able to proliferate with a population doubling time similar to that of trypanosomes during mouse infection, but in its absence they undergo aberrant cell division and growth arrest ***(73)***. Cells cultured for 3 mo in vitro were still able to differentiate normally during mouse infection, suggesting that they had retained their pleomorphic phenotype. Trypanosomes undergo efficient

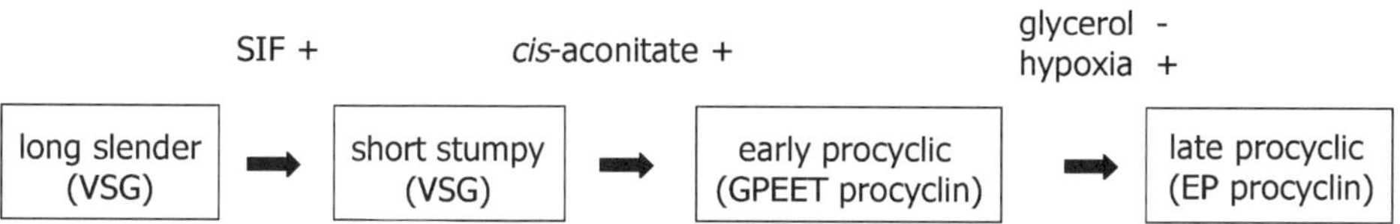

Fig. 4. Three successive differentiations of *T. brucei* life-cycle stages induced in vitro by SIF (stumpy induction factor), *cis*-aconitate, or glycerol withdrawal/hypoxia. Long slender and short stumpy forms of the bloodstream stage express a coat of variant surface glycoprotein (VSG); early procyclic cells have a coat composed mainly of GPEET procyclin; and late procyclic cells have a coat mainly of EP procyclin.

differentiation to the short stumpy form at high cell densities in the culture systems, and this is induced by a trypanosome-released factor (stumpy induction factor [SIF]), which accumulates in the "conditioned" medium ***(61,62)***. The molecular identity of SIF remains mysterious, but the same work demonstrated that the inability of monomorphic Lister 427 bloodstream-stage cells to undergo differentiation is a result of their inability to respond to SIF, even though they still release it.

An established method for inducing synchronous differentiation of bloodstream-stage short stumpy cells to tsetse procyclic-stage cells has been described: a drop in temperature to 27°C and the addition of 6 m*M cis*-aconitate to the medium ***(75,76)***. Note that the same method can be used to trigger differentiation of monomorphic bloodstream-stage cells ***(77,78)***, but the absence of the short stumpy form leads to altered kinetics of differentiation ***(58)***. Recently, we have also established conditions for differentiation of "early" procyclic-stage cells to "late" procyclic-stage, which can be modulated by glycerol or hypoxia and relates to changes in surface antigen that occur in the tsetse fly ***(65,79)***. Thus, it is now possible to trigger differentiation of three successive developmental stages in vitro (**Fig. 4**). A whole palette of differentiation markers is also available to monitor this developmental program ***(58,62,80–82)***.

3.2.2. Growth of Monomorphic Bloodstream-Stage T. brucei

Monomorphic bloodstream stage *T. brucei*, such as stock Lister 427, grow efficiently as the replicative long slender form in liquid HMI-9 medium ***(83)*** (*see* **Note 1**) or semisolid agarose plates derived from this ***(84)*** (*see* **Subheading 3.2.4.**). The parasites can be readily cloned from a single cell with efficiencies of up to 100% ***(85,86)***, although this is likely to be clone- or strain-dependent. They require a humidified, 37°C growth environment with 5% CO_2 and can reach maximal cell densities in liquid media in the range 2–5 × 10^6 cells/mL. The parasites are nonadherent and can grow in most types of standard tissue culture container, including flasks, multiwell plates, and Petri dishes. A total of 5×10^7 cells are used for a single transformation experiments in most published reports, although transformation is possible with as few as 1×10^7 cells in our laboratories. We find that cells at maximal culture density yield lower numbers of transformants than cells that are one or two population doublings from maximum (unpublished observations). Similarly, we find that the cloning efficiency of cells at maximum density is lower than that of cells in midlogarithmic growth. In contrast to other systems, *T. brucei* cells at maximal density are probably not in stationary phase but rather are damaged owing to starvation as the culture medium is exhausted and/or toxic catabolites accumulate. To obtain the required number of cells for transformation, in vitro cultured cells are often sufficient. However, for some experiments (for example, an approach requiring multiple duplicate individual transformations), this may not be practical. The alternative, therefore, is to grow the bloodstream-stage cells in a suitable animal host, normally either a mouse or rat, in which they reach much greater cell densities (maximally around 10^9 cells/mL of blood). In general, we find no difference in transformation efficiency using monomorphic bloodstream-stage *T. brucei* derived from the two sources, although some specific clones or strains may differ.

3.2.3. Growth of Bloodstream-Stage T. brucei *in 1.1% Methylcellulose Medium*

This method is very suitable for in vitro culture of a relatively large number of bloodstream-stage cells, as well as for antibiotic selection of stably transformed cells (*see* **Subheading 3.3.3.**).

Preparation of 500 mL of methylcellulose HMI-9:

1. Prepare 1.6× concentrated HMI-9 medium by supplementing 365 mL of 2× Iscove's minimum essential medium (MEM) (34.06 g/L; can be purchased as a powder from Invitrogen) with 10 mL of each of the following stock solutions: 100 m*M* hypoxanthine in 0.1 *N* NaOH, 5 m*M* bathocuproine disulphonic acid, 20 m*M* 2-mercaptoethanol, 16 m*M* thymidine, 100 m*M* pyruvate, 100 m*M* cysteine, and 150 mL heat-inactivated fetal bovine serum.
2. To 183 mL water, slowly add 5.5 g methylcellulose or Methocel powder (Sigma) with vigorous stirring using a magnetic stirring bar. Next, sterilize by autoclaving, and continue the stirring in a cold room until the viscous solution becomes transparent.
3. To 317 mL 1.6× HMI-9, add 183 mL of the methylcellulose solution and mix well. The resulting methylcellulose HMI-9 medium is very viscous and difficult to pipet.

Cultures of pleomorphic bloodstream-stage *T. brucei* can be established in vitro by inoculating methylcellulose HMI-9 in culture flasks with long slender form cells from a blood stabilate or directly harvested from rodent (or another suitable host) blood. They require the same growth conditions as monomorphic bloodstream-stage cells (*see* **Subheading 3.2.2.**), and the AnTat pleomorphic lines grow logarithmically at low parasite density (population doubling time approx 6 h), but above 1×10^6 cells/mL they undergo differentiation to the nondividing stumpy form. Note that the in vitro population doubling time and the cell density at which the stumpy form is induced is most likely strain- or clone-dependent (*see* Table 2 in ref. ***73***). To recover bloodstream-stage cells from viscous methylcellulose HMI-9 medium, dilute the cell culture with 5–10 vol of phosphate-buffered saline (PBS) containing 10 m*M* glucose, and collect the cells by centrifugation for 10 min at 600–1400*g*.

3.2.4. Growth of Bloodstream-Stage T. brucei *on Agarose Plates*

Bloodstream-stage cells can also be cultured on semisolid agarose HMI-9 plates ***(84)***, on which they form colonies. This method is suitable for isolating clones, although limiting dilution in multiwell plates can also be used. Pleomorphic bloodstream-stage trypanosomes are also able to undergo differentiation to the stumpy form on the surface of semisolid plates, on which colonies consist of around 500 cells ***(62)***. Differentiation on plates occurs at cell densities that are significantly lower than in liquid methylcellulose HMI-9 medium, presumably owing to local accumulation of SIF within the colony. At maximum cell density, greater than 99% of the cells in a culture of AnTat *T. brucei* on agarose plates consist of the stumpy form ***(61)***. In contrast, cell populations of the same strain harvested from mouse infections at maximum cell densities usually contain 10–30% of the long slender form. Thus, the plate method is a powerful tool to highly enrich for stumpy form cells. Up to 2×10^7 cells can be harvested from a single 20-mL plate. These in vitro-derived stumpy form cells are able to differentiate synchronously to the procyclic form and are infective for the tsetse fly ***(61,62,65)***.

Preparation of semisolid HMI-9 agarose plates (adapted from Carruthers and Cross ***[84]***):

1. Add 1.3 g of low melting point agarose (Seaplaque agarose, FMC) to 73 mL H_2O in a 200-mL bottle. Autoclave and equilibrate at 37°C.
2. Transfer 127 mL of 1.6× HMI-9 medium (*see* **Subheading 3.2.3.**), equilibrated at 37°C, to the bottle containing the molten LMP agarose and mix well.
3. Immediately pour plates by dispensing 20 mL of the agarose HMI-9 mix into 100-mm-diameter Petri dishes. Allow 1 h at room temperature for the agarose to solidify, and store inverted at room temperature overnight.
4. Remove condensation from the Petri dish lids, and allow the semisolid agarose plates to dry in a sterile laminar flow hood with the lids off. The length of the drying period is critical for optimal *T. brucei* growth and should be optimized in

each laboratory and for the type of laminar flow hood used; it normally takes 15–30 min.

To grow pleomorphic bloodstream-stage parasites, dilute long slender form cells (either from a midlogarithmic growth phase culture, from a blood stabilate, or directly harvested from mouse blood) to 1×10^5–1×10^6 cells/mL in HMI-9, and gently spread 100 μL of the cell suspension per plate using a bent Pasteur pipet. Incubate at 37°C with 5% CO_2. Cells can be harvested from the agarose plates by washing twice with 2 mL HMI-9 and collected by centrifugation for 10 min at 600–1400*g*. On agarose plates, at these cell densities, the parasites proliferate as the long slender form and after 3–4 d undergo efficient differentiation to the stumpy form. The kinetics of pleomorphic bloodstream-stage cell differentiation can be accelerated by supplementing fresh medium with 50% sterile-filtered conditioned medium harvested from a monomorphic or pleomorphic bloodstream-stage culture at maximal parasite density ***(62)***. Alternatively, differentiation can be induced by the addition of 250 μ*M* 8-(4-chlorophenylthio)-cyclic AMP to the culture medium. Under these conditions, cells undergo complete differentiation within 24–36 h ***(62)***.

3.2.5. Differentiation of Bloodstream-Stage T. brucei *to the Procyclic Stage*

To initiate differentiation of bloodstream-stage trypanosomes to the procyclic stage, the cells should be harvested from an in vitro culture at high cell density (by centrifugation at 600–1400*g* for 10 min), or from an infected host at high parasitemia, and resuspended in either DTM *(77)* or SDM-79 media *(87)* at a density of 2×10^6 cells/mL. (Cunningham's medium *[88]* has also been used in some reports; see, e.g., refs. ***89*** and **Note 2**.) Differentiation is induced by addition of *cis*-aconitate at 6 m*M* and by lowering the incubation temperature to 27°C ***(75,76)***. During the course of differentiation, the parasite's VSG coat is released and replaced by a different coat composed of EP and GPEET procyclins that are characterized by internal tandem dipeptide repeats and internal pentapeptide repeats, respectively ***(65,82)***. Short stumpy form cells are able to differentiate synchronously to the procyclic stage and show maximal levels of EP procyclin expression within 6 h of exposure to the differentiation signal. In contrast, long slender form pleomorphic cells or monomorphic bloodstream-stages differentiate asynchronously to the procyclic form, in which case differentiation is completed within 36–48 h.

3.2.6. Growth of Procyclic-Stage T. brucei

Procyclic-stage *T. brucei* can be cultured in DTM (containing 10 m*M* glycerol) or SDM-79 media in the presence or absence of 10 m*M* glycerol ***(65)***.

In the presence of glycerol, AnTat1.1 parasites grow with a constant population doubling time (around 19 h) and maintain GPEET procyclin expression. In the absence of glycerol, however, they undergo transient cell cycle arrest and differentiation to a late procyclic stage that is GPEET-negative. GPEET expression is repressed completely approx 9 d after exposure of the cells to the differentiation signals (*see* **Subheading 3.2.5.**). *T. brucei* that are cultured under hypoxic conditions (<1% oxygen) in the presence of glycerol are also GPEET-negative, suggesting that hypoxia can override the effect of glycerol ***(65)***. Procyclic-stage cells that are cultured for a few months in the presence of glycerol are still transmissible through the tsetse fly, but they gradually lose this transmissibility upon longer term cultivation. Highly established (GPEET-negative) procyclic-stage cells remain useful for many forms of analysis, however.

Recently established procyclic-stage *T. brucei* proliferate logarithmically to reach maximal cell densities of around $2–3 \times 10^7$ cells/mL but when diluted below approx 1×10^6 cells/mL are normally unable to grow. The exact density at which this loss of growth potential occurs is likely to be strain- or clone-dependent but appears to be a general phenomenon, since even highly established procyclic-stage Lister 427 cells cannot be diluted below approx $1–10 \times 10^4$ cells/mL. This distinction from bloodstream-stage cells in their growth characteristics has important consequences for selection of transformants following electroporation, and also for cloning. One way to circumvent this is, when growing or cloning transformants on antibiotic selection, to supplement the culture with sufficient nontransformed procyclic-stage *T. brucei* to give the required cell densities for efficient outgrowth. During the time that the nontransformed cells are killed by drug selection, the transformants should reach sufficient densities to allow outgrowth. Alternatively, addition of 10–30% "conditioned" medium (harvested from a dense procyclic-stage culture and filter-sterilized) to the procyclic-stage culture allows them to divide at low parasite density. Starting a new culture from a stabilate, or completely replacing the medium of a dense culture with fresh medium, often requires the addition of conditioned medium to the culture medium (*see* **Subheading 5.4.**). Sterile-filtered conditioned medium can be stored for several weeks at 4°C without apparent loss of activity.

Most reports use $2.5–5.0 \times 10^7$ procyclic-stage *T. brucei* in a single transformation experiment. As for bloodstream-stage cells, we find that the transformation efficiency is greater when the cells are harvested for electroporation one or two population doublings from maximum density. In general, however, procyclic-stage transformation is significantly more efficient (normally in the range $1 \times 10^{-3}–10^{-6}$ transformants/cells on selection) than that of the bloodstream stage (normally in the range $1 \times 10^{-5}–10^{-7}$), so this limitation might not be so important, at least for some experiments.

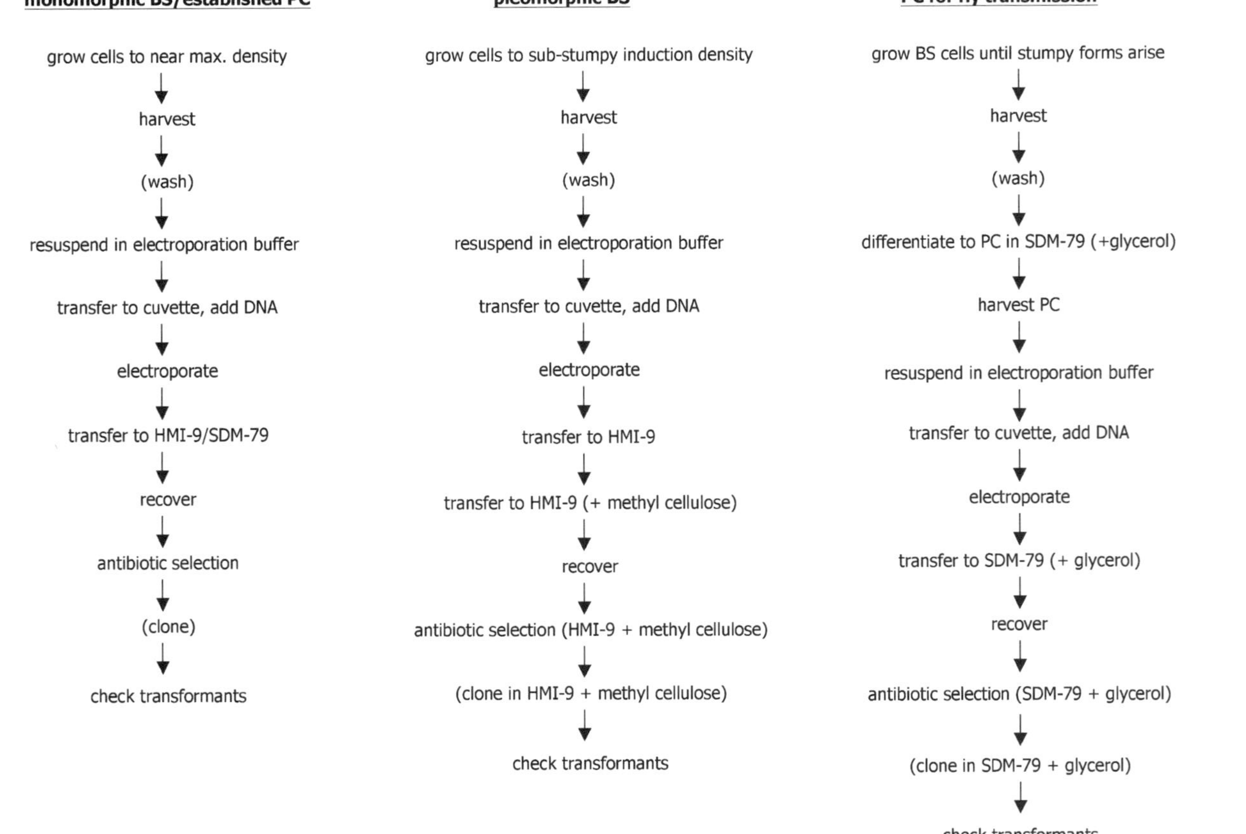

Fig. 5. A road map of the steps involved in *T. brucei* transformation. Bloodstream-stage cells are represented by BS and procyclic-stage by PC. HMI-9 and SDM-79 are culture media discussed in the text. All variables are described in the text.

3.3. Stable Transformation of T. brucei

A road-map of the steps involved in transformation of *T. brucei* is shown in **Fig. 5**.

3.3.1. Transformation of Monomorphic Bloodstream-Stage T. brucei

1. For a single transformation, harvest 5.0×10^7 cells from an HMI-9 culture by centrifugation at 600–1400*g* at 37°C or room temperature for 10 min. Alternatively, the parasites may be recovered from an animal host, as described below for pleomorphic cells (*see* **Subheading 3.3.2.**), although there is normally no need to immunosuppress the host prior to growth of monomorphic cells, and they can be recovered at higher cell densities: up to 1×10^9 cells/mL of blood.
2. Discard supernatant from cell pellet, and resuspend at 1×10^8 cells/mL in 0.5 mL of ZMG (*see* **Note 3**) buffer equilibrated to 37°C or room temperature. As an alternative to ZMG, Cytomix buffer ***(90)*** may be used. Cells may be resuspended in a larger volume of prewarmed ZMG or Cytomix as a wash step to remove serum nucleases prior to final resuspension in 0.5 mL, but we find this usually does not alter transformation efficiency, at least for cells taken from HMI-9 medium.
3. Transfer to an electroporation cuvet (0.1–0.4-cm gap length) and add linearized plasmid DNA (normally 1–50 μg) in 10–40 μL of either sterile, distilled dH_2O or 1 m*M* Tris-HCl, pH 8.0, 0.1 m*M* EDTA.
4. Subject cells to electroporation (one discharge). For a Bio-Rad gene pulser II, use a 1.4–1.5-kV, 25-μF setting; for BTX ECM 630, use 1.5 kV, 25 μF, 175 Ω.
5. Transfer cells to a culture flask containing 10–50 mL HMI-9 medium and incubate for 8–24 h at 37°C, 5% CO_2 to recover from electroporation.
6. Bloodstream-stage transformants can be clonally selected immediately. To ensure clonality, transformants may be selected on semisolid agarose plates (*see* **Subheading 4.4.**) containing the appropriate antibiotic(s): harvest $1–2 \times 10^7$ cells by centrifugation at 600–1400*g* at room temperature for 10 min, resuspend in 0.5 mL of HMI-9, and spread on a single plate. For more reproducible estimates of transformation frequency, the surviving cells should be counted, harvested by centrifugation, and resuspended in 18 (or 36) mL HMI-9 containing antibiotic(s) and then spread in 1.5-mL aliquots over 12 (or 24) wells of a 24-well culture dish (*see* **Note 5**).
7. Antibiotic-resistant transformants should arise as visible colonies on plates, or as dividing populations in dishes, in 5–10 d. If colonies are well separated on dishes, they can be considered clonal; clonality can be ascertained for liquid selection by limiting dilution in multiwell dishes.

3.3.2. Isolation of Pleomorphic Bloodstream-Stage T. brucei *for Transformation*

Pleomorphic bloodstream-stage cells harvested from the blood of a rat or mouse have a more reliable growth potential than cultured parasites and may therefore be more suitable for stable transformation. (If you are using cells grown in vitro in HMI-9 methylcellulose, proceed from **step 5**.)

1. To immunosuppress the host animal, irradiate either rats or mice with 600 rad (total body) from a therapeutic X- or γ-ray source, or inject ip with 250 mg cyclophosphamide per kg body weight.
2. Next day, inject ip with $1–2 \times 10^6$ (mice) or up to 10^7 (rats) parasites.
3. Monitor parasitemia from the tail blood. When it reaches $2 \times 10^7–8 \times 10^7$ cells/mL of blood, harvest the blood from the animal and distribute it between two 15-mL Falcon tubes prelayered with 5 mL CGA (40 m*M* glucose and 100 m*M* sodium citrate, pH 7.7) as anticoagulant.
4. Centrifuge the blood/parasite mix for 10 min at 600–1400*g*, 37°C or room temperature. Parasites and white blood cells form an interface between the red blood cells and the plasma. Take off the plasma, leaving 1–2 mL, and carefully resuspend the cells from the interface in the residual volume of the supernatant by swirling with a Pasteur pipet. Transfer cells to a new Falcon tube containing 40 mL ZMG or Cytomix.
5. Collect cells by centrifugation (600–1400*g* for 10 min at 37°C or room temperature) and resuspend the pellet in ZMG or Cytomix at a density of 4.4×10^7 cells/mL.

3.3.3. Electroporation, Selection, and Cloning of Pleomorphic Bloodstream-Stage T. brucei

1. Mix 0.45 mL of the cell suspension in ZMG or Cytomix (2×10^7 cells) with linearized plasmid DNA (normally 1–50 μg) in 10–40 μL of either sterile, distilled H_2O or 1 m*M* Tris-HCl, pH 8.0, 0.1 m*M* EDTA.
2. Subject cells to electroporation (*see* **Subheading 3.3.1.**) and transfer to a 15-mL Falcon tube containing 10 mL HMI-9. Only approx 5% of the cells will survive electroporation.
2. Incubate cells for 30–60 min at 37°C, 5% CO_2 to allow recovery. Collect cells by centrifugation (600–1400*g*, 37°C or room temperature, 10 min), take off 8 mL of the supernatant, and resuspend the cells in the residual medium in the tube.
3. Transfer cells to a 175-cm² culture flask containing 150 mL HMI-9 methylcellulose medium and incubate for approx 12–18 h at 37°C, 5% CO_2.
4. Dissolve antibiotic(s) in 5 mL HMI-9, transfer to the culture flask, and continue incubation at 37°C, 5% CO_2. The cell density should be monitored so as not to exceed $4–5 \times 10^5$ cells/mL during the incubation period.
5. Antibiotic-resistant transformants should appear as a dividing population after 5–8 d. Harvest the parasites from 5–10 mL methylcellulose medium and inject them into immunosuppressed mice for cell amplification.
6. For cloning, spread 200 long slender cells from the tail blood on semisolid HMI-9 agarose plates (*see* **Subheading 3.2.4.**). After 2 d, pick single colonies and inject into immunosuppressed mice for amplification. Alternatively, perform limiting dilution in liquid methylcellulose HMI-9.

Bloodstream-stage pleomorphic *T. brucei* that have been subjected to three consecutive rounds of transformation, selection, and cloning are still able to grow and differentiate normally in mice; they can be transmitted cyclically through tsetse flies with nearly the same efficiency as the parental line ***(74)*** (P. Blundell, B. Fast, J.D. Barry, and M. Boshart, ms. in preparation). However, one note of caution should be stated. Some pleomorphic lines may be incapable of growing clonally in vitro as they differentiate to the stumpy form at low density (P. Blundell and J.D. Barry, unpublished observations). In such cases, selection for transformants can be undertaken in vivo, in immunosuppressed mice. Two antibiotics have been approved for use in vivo. Bleomycin can be administered at 7 mg/kg bodyweight, a dose found to have no side effects for the 60 d the animals were observed ***(91)***; phleomycin was successfully used at 3 mg/kg ***(92)***. Similarly, G418 has been used at 40 mg/kg bodyweight without significant deleterious side effects ***(93)***.

3.3.4. *Transformation of Procyclic-Stage* T. brucei

1. For a single transformation, harvest $2.5–5.0 \times 10^7$ cells from an SDM-7 or DTM culture by centrifugation at 600–1400*g* at room temperature for 10 min.
2. Discard supernatant from cell pellet, and resuspend at 1×10^8 cells/mL in 0.5 mL of ZM (*see* **Note 3**) or Cytomix buffer equilibrated to room temperature. Cells may be resuspended in a larger volume of ZM or Cytomix as a wash step (*see* **Subheading 3.3.1.**).
3. Transfer to an electroporation cuvet (0.1–0.4-cm gap length) and add linearized plasmid DNA (normally 1–50 µg) in 10–40 µL of either sterile, distilled dH_2O or 1 m*M* Tris-HCl, pH 8.0, 0.1 m*M* EDTA.
4. Subject cells to electroporation with two discharge pulses, using the settings described in **Subheading 3.3.1.**
5. Transfer cells to a culture flask containing 10 mL SDM-79/DTM medium, and incubate for 8–24 h at 27°C to recover from electroporation.
6. Select for transformants by harvesting a range of cell numbers (e.g., 0.1, 0.5, 1, 5, and 10×10^6 cells) by centrifugation at 600–1400*g* at room temperature for 10 min, and resuspend in 10 mL of SDM-79/DTM media containing antibiotic(s). Below final cell densities of 1×10^6 cells/mL, the cultures should either be supplemented with sufficient untransformed cells to give a density of 1×10^6 cells/mL, or 10–30% conditioned medium (*see* **Subheading 3.2.6.**) should be added. If the transformation efficiency of the electroporated DNA is known, then the transformed cells may be diluted in 20 mL of conditioned SDM-79/DTM (supplemented with 2×10^7 untransformed cells if not using conditioned medium) to an appropriate cell density to get 0.5–1 transformant/200 µL and spread over a 96-well plate in 200-µL aliquots. Incubate at 27°C (*see* **Note 5**).
7. Antibiotic-resistant transformants should arise as growing populations by the polyclonal tissue flask method or as clonal populations by the 96-well method after 7–14 d. From the polyclonal population, follow the cloning procedure

described in **Subheading 3.3.4.**, **step 6**, but dilute the transformants to 0.5–1 cells/200 µL of medium.

3.3.5. Transformation of Fly-Transmissible Procyclic-Stage T. brucei

Procyclic-stage cells lose their ability to be transmitted cyclically through tsetse flies upon long-term cultivation. To obtain fly-transmissible transformed clones, we electroporate freshly differentiated trypanosomes and perform subsequent dilutions of the cultures for antibiotic selection of clonal cell lines on 24-well plates. Parasites subjected to two consecutive rounds of stable transformation and selection/cloning are still fly-transmissible (E. Vassella, unpublished results). Thus, this method has the potential to generate fly-transmissible null mutants without intermittent fly transmission between the two transformation steps, the procedure used thus far.

1. Trigger bloodstream-stage cells (from a blood stabilate or directly harvested from mouse blood) to differentiate to procyclic-stage (*see* **Subheading 3.2.5.**), and incubate the cells for several days in SDM-79 in the presence of 10 m*M* glycerol at 27°C until approx 20 mL of a dense culture of 6–8 × 10^6 cells/mL is obtained.
2. Collect cells by centrifugation (600–1400*g*, 10 min, room temperature), transfer the culture supernatant to a fresh Falcon tube, and wash the cell pellet with 10 mL ZM buffer (or Cytomix).
3. Collect cells by centrifugation (as above, **step 2**), and resuspend in ZM or Cytomix buffer at a density of 4.4 × 10^7 cells/mL. Subject cells to electroporation as described in **Subheading 3.3.4.** and transfer to a culture flask containing 10 mL SDM-79 supplemented with 10 m*M* glycerol and 30% conditioned SDM-79 medium (*see* **Subheading 3.2.6.**).
4. Perform subsequent dilutions of 1:10 and 1:100 in SDM-79 supplemented with 10 m*M* glycerol and 30% culture supernatant to obtain a final volume of 24 mL for each dilution. Add untransformed cells to the culture flasks at a density of 5 × 10^5 cells/mL.
5. Transfer 1-mL aliquots to each well of 24-well plates. After 18 h of incubation to allow recovery, add 0.5 mL SDM-79 medium (supplemented with 10 m*M* glycerol and 30% conditioned medium) containing the antibiotic to each well. Incubate at 27°C in a humid chamber. Drug-resistant transformed cells are obtained after 10–12 d.

3.3.6. DNA Transformation as an Assay for T. brucei Recombination

Homologous recombination is a fundamental biological process in organisms from all kingdoms of life. It is critical for the repair of DNA double-strand breaks, generation of genetic diversity, genome stability, and the restart of collapsed replication forks (*see* refs. ***94–96*** for reviews). Even so, we still know relatively little about how homologous recombination acts in African trypanosomes, or other early diverging eukaryotes; it is often not appreciated that the evolutionary distance between *T. brucei* (or other members of the order Kinetoplastida) and members of the "crown" metazoan lineages that contain plants, animals, and fungi is somewhat greater than the distance between species within this crown ***(97)***. Judging from earlier examples of unusual organization and function of its genome, it is possible that *T. brucei* may have novelties in its recombination pathways. One emerging example is that homologous recombination appears to have been co-opted, and perhaps adapted, for a purpose that is critical to survival of this extracellular parasite: periodic changes in the composition of its VSG coat to evade the mammalian immune system. The evidence that this occurs by general homologous recombination, rather than by a more specialized recombination pathway, now appears convincing ***(7,26,71,85)***.

As a main route to understanding how homologous recombination operates in *T. brucei*, determining the rate and pathway of exogenous DNA integration into the parasite's genome following transformation should be a simple and robust assay. So far, work with the monomorphic Lister 427 strain, primarily in the bloodstream stage, suggests that this is so ***(26,53,85,86,98)***. By carefully matching the growth stage at which cells are used for electroporation, the number of cells transformed and put on selection, and the amount of DNA used (all potential sources of variation), transformation efficiency has been used as an indirect measure of recombination efficiency. This has allowed the function of putative recombination genes to be examined and pathways of recombination to be detailed (*see* next paragraph). This is a simple but versatile assay that can be used to compare recombination in different strains or mutants, or to examine the influence of other factors such as transcription levels and genomic location. Thus far, only ends-out recombination events have been examined, and it remains to be seen how ends-in recombination compares. In addition, it remains to be demonstrated that the assay is as robust a measure of recombination in procyclic-stage cells or in the less tractable pleomorphic bloodstream stages.

Until recently, all reported integrations of transformed DNA in *T. brucei* have been mediated by sequence homology ***(15,17)***, suggesting that, as in yeast ***(38)***, nonhomologous recombination is a minor component of DSB repair in this organism. For any kind of homologous recombination event to occur, homologous sequences from two separate DNA molecules must pair to form a heteroduplex intermediate, a process termed *strand invasion* ***(37)***. The integration of exogenous DNA is no exception. Strand invasion of the end of a linearized plasmid or PCR product into a homologous region of the genome can initiate DNA synthesis or allow the formation of Holliday junctions, leading to the integration of DNA (*see* **Figs. 1–3**). Proteins involved in the process of strand invasion therefore should severely affect the efficiency of integration if they are mutated. RAD51, aided by cofactors, forms a nucleoprotein filament upon DNA and catalyzes the invasion of this DNA into a homologous duplex. *T. brucei* null mutants of *RAD51* display a 10-fold reduction in transformation efficiency ***(26,53)***, consistent with homologous recombination mediating DNA integration. The fact that integrations still occur, however, indicates that other pathways of recombination are available.

A glimpse of what such pathway(s) may be is revealed by the fact that *rad51* mutants occasionally integrate DNA into chromosomal regions not dictated by the homologous flanks of transformed DNA. These integrations appear to be mediated by short (7–13-bp) stretches of homology, which often contain mismatches. The frequency at which these aberrant integrations occur appears to be somewhat locus-specific and influenced by the number of homologous targets available, perhaps indicating differences between RAD51-dependent and RAD51-independent recombination to access some regions. Preliminary examination of the substrate requirements of the *T. brucei* RAD51-dependent and -independent pathways suggests that the former favors long substrates and the latter short substrates, which is reassuringly consistent with findings in *Escherichia coli* ***(99)*** and *Saccharomyces cerevisiae* ***(100)***.

Although these broad findings suggest a continuity of recombination in *T. brucei* compared with that of other organisms, a more thorough dissection of the substrate requirements of recombination pathways is required. For example, although mutation of *KU70* and *KU80* (major factors of nonhomologous recombination in other organisms ***[101,102]***) appeared to play no significant role in DNA repair in *T. brucei* ***(85)***, we cannot exclude the possibility that some of the integration reactions involving short substrates thus far described as being homology-based may in fact be random and driven by *KU70/80*. Nevertheless, it is tempting to speculate that the integration of PCR-amplified selectable markers (*see* **Subheading 3.1.3.**), utilizing short targeting sequences, is mediated by RAD51-independent recombination. If this is the case, the detailed mechanism of integration using short stretches of homology may sometimes be unusual. In fact, there may be some indirect evidence for this: aberrant integrations in *rad51* mutants often appear to be associated with chromosomal rearrangements. Indeed, in *S. cerevisiae*, RAD51-independent reactions using short substrates were associated with long stretches of DNA synthesis, a phenomenon known as *break-induced replication* ***(37,100)***.

Little is known not only about the substrate length requirements of integration and recombination in *T. brucei*, but also the effect of base mismatches. Transformation experiments in *T. brucei* demonstrated that integration occurs preferentially into perfectly homologous sequences when other, slightly less, homologous sequences are available within the genome ***(50)***. More recently, we have analyzed mutants of two genes involved in mismatch repair and confirmed that this repair system regulates the integration of DNA into the genome of *T. brucei* in a manner similar to the regulation of homologous recombination in other organisms (J. Bell and R. McCulloch, ms. submitted).

The only other gene product that has thus far been demonstrated to have a role in transformation in *T. brucei* is *MRE11* ***(86,98)***. Here, mutation causes a reduction in transformation efficiency that ranges, depending on the study, from 2- to 10-fold. Mutation of the gene also influences the DNA integration pattern, in a similar manner to *RAD51*, when a choice of target loci is present. Aberrant integrations have not been detected, however. The role of MRE11 in homologous recombination is not fully understood. It appears to be a multifaceted protein acting in a complex with two other proteins (RAD50 and XRS2 in yeast, or RAD50 and NBS1 in mammalians cells) and may play a role in the resection of DNA DSBs, as well as functioning in the DNA damage checkpoint response and maintaining genome integrity ***(103)***. A measure of the importance of *T. brucei MRE11* to genome integrity is seen in the huge attrition in chromosome sizes seen in mutants ***(86)***. Interestingly, in yeast, RAD51-independent recombination is dependent on RAD50 ***(100)***, and presumably MRE11 and XRS2, so MRE11 may play a role in this pathway in *T. brucei*.

3.4. Conclusions and Perspectives

Genetic transformation of *T. brucei* is the cornerstone of molecular genetic analysis in this parasite. Its importance is unlikely to diminish, since the imminent unveiling of the complete genome sequence of *Trypanosoma* species will undoubtedly reveal novel genes whose function must be analyzed ultimately by transformation techniques. Even in utilizing other genetic approaches, DNA transformation is essential. Double-stranded RNA interference ***(4)*** is currently only possible in *T. brucei* by stably transforming constructs that controllably express siRNAs as a result of other genetic components, themselves stably integrated into the genome. This complementary approach to conventional gene knockout analysis has yielded a spectacularly successful library to screen for novel gene functions ***(104)***, and a systematic, genome-wide siRNA-based phenotypic screen is under development (*see* http://www.trypanofan.org/cgi-bin/WebObjects/trypanofan).

Complementation analysis, so far only sporadically employed, is sure to be used more often as our wish to examine genome functions progresses. Even proteomic approaches, such as epitope-tagging of known genes ***(51)*** and examination of functionally interacting proteins by, for instance, tandem affinity purification ***(105,106)***, rely on stable transformation. Nevertheless, the absolute efficiency of transformation remains low, at least in the bloodstream stage, and remains a limiting factor for such desirable approaches as insertional mutagenesis or transposon tagging. Currently, it is unclear which step in the process of transformation (e.g., entry of DNA into the cell, routing to the nucleus, nucleolytic degradation, or genome integration) is limiting. A detailed analysis of the recombination machinery might provide us with mutant lines with enhanced intracellular stability of transformed DNA or enhanced recombinational integration and thus stimulate a further leap in transformation technology. We have suggested that DNA transformation is, in itself, a valuable assay to examine DNA recombination in *T. brucei*. Nevertheless, as we wish to design more complex assays in the future, perhaps to ask more global questions of recombination and antigenic variation, there is little doubt that DNA transformation will be required to establish them in the trypanosome.

4. Notes

1. The composition of HMI-9 medium is described in Hirumi and Hirumi ***(83)***. The base of this medium is Iscove's modified Dulbecco's medium, which is available commercially (e.g., from Invitrogen), as are all the necessary supplements. Indeed, many companies will be happy to make the complete medium to order. (This is also true for other media, such as SDM-79; *see* **Note 2**.)
2. The composition of SDM-79 is described in Brun and Schonenberger ***(87)***. For procyclic-stage cells, alternative growth media may also be used. One alternative is Cunningham's medium ***(88)***, and another is DTM (e.g., ref. ***65***).
3. The composition of Zimmerman's postfusion medium that we describe is used for the transformation of procyclic-stage *T. brucei*, and we refer to it in the text as ZM buffer. For bloodstream-stage cells, this should be supplemented with 1% glucose. To distinguish this formulation from ZM buffer, it is referred to as ZMG buffer. Cytomix is an alternative buffer used for both life cycle stages in some laboratories. *T. brucei* does not survive in any of these buffers for prolonged periods, so it is advisable that electroporation be conducted rapidly.
4. Sequencing of the genome of *T. brucei* is being undertaken at two institutions: the Wellcome Trust Sanger Institute (United Kingdom) and the Institute for Genomic Research (United States). Both have databases of the available sequence that can be searched via the internet and are annotating features of the sequence as it emerges.
5. The efficiency of monomorphic bloodstream-stage transformation is generally lower than that of procyclic-stage cells, and for both cell types the frequency is somewhat variable. In general, however, transformants arise in bloodstream-stage cells at around 1 transformant in 10^5–10^7 cells put on selection. In contrast, procyclic-stage transformants arise at frequencies of around 1 in 10^3–10^6. Given this variation, it is advisable at the outset of a study to put a range of parasite numbers on selection following electroporation.

Acknowledgments

Work in the laboratories of J.D.B. and R.McC. is funded by grants from the Wellcome Trust, Royal Society, and Medical Research Council; J.D.B. is a Wellcome Trust Principal Research Fellow, and R.McC. is a Royal Society University Research Fellow. Work in the laboratory of M.B. is funded by grants from Deutsche Forschungsgemeinschaft and Fonds der Chemischen Industrie. E.V. is supported by the Hans Sigrist Foundation. Dr. M. Engstler is thanked for critical reading of the manuscript.

References

1. Li, F., Hua, S. B., Wang, C. C., and Gottesdiener, K. M. (1996) Procyclic *Trypanosoma brucei* cell lines deficient in ornithine decarboxylase activity. *Mol. Biochem. Parasitol.* **78,** 227–236.
2. Sommer, J. M., Hua, S., Li, F., Gottesdiener, K. M., and Wang, C. C. (1996) Cloning by functional complementation in *Trypanosoma brucei. Mol. Biochem. Parasitol.* **76,** 83–89.
3. Wirtz, E., Leal, S., Ochatt, C., and Cross, G. A. (1999) A tightly regulated inducible expression system for conditional gene knock-outs and dominant-negative genetics in *Trypanosoma brucei. Mol. Biochem. Parasitol.* **99,** 89–101.
4. Ngo, H., Tschudi, C., Gull, K., and Ullu, E. (1998) Double-stranded RNA induces mRNA degradation in *Trypanosoma brucei. Proc. Natl. Acad. Sci. USA* **95,** 14,687–14,692.
5. Wirtz, E. and Clayton, C. (1995) Inducible gene expression in trypanosomes mediated by a prokaryotic repressor. *Science* **268,** 1179–1183.
6. Ulbert, S., Cross, M., Boorstein, R. J., Teebor, G. W., and Borst, P. (2002) Expression of the human DNA glycosylase hSMUG1 in *Trypanosoma brucei* causes DNA damage and interferes with biosynthesis. *Nucleic Acids Res.* **30,** 3919–3926.
7. Barry, J. D. and McCulloch, R. (2001) Antigenic variation in trypanosomes: enhanced phenotypic variation in a eukaryotic parasite. *Adv. Parasitol.* **49,** 1–70.
8. Gray, M. A., Ross, C. A., Taylor, A. M., and Luckins, A. G. (1984) In vitro cultivation of *Trypanosoma congolense*: the production of infective metacyclic trypanosomes in cultures initiated from cloned stocks. *Acta Trop.* **41,** 343–353.
9. Ross, C. A., Gray, M. A., Taylor, A. M., and Luckins, A. G. (1985) In vitro cultivation of *Trypanosoma congolense*: establishment of infective mammalian forms in continuous culture after isolation from the blood of infected mice. *Acta Trop.* **42,** 113–122.
10. Downey, N. and Donelson, J. E. (1999) Expression of foreign proteins in *Trypanosoma congolense. Mol. Biochem. Parasitol.* **104,** 39–53.
11. Inoue, N., Otsu, K., Ferraro, D. M., and Donelson, J. E. (2002) Tetracycline-regulated RNA interference in *Trypanosoma congolense. Mol. Biochem. Parasitol.* **120,** 309–313.
12. Zomerdijk, J. C., Ouellette, M., Ten Asbroek, A. L., et al. (1990) The promoter for a variant surface glycoprotein gene expression site in *Trypanosoma brucei. EMBO J.* **9,** 2791–2801.

13. Clayton, C. E., Fueri, J. P., Itzhaki, J. E., et al. (1990) Transcription of the procyclic acidic repetitive protein genes of *Trypanosoma brucei. Mol. Cell Biol.* **10,** 3036–3047.
14. Rudenko, G., Le Blancq, S., Smith, J., Lee, M. G., Rattray, A., and Van der Ploeg, L. H. (1990) Procyclic acidic repetitive protein (PARP) genes located in an unusually small alpha-amanitin-resistant transcription unit: PARP promoter activity assayed by transient DNA transfection of *Trypanosoma brucei. Mol. Cell Biol.* **10,** 3492–3504.
15. Clayton, C. E. (1999) Genetic manipulation of kinetoplastida. *Parasitol. Today* **15,** 372–378.
16. Ginger, M. L., Blundell, P. A., Lewis, A. M., Browitt, A., Gunzl, A., and Barry, J. D. (2002) Ex vivo and in vitro identification of a consensus promoter for VSG genes expressed by metacyclic-stage trypanosomes in the tsetse fly. *Eukaryot. Cell* **1,** 1000–1009.
17. Beverley, S. M. (2003) Protozomics: trypanosomatid parasite genetics comes of age. *Nat. Rev. Genet.* **4,** 11–19.
18. Ten Asbroek, A. L., Ouellette, M., and Borst, P. (1990) Targeted insertion of the neomycin phosphotransferase gene into the tubulin gene cluster of *Trypanosoma brucei. Nature* **348,** 174–175.
19. Ten Asbroek, A. L., Mol, C. A., Kieft, R., and Borst, P. (1993) Stable transformation of *Trypanosoma brucei. Mol. Biochem. Parasitol.* **59,** 133–142.
20. Patnaik, P. K., Kulkarni, S. K., and Cross, G. A. (1993) Autonomously replicating single-copy episomes in *Trypanosoma brucei* show unusual stability. *EMBO J.* **12,** 2529–2538.
21. Metzenberg, S. and Agabian, N. (1994) Mitochondrial minicircle DNA supports plasmid replication and maintenance in nuclei of *Trypanosoma brucei. Proc. Natl. Acad. Sci. USA* **91,** 5962–5966.
22. Nagamune, K., Nozaki, T., Maeda, Y., et al. (2000) Critical roles of glycosylphosphatidylinositol for *Trypanosoma brucei. Proc. Natl. Acad. Sci. USA* **97,** 10,336–10,341.
23. Lee, M. G. (1995) A foreign transcription unit in the inactivated VSG gene expression site of the procyclic form of *Trypanosoma brucei* and formation of large episomes in stably transformed trypanosomes. *Mol. Biochem. Parasitol.* **69,** 223–238.
24. Patnaik, P. K., Axelrod, N., Van der Ploeg, L. H., and Cross, G. A. (1996) Artificial linear mini-chromosomes for *Trypanosoma brucei. Nucleic Acids Res.* **24,** 668–675.
25. Alsford, N. S., Navarro, M., Jamnadass, H. R., et al. (2003) The identification of circular extrachromosomal DNA in the nuclear genome of *Trypanosoma brucei. Mol. Microbiol.* **47,** 277–289.
26. McCulloch, R. and Barry, J. D. (1999) A role for RAD51 and homologous recombination in *Trypanosoma brucei* antigenic variation. *Genes Dev.* **13,** 2875–2888.

27. Chaves, I., Rudenko, G., Dirks-Mulder, A., Cross, M., and Borst, P. (1999) Control of variant surface glycoprotein gene-expression sites in *Trypanosoma brucei*. *EMBO J.* 18 , 4846–4855.
28. Berriman, M., Hall, N., Sheader, K., et al. (2002) The architecture of variant surface glycoprotein gene expression sites in *Trypanosoma brucei*. *Mol. Biochem. Parasitol.* **122,** 131–140.
29. Wickstead, B., Ersfeld, K., and Gull, K. (2002) Targeting of a tetracycline-inducible expression system to the transcriptionally silent minichromosomes of *Trypanosoma brucei*. *Mol. Biochem. Parasitol.* **125,** 211–216.
30. Ersfeld, K., Melville, S. E., and Gull, K. (1999) Nuclear and genome organization of *Trypanosoma brucei*. *Parasitol. Today* **15,** 58–63.
31. Melville, S. E., Leech, V., Gerrard, C. S., Tait, A., and Blackwell, J. M. (1998) The molecular karyotype of the megabase chromosomes of *Trypanosoma brucei* and the assignment of chromosome markers. *Mol. Biochem. Parasitol.* **94,** 155–173.
32. Rudenko, G., Blundell, P. A., Dirks-Mulder, A., Kieft, R., and Borst, P. (1995) A ribosomal DNA promoter replacing the promoter of a telomeric VSG gene expression site can be efficiently switched on and off in *T. brucei*. *Cell* **83,** 547–553.
33. McCulloch, R., Rudenko, G., and Borst, P. (1997) Gene conversions mediating antigenic variation in *Trypanosoma brucei* can occur in variant surface glycoprotein expression sites lacking 70- base-pair repeat sequences. *Mol. Cell Biol.* **17,** 833–843.
34. Horn, D. and Cross, G. A. (1997) Analysis of *Trypanosoma brucei* vsg expression site switching in vitro. *Mol. Biochem. Parasitol.* **84,** 189–201.
35. Horn, D. and Cross, G. A. (1997) Position-dependent and promoter-specific regulation of gene expression in *Trypanosoma brucei*. *EMBO J.* **16,** 7422–7431.
36. Wirtz, E., Hartmann, C., and Clayton, C. (1994) Gene expression mediated by bacteriophage T3 and T7 RNA polymerases in transgenic trypanosomes. *Nucleic Acids Res.* **22,** 3887–3894.
37. Paques, F. and Haber, J. E. (1999) Multiple pathways of recombination induced by double-strand breaks in *Saccharomyces cerevisiae*. *Microbiol. Mol. Biol. Rev.* **63,** 349–404.
38. Symington, L. S. (2002) Role of RAD52 epistasis group genes in homologous recombination and double-strand break repair. *Microbiol. Mol. Biol. Rev.* **66,** 630–670, table.
39. Rudenko, G., Blundell, P. A., Taylor, M. C., Kieft, R., and Borst, P. (1994) VSG gene expression site control in insect form *Trypanosoma brucei*. *EMBO J.* **13,** 5470–5482.
40. Wirtz, E., Hoek, M., and Cross, G. A. (1998) Regulated processive transcription of chromatin by T7 RNA polymerase in *Trypanosoma brucei*. *Nucleic Acids Res.* **26,** 4626–4634.
41. Leung, W., Malkova, A., and Haber, J. E. (1997) Gene targeting by linear duplex DNA frequently occurs by assimilation of a single strand that is subject to preferential mismatch correction. *Proc. Natl. Acad. Sci. USA* **94,** 6851–6856.

42. Cruz, A. K., Titus, R., and Beverley, S. M. (1993) Plasticity in chromosome number and testing of essential genes in *Leishmania* by targeting. *Proc. Natl. Acad. Sci. USA* **90,** 1599–1603.
43. Mottram, J. C., McCready, B. P., Brown, K. G., and Grant, K. M. (1996) Gene disruptions indicate an essential function for the LmmCRK1 cdc2-related kinase of *Leishmania mexicana. Mol. Microbiol.* **22,** 573–583.
44. Tovar, J., Wilkinson, S., Mottram, J. C., and Fairlamb, A. H. (1998) Evidence that trypanothione reductase is an essential enzyme in *Leishmania* by targeted replacement of the tryA gene locus. *Mol. Microbiol.* **29,** 653–660.
45. Valdes, J., Taylor, M. C., Cross, M. A., Ligtenberg, M. J., Rudenko, G., and Borst, P. (1996) The viral thymidine kinase gene as a tool for the study of mutagenesis in *Trypanosoma brucei. Nucleic Acids Res.* **24,** 1809–1815.
46. Ha, D. S., Schwarz, J. K., Turco, S. J., and Beverley, S. M. (1996) Use of the green fluorescent protein as a marker in transfected *Leishmania. Mol. Biochem. Parasitol.* **77,** 57–64.
47. Das, A. and Bellofatto, V. (2003) RNA polymerase II-dependent transcription in trypanosomes is associated with a SNAP complex-like transcription factor. *Proc. Natl. Acad. Sci. USA* **100,** 80–85.
48. Gilinger, G. and Bellofatto, V. (2001) Trypanosome spliced leader RNA genes contain the first identified RNA polymerase II gene promoter in these organisms. *Nucleic Acids Res.* **29,** 1556–1564.
49. Papadopoulou, B. and Dumas, C. (1997) Parameters controlling the rate of gene targeting frequency in the protozoan parasite *Leishmania. Nucleic Acids Res.* **25,** 4278–4286.
50. Blundell, P. A., Rudenko, G., and Borst, P. (1996) Targeting of exogenous DNA into *Trypanosoma brucei* requires a high degree of homology between donor and target DNA. *Mol. Biochem. Parasitol.* **76,** 215–229.
51. Shen, S., Arhin, G. K., Ullu, E., and Tschudi, C. (2001) In vivo epitope tagging of *Trypanosoma brucei* genes using a one step PCR-based strategy. *Mol. Biochem. Parasitol.* **113,** 171–173.
52. Gaud, A., Carrington, M., Deshusses, J., and Schaller, D. R. (1997) Polymerase chain reaction-based gene disruption in *Trypanosoma brucei. Mol. Biochem. Parasitol.* **87,** 113–115.
53. Conway, C., Proudfoot, C., Burton, P., Barry, J. D., and McCulloch, R. (2002) Two pathways of homologous recombination in *Trypanosoma brucei. Mol. Microbiol.* **45,** 1687–1700.
54. Krawchuk, M. D. and Wahls, W. P. (1999) High-efficiency gene targeting in *Schizosaccharomyces pombe* using a modular, PCR-based approach with long tracts of flanking homology. *Yeast* **15,** 1419–1427.
55. Lee, M. G. and Van der Ploeg, L. H. (1990) Homologous recombination and stable transfection in the parasitic protozoan *Trypanosoma brucei. Science* **250,** 1583–1587.
56. Eid, J. and Sollner-Webb, B. (1991) Stable integrative transformation of *Trypanosoma brucei* that occurs exclusively by homologous recombination. *Proc. Natl. Acad. Sci. USA* **88,** 2118–2121.

57. Melville, S. E., Leech, V., Navarro, M., and Cross, G. A. (2000) The molecular karyotype of the megabase chromosomes of *Trypanosoma brucei* stock 427. *Mol. Biochem. Parasitol.* **111,** 261–273.
58. Matthews, K. R. and Gull, K. (1994) Evidence for an interplay between cell cycle progression and the initiation of differentiation between life cycle forms of African trypanosomes. *J. Cell Biol.* **125,** 1147–1156.
59. Blundell, P. A., van Leeuwen, F., Brun, R., and Borst, P. (1998) Changes in expression site control and DNA modification in *Trypanosoma brucei* during differentiation of the bloodstream form to the procyclic form. *Mol. Biochem. Parasitol.* **93,** 115–130.
60. Muller, I. B., Domenicali-Pfister, D., Roditi, I., and Vassella, E. (2002) Stage-specific requirement of a mitogen-activated protein kinase by *Trypanosoma brucei. Mol. Biol. Cell* **13,** 3787–3799.
61. Reuner, B., Vassella, E., Yutzy, B., and Boshart, M. (1997) Cell density triggers slender to stumpy differentiation of *Trypanosoma brucei* bloodstream forms in culture. *Mol. Biochem. Parasitol.* **90,** 269–280.
62. Vassella, E., Reuner, B., Yutzy, B., and Boshart, M. (1997) Differentiation of African trypanosomes is controlled by a density sensing mechanism which signals cell cycle arrest via the cAMP pathway. *J. Cell Sci.* **110,** 2661–2671.
63. Matthews, K. R. (1999) Developments in the differentiation of *Trypanosoma brucei. Parasitol. Today* **15,** 76–80.
64. Hendriks, E., van Deursen, F. J., Wilson, J., Sarkar, M., Timms, M., and Matthews, K. R. (2000) Life-cycle differentiation in *Trypanosoma brucei*: molecules and mutants. *Biochem. Soc. Trans.* **28,** 531–536.
65. Vassella, E., Den Abbeele, J. V., Butikofer, P., et al. (2000) A major surface glycoprotein of *Trypanosoma brucei* is expressed transiently during development and can be regulated post-transcriptionally by glycerol or hypoxia. *Genes Dev.* **14,** 615–626.
66. Butikofer, P., Vassella, E., Boschung, M., et al. (2002) Glycosylphosphatidylinositol-anchored surface molecules of *Trypanosoma congolense* insect forms are developmentally regulated in the tsetse fly. *Mol. Biochem. Parasitol.* **119,** 7–16.
67. Butikofer, P., Vassella, E., Mehlert, A., Ferguson, M. A., and Roditi, I. (2002) Characterisation and cellular localisation of a GPEET procyclin precursor in *Trypanosoma brucei* insect forms. *Mol. Biochem. Parasitol.* **119,** 87–95.
68. Robinson, N. P., Burman, N., Melville, S. E., and Barry, J. D. (1999) Predominance of duplicative VSG gene conversion in antigenic variation in African trypanosomes. *Mol. Cell Biol.* **19,** 5839–5846.
69. Borst, P. and Cross, G. A. (1982) Molecular basis for trypanosome antigenic variation. *Cell* **29,** 291–303.
70. Cross, G. A., Wirtz, L. E., and Navarro, M. (1998) Regulation of vsg expression site transcription and switching in *Trypanosoma brucei. Mol. Biochem. Parasitol.* **91,** 77–91.

71. Vanhamme, L., Pays, E., McCulloch, R., and Barry, J. D. (2001) An update on antigenic variation in African trypanosomes. *Trends Parasitol.* **17,** 338–343.
72. Borst, P. (2002) Antigenic variation and allelic exclusion. *Cell* **109,** 5–8.
73. Vassella, E. and Boshart, M. (1996) High molecular mass agarose matrix supports growth of bloodstream forms of pleomorphic *Trypanosoma brucei* strains in axenic culture. *Mol. Biochem. Parasitol.* **82,** 91–105.
74. Vassella, E., Kramer, R., Turner, C. M., et al. (2001) Deletion of a novel protein kinase with PX and FYVE-related domains increases the rate of differentiation of *Trypanosoma brucei. Mol. Microbiol.* **41,** 33–46.
75. Brun, R. and Schonenberger, M. (1981) Stimulating effect of citrate and cis-Aconitate on the transformation of *Trypanosoma brucei* bloodstream forms to procyclic forms in vitro. *Z. Parasitenkd.* **66,** 17–24.
76. Ziegelbauer, K., Quinten, M., Schwarz, H., Pearson, T. W., and Overath, P. (1990) Synchronous differentiation of *Trypanosoma brucei* from bloodstream to procyclic forms in vitro. *Eur. J. Biochem.* **192,** 373–378.
77. Overath, P., Czichos, J., and Haas, C. (1986) The effect of citrate/cis-aconitate on oxidative metabolism during transformation of *Trypanosoma brucei. Eur. J. Biochem.* **160,** 175–182.
78. Overath, P., Czichos, J., Stock, U., and Nonnengaesser, C. (1983) Repression of glycoprotein synthesis and release of surface coat during transformation of *Trypanosoma brucei. EMBO J.* **2,** 1721–1728.
79. Vassella, E., Acosta-Serrano, A., Studer, E., Lee, S. H., Englund, P. T., and Roditi, I. (2001) Multiple procyclin isoforms are expressed differentially during the development of insect forms of *Trypanosoma brucei. J. Mol. Biol.* **312,** 597–607.
80. Tyler, K. M., Matthews, K. R., and Gull, K. (1997) The bloodstream differentiation-division of *Trypanosoma brucei* studied using mitochondrial markers. *Proc. R. Soc. Lond B Biol. Sci.* **264,** 1481–1490.
81. Vassella, E., Straesser, K., and Boshart, M. (1997) A mitochondrion-specific dye for multicolour fluorescent imaging of *Trypanosoma brucei. Mol. Biochem. Parasitol.* **90,** 381–385.
82. Roditi, I., Schwarz, H., Pearson, T. W., et al. (1989) Procyclin gene expression and loss of the variant surface glycoprotein during differentiation of *Trypanosoma brucei. J. Cell Biol.* **108,** 737–746.
83. Hirumi, H. and Hirumi, K. (1989) Continuous cultivation of *Trypanosoma brucei* blood stream forms in a medium containing a low concentration of serum protein without feeder cell layers. *J. Parasitol.* **75,** 985–989.
84. Carruthers, V. B. and Cross, G. A. (1992) High-efficiency clonal growth of bloodstream- and insect-form *Trypanosoma brucei* on agarose plates. *Proc. Natl. Acad. Sci. USA* **89,** 8818–8821.
85. Conway, C., McCulloch, R., Ginger, M. L., Robinson, N. P., Browitt, A., and Barry, J. D. (2002) Ku is important for telomere maintenance, but not for differential expression of telomeric VSG genes, in African trypanosomes. *J. Biol. Chem.* **277,** 21,269–21,277.

86. Robinson, N. P., McCulloch, R., Conway, C., Browitt, A., and Barry, J. D. (2002) Inactivation of Mre11 does not affect VSG gene duplication mediated by homologous recombination in *Trypanosoma brucei*. *J. Biol. Chem.* **277,** 26,185–26,193.
87. Brun, R. and Schonenberger (1979) Cultivation and in vitro cloning or procyclic culture forms of *Trypanosoma brucei* in a semi-defined medium. Short communication. *Acta Trop.* **36,** 289–292.
88. Cunningham, I. (1977) New culture medium for maintenance of tsetse tissues and growth of trypanosomatids. *J. Protozool.* **24,** 325–329.
89. Bingle, L. E., Eastlake, J. L., Bailey, M., and Gibson, W. C. (2001) A novel GFP approach for the analysis of genetic exchange in trypanosomes allowing the in situ detection of mating events. *Microbiology* **147,** 3231–3240.
90. Van Den Hoff, M. J., Moorman, A. F., and Lamers, W. H. (1992) Electroporation in 'intracellular' buffer increases cell survival. *Nucleic Acids Res.* **20,** 2902.
91. Nathan, H. C., Bacchi, C. J., Sakai, T. T., Rescigno, D., Stumpf, D., and Hutner, S. H. (1981) Bleomycin-induced life prolongation of mice infected with *Trypanosoma brucei brucei* EATRO 110. *Trans. R. Soc. Trop. Med. Hyg.* **75,** 394–398.
92. Jefferies, D., Tebabi, P., Le Ray, D., and Pays, E. (1993) The ble resistance gene as a new selectable marker for *Trypanosoma brucei*: fly transmission of stable procyclic transformants to produce antibiotic resistant bloodstream forms. *Nucleic Acids Res.* **21,** 191–195.
93. Murphy, N. B., Muthiani, A. M., and Peregrine, A. S. (1993) Use of an in vivo system to determine the G418 resistance phenotype of bloodstream-form *Trypanosoma brucei brucei* transfectants. *Antimicrob. Agents Chemother.* **37,** 1167–1170.
94. van Gent, D. C., Hoeijmakers, J. H., and Kanaar, R. (2001) Chromosomal stability and the DNA double-stranded break connection. *Nat. Rev. Genet.* **2,** 196–206.
95. Cox, M. M. (2001) Recombinational DNA repair of damaged replication forks in *Escherichia coli*: questions. *Annu. Rev. Genet.* **35,** 53–82.
96. Masson, J. Y. and West, S. C. (2001) The Rad51 and Dmc1 recombinases: a non-identical twin relationship. *Trends Biochem. Sci.* **26,** 131–136.
97. Sogin, M. L., Hinkle, G., and Leipe, D. D. (1993) Universal tree of life. *Nature* **362,** 795.
98. Tan, K. S., Leal, S. T., and Cross, G. A. (2002) *Trypanosoma brucei* MRE11 is non-essential but influences growth, homologous recombination and DNA double-strand break repair. *Mol. Biochem. Parasitol.* **125,** 11–21.
99. Lovett, S. T., Hurley, R. L., Sutera, V. A., Jr., Aubuchon, R. H., and Lebedeva, M. A. (2002) Crossing over between regions of limited homology in *Escherichia coli*. RecA-dependent and RecA-independent pathways. *Genetics* **160,** 851–859.
100. Ira, G. and Haber, J. E. (2002) Characterization of RAD51-independent break-induced replication that acts preferentially with short homologous sequences. *Mol. Cell Biol.* **22,** 6384–6392.
101. Barnes, D. E. (2001) Non-homologous end joining as a mechanism of DNA repair. *Curr. Biol.* **11,** R455–R457.

102. Critchlow, S. E. and Jackson, S. P. (1998) DNA end-joining: from yeast to man. *Trends Biochem. Sci.* **23,** 394–398.
103. D'Amours, D. and Jackson, S. P. (2002) The Mre11 complex: at the crossroads of DNA repair and checkpoint signalling. *Nat. Rev. Mol. Cell Biol.* **3,** 317–327.
104. Morris, J. C., Wang, Z., Drew, M. E., and Englund, P. T. (2002) Glycolysis modulates trypanosome glycoprotein expression as revealed by an RNAi library. *EMBO J.* **21,** 4429–4438.
105. Panigrahi, A. K., Schnaufer, A., Ernst, N. L., Wang, B., Carmean, N., Salavati, R., and Stuart, K. (2003) Identification of novel components of *Trypanosoma brucei* editosomes. *RNA* **9,** 484–492.
106. Aphasizhev, R., Aphasizheva, I., Nelson, R. E., et al. (2003) Isolation of a U-insertion/deletion editing complex from *Leishmania tarentolae* mitochondria. *EMBO J.* **22,** 913–924.
107. Lee, M. G. and Van der Ploeg, L. H. (1991) The hygromycin B-resistance-encoding gene as a selectable marker for stable transformation of *Trypanosoma brucei. Gene* **105,** 255–257.
108. Lorenz, P. Maier, A. G., Baumgart, E., Erdmann, R., and Clayton, C. (1998) Elongation and clustering of glycosomes in *Trypanosoma brucei* overexpressing the glycosomal Pex11p. *EMBO J.* **17,** 3542–3555.
109. Brooks, D. R., McCulloch, R., Coombs, G. H., and Mottram, J. C. (2000) Stable transformation of trypanosomatids through targeted chromosomal integration of the selectable marker gene encoding blasticidin S deaminase. *FEMS Microbiol. Lett.* **186,** 287–291.
110. Ruepp, S., Furger, A., Kurath, U., et al. (1997) Survival of *Trypanosoma brucei* in the tsetse fly is enhanced by the expression of specific forms of procyclin. *J. Cell Biol.* **137,** 1369–1379.

6

Forward Genetic Screens for Meiotic and Mitotic Recombination-Defective Mutants in Mice

Laura Reinholdt, Terry Ashley, John Schimenti, and Naoko Shima

Summary

The goal of understanding the function of all mammalian genes is best accomplished through mutational analyses. Although the sequence of the mouse genome is now available and many genes have been identified, it is not possible to ascribe functions accurately to these genes *in silico*. Gene targeting using embryonic stem cells is ideal for analysis of individual genes selected on the basis of sequence features, but it is impractical for identifying novel genes involved in particular biological processes. Phenotype-based random mutagenesis of the genome is well suited for this goal. In the mouse, *N*-ethyl-*N*-nitrosourea (ENU) induces point mutations at a high frequency in the mouse germline. In this chapter, we describe methods for detecting and characterizing recombination mutations in mice produced by ENU mutagenesis. Potential meiotic recombination mutants are identified in a hierarchical fashion, by performing a screen for infertility, then gonad histology to determine whether meiotic arrest occurs, and finally by immunohistochemical analysis of meiotic chromosome with a battery of antibody markers. Screening for mutations potentially required for recombinational repair of DNA damage in somatic cells is performed using a flow cytometry-based micronucleus assay. Both strategies have proved effective in identifying desired classes of mutations.

Key Words: mouse, ENU mutagenesis, recombination, meiosis, micronucleus assay, genomic instability, Rad51, chromosome synapsis

1. Introduction

1.1. ENU Mutagenesis in Mice

Spurred by the genomic resources created by the Human Genomic Project, which has greatly facilitated the positional cloning of mutations, there has been a surge in forward genetic approaches to gene identification in mice. The concept of forward genetics has long been recognized in model organisms such as yeast, worms, and fruitflies and has driven the remarkable advances made in

From: *Methods in Molecular Biology, vol. 262, Genetic Recombination: Reviews and Protocols*
Edited by: A. S. Waldman © Humana Press Inc., Totowa, NJ

understanding the developmental genetics of those organisms. The most effective way to induce random germline mutations in mice is with the chemical *N*-ethyl-*N*-nitrosourea (ENU) ***(1,2)***. It primarily induces point mutations and therefore results in a variety of alleles. These include hypomorphs, recessive nulls, and dominant mutations. Multiple alleles are extremely useful in discerning the function of genes and in positional cloning. ENU induces mutations in mice at a frequency of greater than 1/750/locus/gamete, which enables highly efficient screening of mutagenized animals for aberrant phenotypes ***(3)***. Another feature of ENU is that it mutagenizes stem cell spermatogonia. Therefore, treated males will produce mutagenized gametes throughout the remainder of their reproductive lives.

In the past several years, several large-scale mutagenesis programs have been initiated around the world. The focus of these screens has included behavior, clinical chemistry, cardiovascular function, neurological phenotypes, obesity, immunology, and embryonic development. Most of these screens seek dominant mutations, which has the advantage of being logistically simple and of simplifying stock maintenance and genetic mapping ***(4,5)***. The identification of recessive mutations can be done either on a genome-wide scale, requiring three-generation breeding schemes followed by classical linkage analysis to identify the location of the mutation, or by region-directed screens that predetermine the chromosomal linkage of mutations (for reviews, *see* refs. ***6*** and ***7***).

For the purposes of this chapter, we'll assume that readers considering a mutagenesis program will be familiar with mouse husbandry and the logistical issues associated with various mating paradigms. The technique of ENU mutagenesis is not very complicated, and protocols will not be delineated here. However, the following points could prove helpful to first-time users. First, the preparation of ENU and accurate quantitation is critical. This and safety issues are well discussed by Nolan et al. ***(8)***. Second, different mouse strains respond quite differently to ENU doses, so one should choose a regimen that has been established in the literature for a given strain ***(9)***. Even so, we have found that the dosages may need to be altered to suit unknown factors associated with different environments. Third, don't take shortcuts when measuring the appropriate dosage. Weigh each animal separately for each treatment, and use an accurate pipetor to measure out the correct amount of ENU. If all goes well, the weight of the animals should decline slightly or be maintained over a multiweek regimen. Fourth, the worst thing that can happen is that the mice survive the treatments but never regain fertility. Since this takes many weeks to realize, one can easily lose months of time before finding that the proper dosage should have been lower. Consequently, it is worth doing pilot studies including dosages that are slightly lower and higher than the published optimum.

1.2. Screens for Meiotic Recombination Mutants

The likely phenotype of mutants defective in meiotic recombination—infertility—mandates that such mutations be recessive in order to be propagated and studied. Mutations that are essential for meiotic chromosome behavior and recombination in yeast typically result in inviable spores. Furthermore, knockouts of mouse homologs of yeast recombination genes, such as *Dmc*1, *Spo1*1, *Msh*4, and *Msh*5, cause infertility ***(10–14,16)***. Thus, the lab of J.S. initiated phenotype-driven mutagenesis screens for mouse infertility mutants derived from chemical mutagenesis of ES cells or classical ENU mutagenesis ***(15)***.

As infertility can result from any number of defects not involving recombination, a "triage" of infertile mice is conducted to identify those that may indeed have recombination defects. A common characteristic of male mice bearing mutations in the four genes mentioned in the preceding paragraph is that none produce sperm, and their testes are very small. This is a result of an arrest of spermatogenesis at meiotic prophase or at the first meiotic division, depending on the gene involved. Recombinational repair during meiosis is a complex process, which occurs in the context of dynamic changes in nuclear architecture. Therefore, genes that are required for meiotic recombination may be either directly involved in the enzymology of recombinational repair or indirectly involved via meiotic chromosome mechanics. The phenotypes of these mutants provide the framework for the histological criteria we use to identify putative recombinational repair mutants. These criteria include meiotic arrest at zygotene and/or pachytene stages (as seen in *Spo1*1 ***[16]***, *Dmc*1 ***[17,18]***, *Atm* ***[19]***) or meiotic arrest at metaphase I (as seen in *Mlh*1 knockouts ***[20,21]***). In some cases, mutations that disrupt meiotic chromosome mechanics may also, indirectly, disrupt recombination and will cause sterility and meiotic arrest in males, but not females. Examples include SCP3 ***(22)*** and γH2AX deficiencies ***(22,23)***. However, in these cases, fertile females, are not entirely unaffected, as they often produce small litters and have shorter reproductive life spans ***(24)***. Therefore, if a chemically induced recessive mutation causes meiotic arrest and sterility in males but not females, a possible indirect meiotic recombination defect should not be excluded.

When a putative meiotic recombination mutant has been identified based on the criteria just described, microspreading of meiotic nuclei and subsequent immunolabeling are used to define precisely the appearance and distribution of meiotic chromosome cores and accompanying meiotic recombination complexes at progressive meiotic stages. Both light and electron microscopy can be used to evaluate immunolabeled, microspread meiotic nuclei; however, the methods we describe here refer only to the less expensive and less technically challenging light microscopy procedures. To determine whether early or late meiotic recombination steps have been disrupted, synapsis and chiasmata maintenance are examined. Primary antibodies to synaptonemal complex proteins can be used to visualize the meiotic chromosome cores, which undergo a series of cytologically defined stages as homologous chromosomes pair, synapse, and form chiasmata during prophase I. These primary antibodies include anti-COR1 (hamster) ***(25)*** or anti-SCP3 (rat) ***(26)***, which localize to the axial elements, and anti-SYN1 (hamster) ***(25)*** or anti-SCP1 (rat) ***(26)***, which localize to the central element. To examine the appearance and distribution of known meiotic recombination proteins, there are currently many antibodies available either commercially or through collaboration. For a list of these antibodies, *see* **Table 1**, which is discussed in **Note 1**. For comprehensive reviews of the meiotic chromosome core morphogenesis and distribution of meiotic recombination proteins during prophase I in normal and mutant mice, the reader is referred to refs. ***27*** and ***28*** and to the references therein.

1.3. Screen for Mouse Mutants Potentially Defective in Recombinational Repair by a Flow Cytometric Peripheral Blood Micronucleus Assay

The identification of genes involved in recombinational repair of DNA damage in somatic cells is more problematic. There is no obvious phenotypic consequence of such mutations, if homozygous mutant mice are viable (unlike null mutations in *Rad51* homologs, for example). Such mutations may predispose to cancer, but such a phenotype, which is typically a late-onset phenomenon, is not practical for large-scale, high-throughput mutagenesis screens. Accordingly, we adopted a high-throughput micronucleus assay, which provides a quantitative measure of in vivo chromosome damage ***(29)***, as a screening tool for mice bearing mutations affecting double-strand break (DSB) repair. Micronuclei can arise from acentric chromosome fragments or whole chromosomes that have not been incorporated in the main nuclei at cell division ***(30)***. The formation of micronuclei can be stimulated by DNA-damaging agents that induce DNA DSBs or abnormal chromosome segregation, and thus the micronucleus assay has been used as a genetic toxicology tool for quantitative analysis of in vivo chromosome damage induced by potential mutagens ***(31)***. A peripheral blood micronucleus assay has been semiautomated by flow cytometry ***(32,33)***, making it practical for screening large numbers of mice with high statistical power. We have used this assay in preliminary screens for ENU-induced chromosome instability mutations in mice and have reported on the positional cloning of a gene (DNA polymerase theta) using this strategy ***(34)***. Consequently, this is a potentially powerful tool for the identification of new genes involved in DNA repair in mice, while simultaneously providing the animal model.

2. Materials

2.1. Immunohistochemical Analysis of Meiotic Chromosome Spreads

2.1.1. Microspreading Meiotic Nuclei

1. Benchtop microfuge with adjustable speed.
2. Precleaned glass microscope slides.
3. Hydrophobic marker (PAP pen or clear nailpolish).
4. Unsharpened pencils or plastic serological pipets.
5. 245 × 245 mm square bioassay dish.
6. Paper towels.
7. Small carpenter's level.
8. Coplin jars.
9. 15 mL polystyrene conical tubes.
10. 35 × 10 mm petri dishes.
11. 10% stock solution of Triton-X.
12. 20× Protease inhibitor cocktail ("complete," protease inhibitor cocktail tablet, dissolved in water for 20× stock [Roche]; or *see* **Note 2**).
13. Phosphate-buffered saline (PBS).
14. Minimal essential medium, high glucose (i.e., Eagle's MEM).
15. Photo-flo (Kodak).

Solutions to be prepared fresh, on a weekly basis, and stored at 4°C:

16. 0.5 *M* sucrose, sterile.
17. 1% paraformaldehyde in water (can be prepared from reagent or freshly diluted from one ampule of 16%, EM grade paraformaldehyde).

Additional solutions needed for microspreading meiotic nuclei from E17 ovaries:

18. 24-Well dish.
19. Hypo/extraction buffer: 1 mL 600 m*M* Tris-HCl, pH 8.2, 1 mL 500 m*M* sucrose, pH 8.2, 2 mL 170 m*M* sodium citrate, pH 8.2, 0.24 mL 100 m*M* ethylenediaminetetraacetic acid (EDTA), 14.68 mL distilled water, and 1 mL 20× protease inhibitor cocktail (add just before use).
20. 100- and 35-mm dishes.
21. Fine watchmaker's forceps.

2.1.2. Immunolabeling Microspread Meiotic Nuclei

1. Coplin jars.
2. 22 × 50 mm #1 cover glass.
3. 10× PBS, sterile.
4. Amber tubes.
5. Humid chamber, i.e., 245 × 245-mm square bioassay dish, lined with moistened paper towels. Use serological pipets or unsharpened pencils for supports.

6. Antibody dilution buffer (ADB stock): 10 mL serum (goat, donkey, or horse), 30 g bovine serum albumin, Triton-X to 0.05%, PBS to 1 L. Store at 0°C in 25–50-mL aliquots.
7. DAPI (4,6-diamino-2-phenylindole), if desired.
8. SlowFade Antifade reagent (Molecular Probes).
9. Primary and secondary antibodies (*see* **Notes 1** and **3–5**).

2.2. Micronucleus Assay

1. Heparin, 500 USP units/mL saline (Sigma); aliquot 250 µL in 1.5-mL microtubes and store at 4°C.
2. Heparinized microhematocrit capillary tubes.
3. Absolute methanol (Fisher) as fixative; aliquot 2 mL absolute methanol into polypropylene centrifuge tubes and place them in a –80°C freezer 1 d before blood collection.
4. Propidium iodide (PI; Sigma). Make a stock at 1.25 mg/mL in bicarbonate buffer and store at 4°C. For each assay, prepare fresh 1.25 µg/mL working solution from the stock.
5. Bicarbonate buffer: 0.9% NaCl, 5.3 m*M* sodium bicarbonate, pH 7.5; autoclave and store at 4°C.
6. Fluorescein-conjugated anti-CD71 monoclonal antibody (CD71-FITC; BioDesign, Kennebunk, ME; clone no. R17217.1.4).
7. RNaseA (Amresco, Solon, OH).

3. Methods

3.1. Immunohistochemical Analysis of Meiotic Chromosome Spreads

3.1.1. Microspreading Meiotic Nuclei

The success of immunocytology and fluorescence *in situ* hybridization (FISH) is highly dependent on the preservation and accessibility of subnuclear structure and protein content. In 1997, Peters et al. ***(35)*** described a drying down technique, which was not only useful for immunocytology and FISH, but also allowed for the collection of large numbers of nuclei with better representation of prophase I substages. The protocol described here is a derivation of that described by Peters et al. and was developed by Drs. Peter de Boer and Terry Ashley.

3.1.1.1. Protocol for One 3-Wk-Old Male Mouse (*see* **Note 6**)

1. Draw a border around an appropriate number of slides (*see* **Note 7**) using clear nail polish or a hydrophobic pen; air-dry.
2. Line the bottom of each bioassay dish with paper towels, wet the paper towels with water, and then, by rolling a serological pipet, smooth the surface of the paper towels (*see* **Note 8**).
3. Use four unsharpened pencils or serological pipets (broken to fit the dish) to create two rows of support for slides.

4. Place carpenter's level on a slide. Put slide on supports to check that supports are level. Adjust dish as necessary.
5. Place slides prepared in **step 1** on the supports and cover.
6. Adjust the pH of the 1% paraformaldehyde to 9.2. In a 15-mL conical tube, dissolve Triton-X to a concentration of 0.1% in 5 mL of the pH-adjusted, paraformaldehyde, and keep on ice.
7. Prepare 5 mL of 0.1 *M* sucrose in a 15-mL conical tube, and keep on ice.
8. Place 7 mL MEM-high glucose in a 15-mL conical tube, and keep on ice.
9. Add protease inhibitor cocktail to a final concentration of 1×.
10. Remove 3–4 mL of the MEM prepared in **step 8** and place it in a 35-mm culture dish.
11. In a second 35-mm culture dish, place 2–3 mL PBS.
12. Remove both testes and rinse in the PBS-containing dish; remove fat and adherent tissue.
13. Transfer testes to the dish containing MEM and remove the tunicae.
14. Roughly separate the tubules using watchmaker's forceps and use a Pasteur pipet or a P1000 to disperse the tubules by repeated pipeting.
15. Being careful to leave large fragments behind (*see* **Note 9**), pipet the suspension back into the conical tube that contains the remaining MEM.
16. Allow large fragments (if any) to settle.
17. Pipet 1 mL of suspension into each of 6 1.5-mL microcentrifuge tubes (*see* **Note 10**). One milliliter should be left behind.
18. Centrifuge the tubes for 5 min at 5800*g* (or 9000 rpm on benchtop microcentrifuge).
19. While tubes are centrifuging, add protease inhibitor (final concentration 1×) to the 5 mL of 1% paraformaldehyde, 0.1% Triton-X. Then dispense 5–7 drops (or 65 μL) of this solution to each of the slides in the humid chamber. Evenly distribute (by tilting the slide) the solution throughout the surface of the slide, keeping the solution within the hydrophobic boundaries. Replace the lid on the chamber.
20. After centrifugation, decant most of the supernatant from each of the pellets. Be sure to leave a small amount of medium behind to moisten the pellet.
21. Add protease inhibitor cocktail to the 0.1 *M* sucrose solution, to a final concentration of 1×.
22. Working with three centrifuge tubes at a time (see **Note 11**), remove the remaining medium from each pellet and then resuspend each pellet in 40 μL of 0.1 *M* sucrose.
23. Evenly drop 20 μL of the sucrose suspension onto each slide. (Two slides can be made from each pellet.) Repeat with remaining resuspended pellets.
24. Cover the humid chamber and let slides sit for 2–3 h. *Do not move the tray while the nuclei are spreading.*
25. Using a Pasteur pipet, rinse the slides well with 1:250 dilution of PhotoFlo (in water). If nail polish was used, carefully remove the nail polish as the slide is rinsed. Do not aim the pipet directly at the surface of the slide, but rather allow the solution to run down the slide from the frosted end.

26. Allow the slides to air-dry and store at –20°C until ready to use. For best results, slides should be used for immunocytology within 1 wk.

3.1.1.2. Protocol for E17 Ovaries

1. Prepare slides and humid chamber as described in **Subheading 3.1.1., steps 1–5**.
2. In a 24-well dish, fill 6 wells half full with hypo/extraction buffer.
3. Sacrifice pregnant mouse at E17 and remove uterus to a 100-mm dish filled with PBS.
4. Quickly dissect fetuses out of uterus and remove them to another PBS-filled dish. Decapitate each fetus as it is removed from the uterus.
5. Place a fetus in a dissecting dish, and remove ovaries to a 35-mm dish filled with PBS, which is sitting on the dissecting scope stage.
6. Carefully peel back surrounding tissue from each ovary and then carefully move the ovary to a well in a 24-well dish, which is half-filled with hypo/extraction buffer (Two ovaries from each animal/well). Allow to sit for at least 30 min (at room temperature).
7. Apply buffered paraformaldehyde (PFA) to slides. Leave slides in humid chamber.
8. Place a clean microscope slide on the dissecting scope stage. Aliquot 10 µL of 0.1 *M* sucrose solution onto slide. Place one ovary in the drop of sucrose, and, using watchmakers forceps, pull it apart. Add another 10-µL drop of sucrose and further disperse the tissue using a 20-µL pipet.
9. Apply as much of the ovary suspension as can be sucked off the slide to one PFA-coated slide. *Thus, one ovary/slide is applied in small drops in a total volume of 20 µL.*
10. Allow slides to dry for 2 h in closed moist chamber.
11. Using a Pasteur pipet, rinse the slides well with a 1:250 dilution of PhotoFlo (in water). If nail polish was used, carefully remove the nail polish as the slide is rinsed. Do not aim the pipet directly at the surface of the slide, but rather allow the solution to run down the slide from the frosted end.

3.1.2. Immunolabeling Microspread Meiotic Nuclei

To examine the timing and localization of specific recombination proteins during mammalian meiosis, multiple labeling is required. Primary antibodies to most of the proteins currently known or thought to play a role in meiotic recombination are available, either commercially or through collaboration (*see* **Note 1**). Simultaneous visualization of the meiotic cores and of the localization of meiotic recombination proteins can be achieved by labeling with multiple primary antibodies. The major limitations to the number of primary antibodies that can be used simultaneously are that each antibody should be derived from immunization of different species (i.e., rat and rabbit) and that a distinct fluorochrome (and appropriate filter set) be available to detect each primary antibody. When choosing secondary antibodies, several considerations should be made. First, the fluorochromes chosen should correctly match the filter sets available for fluorescent microscopy. Consideration should also be given to wavelength separation of the filters to minimize bleed through from one filter channel to another. Second, care should be taken to avoid crossreactivity between secondary antibodies and antigens present in serum. Crossadsorbed antibodies are available for this purpose.

For detailed information on immunochemistry and indirect immunofluorescent labeling of fixed cytological specimens in general, the reader is directed to Javois ***(36)***. The following protocol for immunolabeling microspread meiotic nuclei is a modification of one described ***(25)***.

1. Prepare 1 L of 10% antibody dilution buffer (ADB) in 1× PBS.
2. Wash slides containing microspread meiotic nuclei 3 × 10 min in 10% ADB.
3. Dilute primary antibody(ies) in ADB (undiluted) (*see* **Note 3**).
4. Place washed slides on supports in a humid chamber.
5. Apply each primary antibody, dropwise, to each slide; do not exceed 50 μL total volume. Add coverslip.
6. Incubate for 2 h at 37°C or overnight at 4°C.
7. Remove coverslip, and wash 3 × 10 min in 10% ADB.
8. Dilute secondary antibody(ies) in ADB in an amber tube (*see* **Note 4**).
9. In the dark, place washed slides in supports in humid chamber.
10. Apply secondary antibody(ies) in a volume not exceeding 50 μL; add coverslip.
11. Incubate *in the dark* for 1 or more h at 37°C.
12. Remove coverslip, wash 3 × 10 min in 1× PBS, and wash 1 × 5 min in water.
13. Optional: incubate for 10 min in 0.02 μg/mL DAPI in 1× PBS.
14. Wash for 10 min in water.
15. Allow slide to dry, mount in Antifade reagent, and add coverslip.

3.2. Micronucleus Assay

3.2.1. Blood Collection and Fixation

1. Collect 50 μL peripheral blood from the retro-orbital sinus using a heparinized tube into a microtube containing 250 μL anticoagulant heparin solution. Mix immediately.
2. Keeping the sample tubes at –80°C, take one tube with cold methanol at a time and transfer a 180-μL aliquot of each blood sample into it with a micropipeter. Hold the top part of the tube so as not to warm the methanol. Disrupt any aggregates by striking the tube several times and returning it to the freezer. Keep the samples at –80°C at least overnight before further processing (*see* **Notes 12** and **13**).

3.2.2. Setting Gates, Parameters, and Compensations for Flow Cytometric Analysis

Originally developed and described by Dertinger et al. ***(32)***, this assay is performed on a flow cytometer with a laser capable of 488-nm emission required for excitation of both fluorescein isothiocyanate (FITC) and PI. A simple

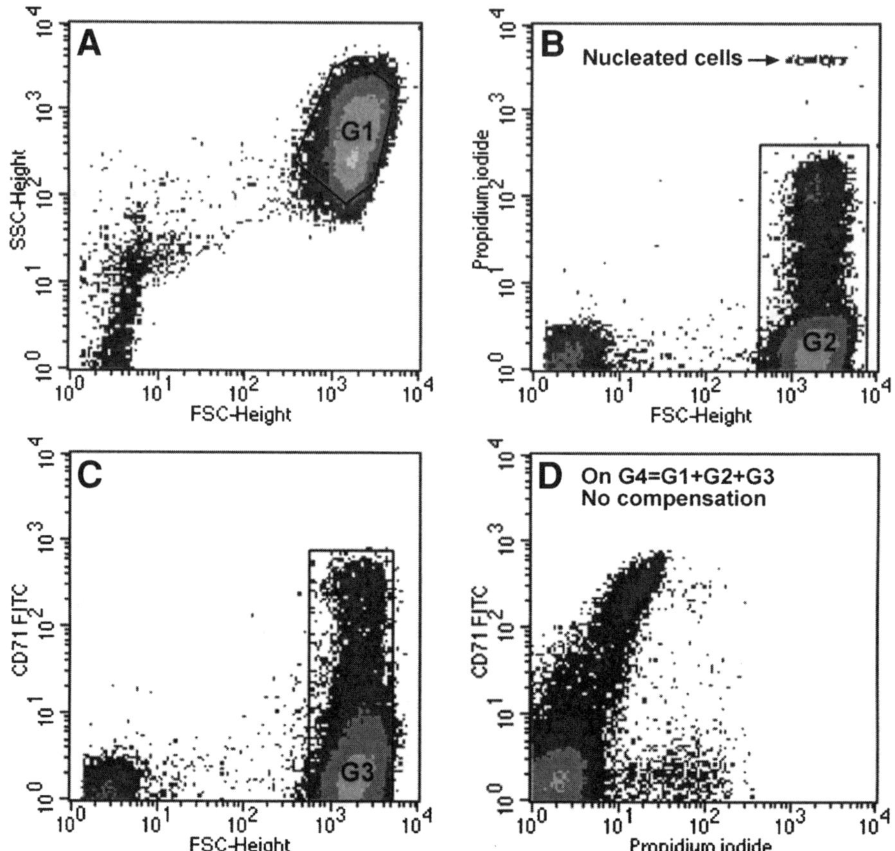

Fig. 1. Gating for flow cytometric analysis. **(A)** Erythrocytes are isolated by Gate 1 (G1) on forward scatter × side scatter (FSC × SSC). **(B)** Since the cell population gated by G1 may contain nucleated cells such as white blood cells, Gate 2 (G2) is placed to eliminate nucleated cells. **(C)** Cell population with CD71-positives is defined with Gate 3 (G3). **(D)** A sample stained with CD71-fluorescein isothiocyanate (FITC) and PI on FL-3-PI and FL-1-FITC using Gate 4 (G4). At this stage, longer emissions from FITC and shorter emissions from PI interfere with FL-3 and FL-1 signals, respectively.

single-laser flow cytometer such as FACScan (Becton Dickinson, San Jose, CA) is sufficient. If there is no access to a flow cytometer, this assay is commercially available from Litron Laboratory (Rochester, NY).

1. Prepare two single-colored samples stained with either CD71-FITC or PI as described below (*see* **Subheading 3.2.5.**).
2. Add 1 mL 1.25 µg/mL PI solution to a PI single-colored sample. Use a flow rate around 5000 events/s.

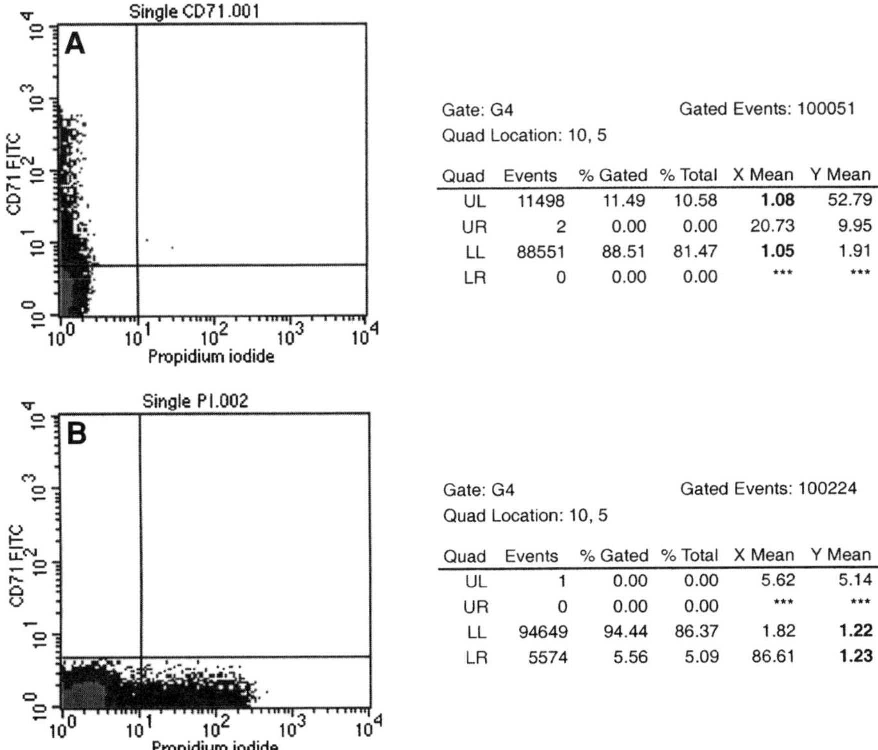

Gate: G4 Gated Events: 100051
Quad Location: 10, 5

Quad	Events	% Gated	% Total	X Mean	Y Mean
UL	11498	11.49	10.58	**1.08**	52.79
UR	2	0.00	0.00	20.73	9.95
LL	88551	88.51	81.47	**1.05**	1.91
LR	0	0.00	0.00	***	***

Gate: G4 Gated Events: 100224
Quad Location: 10, 5

Quad	Events	% Gated	% Total	X Mean	Y Mean
UL	1	0.00	0.00	5.62	5.14
UR	0	0.00	0.00	***	***
LL	94649	94.44	86.37	1.82	**1.22**
LR	5574	5.56	5.09	86.61	**1.23**

Fig. 2. Compensations of overlapping emissions from two dyes. **(A)** CD71-fluorescein isothiocyanate (FITC) single-colored control is used to subtract FITC-signals detected by FL-3, by adjusting that UL and LL have similar X means (indicated as bold). **(B)** PI-single-colored control is used to subtract PI signals from FL-1 (see Y means of LL and LR in bold).

3. The first step is to isolate erythrocytes (possibly including some nucleated cells) by gating on the forward scatter (FSC) and side scatter (SSC) parameters on a log scale, eliminating both smaller and larger events (Gate 1, **Fig. 1A**).
4. By optimizing the photomultiplier tube (PMT) voltage on FL-3 (which detects red fluorescence signals from PI), visualize the brightest PI-staining cells (nucleated cells) on the FSC and FL-3-PI, and gate out this population (Gate 2, **Fig. 1B**).
5. Likewise, using the CD71-FITC single-colored sample (to which 1 mL bicarbonate buffer is added), and find an appropriate PMT voltage on FL-1 for green fluorescence signals to define CD71-FITC-positive cells (reticulocytes) on the FSC and FL-1-FITC (Gate 3, **Fig. 1C**).
6. Using a combination of these three gates (Gate 4 = G1 + G2 + G3), erythrocytes are analyzed for the presence of micronuclei on FL-3-PI and FL-1-FITC (**Fig. 1D**). Save these gates in an acquisition template. To detect micronucleated erythro-

cytes precisely, overlapping emissions from FITC and PI must be electronically compensated.

7. Make quadrants on the FL-3-PI and FL-1-FITC plot to separate erythrocytes into four populations.
8. Flow CD71-FITC single-colored control. Set FL-1-%FL-2 compensation so that upper left and lower left quadrants have similar *X* means. (Use quadrant-stats; **Fig. 2A**).
9. Using the PI single-colored control, set a right FL-3-%FL-2 compensation rate to have similar *Y* means in lower left and lower right quadrants (**Fig. 2B**). All parameters are run on logarithmic detection mode using CellQuest acquisition software (Becton-Dickinson).
10. Save information on the PMT voltages and compensation rates as an instrument setting.

3.2.3. RNaseA Titration

RNaseA is required to eliminate reticulocyte RNA, which would interfere with quantitation of micronuclei. In the original protocol *(32)*, the final concentration of RNaseA is 1 mg/mL. However, since the specific activity of RNaseA varies between lots and manufacturers, it should be titrated for optimal performance. This is particularly important when radiation-induced micronuclei are measured in reticulocytes, as described in **Subheading 3.2.6.**

1. Make a series of RNaseA dilutions including CD71-FITC antibody with bicarbonate buffer in a total volume of 80 μL.
2. Incubate at 4°C for 45 min (*see* **Note 14**).
3. Flow by adding 1 mL PI (1.25 μg/mL).
4. Look carefully at the FSC and SSC plot. If RNaseA concentration is too high, the entire erythrocyte population will not fit into the G1 gate (compare **Fig. 1A** and **3A**). Moreover, there should not be excess events between nucleated cells and the G2 gate (**Fig. 3B**). Find RNaseA concentrations giving an intact erythrocyte population (*see* **Notes 15** and **16**).
5. Pick an RNaseA concentration that also completes digestion of RNA. This can be assessed by the FL-3 and FL-1 plots. There should be no bending of the CD71-positive population to the PI-positive side (**Figs. 3C** and **D**).

3.2.4. γ-Irradiation to Induce DNA Double-Strand Breaks

Many recombination repair mutants traditionally have been isolated on the basis of radiation hypersensitivity in yeast and rodent cell lines *(37–40)*. Radiation sensitivity is assessed here by radiation-induced micronucleus frequencies in individual mice. Mice are exposed to a low dose of γ-rays, which does not affect their survival or reproduction, and potential mutants are identified by elevated γ-ray-induced micronucleus frequencies in CD71-labeled reticulocytes.

1. Expose 6–8-wk-old animals to a 0.7-Gy dose of γ-rays from a ^{137}Cs source.

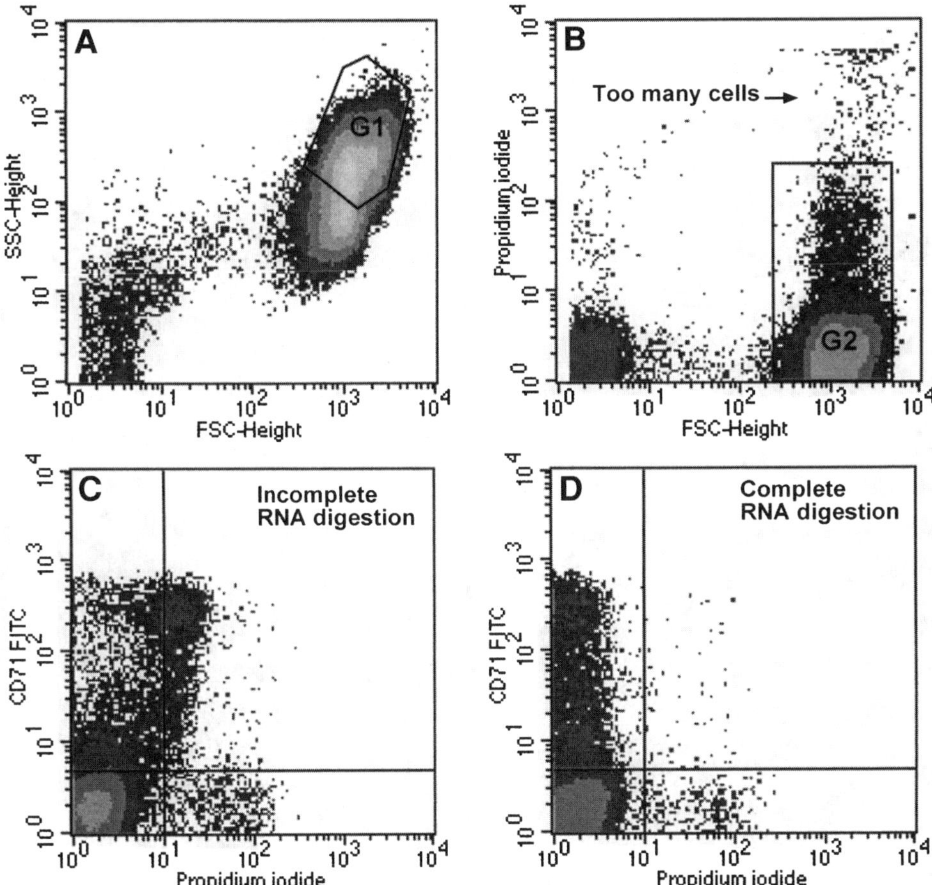

Fig. 3. Examples of inappropriate RNaseA concentrations. **(A)** If RNaseA treatment is too strong, erythrocyte population does not fit to G1 gate. **(B)** Too many events between nucleated cells and G2 gate are indicative of lysis of the cells by RNaseA. **(C)** Incomplete digestion of RNA in reticulocytes. PI also stains RNA; signals are detected by FL-3, limiting the resolution of micronucleated reticulocytes. **(D)** Complete digestion of RNA ensures the accuracy of the micronucleus measurement. FITC, fluorescein isothiocyanate; FSC, forward scatter; SSC, side scatter.

2. Forty-eight hours later, collect and fix peripheral blood as described in **Subheading 3.2.1.** (*see* **Note 17**).

3.2.5. Cell Preparation and Staining

1. Take fixed blood samples out of the freezer, and place them on ice.
2. Wash a fixed blood sample with 12 mL ice-cold bicarbonate buffer and centrifuge at 500*g* for 5 min.

3. Drain the supernatant, and resuspend the blood cells in minimum carry-over solution.
4. Make a chart to record tube number and sample description for each sample including single-colored controls.
5. Aliquot 20 µL blood cells from each sample into a 5-mL polystyrene round tube.
6. For each sample, make an 80-µL solution containing CD71-FITC antibody and RNaseA, and add it to a polystyrene tube containing 20 µL blood cells (*see* **Note 18**).
7. Prepare two single-colored samples stained with either CD71-FITC or PI as controls. Add 80 µL bicarbonate buffer to 20 µL blood cells for PI-single colored sample.
8. Stain the cells at 4°C for 45 min.
9. Wash the cells with cold 2 mL bicarbonate buffer by centrifugation.
10. Keep the samples on ice until analysis.
11. Immediately before flow cytometric analysis, add ice-cold 1 mL PI (1.25 µg/mL) to stain DNA contained in micronuclei. Start with single-colored controls (*see* **Note 19**).

3.2.6. Flow Cytometric Analysis of Spontaneous and γ-Ray-Induced Micronucleus Frequencies

Radiation-induced micronuclei must be analyzed in CD71-positive RETs, the youngest population in erythrocytes, since micronucleus formation requires a mitosis. On the other hand, because mature erythrocytes (normochromatic erythrocytes [NCEs]; CD71-negative) lacked nuclei at the time of irradiation, the frequencies of MN-NCEs would not be affected by irradiation; instead, they represent spontaneous micronucleus frequencies. Therefore, both spontaneous and radiation-induced micronucleus levels can be measured simultaneously in the same sample.

1. Call up the acquisition template and instrument setting. Before analyzing samples, wash the sample line in the flow cytometer with bicarbonate buffer for a few minutes to eliminate any debris (*see* **Note 20**).
2. Analyze samples at a rate around 5000 events/s. Start with single-colored controls.
3. Collect at least 10,000 CD71-positive RETs and 500,000 CD71-negative NCEs per sample.
4. Calculate micronucleus frequencies using the numbers of events from the quadrant-stats as follows (**Fig. 4**):

$$\gamma\text{-Ray-induced micronucleus frequencies (\%) in RET} = UR/(UL + UR) \times 100$$

$$\text{Spontaneous micronucleus frequencies (\%) in NCE} = LR/(LL + LR) \times 100$$

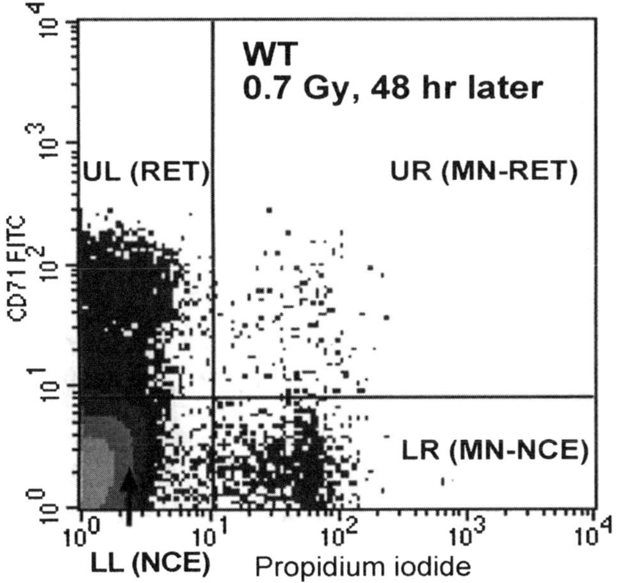

Quad	Events	% Gated	% Total	X Mean	Y Mean
UL	9035	1.19	1.08	1.50	43.06
UR	270	0.04	0.03	62.52	35.68
LL	750105	98.49	89.60	1.24	1.37
LR	2193	0.29	0.26	60.31	1.48

Fig. 4. Flow cytometric measurement of micronucleated erythrocytes. Lower left (LL) represents CD71-negative normochromatic erythrocytes (NCE, mature erythrocytes) without micronuclei. Micronucleated NCEs (MN-NCE) are stained with PI and detected in lower right (LR). CD71-positive cells in upper left (UL) and upper right (UR) are reticulocytes. Radiation-induced MN-RETs appear in UR. FITC, fluorescein isothiocyanate.

3.2.7. Potential Mutants Appear as Outliers

After analyzing some mice, distributions of spontaneous and γ-ray-induced micronucleus frequencies can be established. Obtain the means and standard deviations (SDs); potential mutants can be identified as those with micronucleus frequencies higher than 3 SD of the mean. If potential mutants have higher spontaneous micronucleus frequencies, bleed the animals later, and analyze the fresh blood samples (*see* **Note 21**). This step excludes some false-positives and confirms them as potential mutants. Then analyze the mutant families to confirm genetic inheritance.

4. Notes

1. **Table 1** gives a list of currently available antibodies to proteins with known and/or suspected roles in meiotic recombination. Only those references that describe the localization of each antibody to meiotic nuclei in mice are provided.
2. An alternative to commercial protease inhibitor cocktails is as follows: make stock solutions of pepstatin A 5 mg/mL in dimethylsulfoxide (DMSO), chymotrypsin 5 mg/mL in DMSO, antipain 5 mg/mL in water, leupeptin 5 mg/mL in water, aprotinin 10 mg/mL in water. For a 250× stock solution: 50 mL each of pepstatin A, chymotrypsin, antipain, leupeptin stocks + 250 mL of aprotinin stock + 9.55 mL of water; freeze in small aliquots at –20°C. Note that the protocol described assumes a 20× protease inhibitor cocktail. If this recipe is used, appropriate volume adjustments should be made for a final working concentration of 1×.
3. Antibody dilutions should be followed as recommended by the supplier. If no dilution is recommended or if results are unsatisfactory (i.e., too much background or nonspecific labeling), an appropriate dilution should be determined empirically. In addition, appropriate controls are essential for interpretation of immunocytological results. These controls should include one slide with no primary or secondary antibody, one slide with primary antibody alone, and one slide with secondary antibody alone. If mutant animals are used, then additional slides prepared from normal littermates should be included in each experiment.
4. If multiple secondary antibodies are used, they can be diluted together. For example, if two secondary antibodies are needed and the recommended dilution for both is 1:000, 1 μL of each secondary antibody can be diluted in 998 μL of ADB.
5. Two excellent sources for fluorophore-conjugated secondary antibodies are Molecular Probes (www.probes.com) and Jackson ImmunoResearch (www.jacksonimmuno.com).
6. The testes of a 3-wk-old mouse will make 12 slides. Older mice can be used, but they will not yield ideal results, because maturing spermatids and spermatozoa interfere with spreading. In addition, many antibodies will bind nonspecifically to the highly proteinaceous acrosome of developing, postmeiotic spermatids.
7. The microspreading protocol described in **Subheading 3.1.1.** is designed for one 3-wk-old male mouse. If more than one male mouse is required, then make the following changes: In **step 1**, prepare 12 slides for each animal. In **step 8**, prepare one 15-mL Falcon tube with 7 mL MEM-high glucose for each animal. If the animal has small testes, then a volume adjustment should be made and the number of slides reduced. The appropriate volume should be determined empirically, but in our experience 4–5 mL MEM-high glucose is appropriate for mutant animals with smaller testes.
8. The amount of humidity in the chamber is very important to the quality of spreading. If there is too little moisture, the slides will dry too quickly. It is important to use just enough moisture to keep the slides from drying. If too many paper towels are used, it will be difficult to keep the liquid on the slides level.
9. To avoid large fragments at this step, prop the dish at an angle and allow the fragments to settle. It is best to sacrifice some of the suspension, if necessary.

Table 1
Currently Available Antibodies

Designation	Description	Ref.	Source
SPO11	Required for double strand break formation	*38*	By collaboration
RAD51/DMC1	RecA homolog, single-stranded binding protein, early recombination protein	*39,40*	Oncogene Research Products
γ-H2AX	Histone H2AX, phosphorylated at Ser129, marker of double-strand breaks and chromatin dynamics	*20,41*	Upstate Biotechnology
RPA	Replication protein A, single-stranded DNA binding protein	*42,43*	By collaboration, Oncogene Research Products
MLH1	MutL homolog 1, mismatch repair, required for meiotic recombination	*17,44,45*	Pharmingen
MLH3	MutL homolog 3, mismatch repair, required for meiotic recombination	*46*	By collaboration
MRE11	Required for telomere maintenance	*47*	Oncogene Research
BLM	Bloom's syndrome, required for chromosome stability and recombination	*48–50*	By collaboration, Genetex
ATR	Ataxia telangiectasia and Rad 3 related	*51*	By collaboration, Genetex
ATM	Ataxia telangiectasia mutated	*43,52*	By collaboration, Novus
MSH4	MutS homolog 4, chromosome pairing	*13,53*	By collaboration

10. Generally, 1 mL of suspension per tube is appropriate, but if the cell density is higher or lower, more or less suspension may be required at this step. If desired, cells can be counted and optimum density can be determined empirically.
11. Once resuspended in sucrose, the cells should be allowed to sit no longer than 1–2 min before being dispensed onto fixative. If hypotonic treatment is prolonged and the nuclei become too swollen, they may burst. Conversely, if the nuclei do not swell sufficiently, they will not flatten as they adhere to the slide.
12. The heparinized blood should be fixed in absolute methanol kept at –80°C within 2 h.
13. Fixation is the most critical step in this assay. To maintain the temperature of methanol and the inside of the freezer, this step must be done as quickly as possible.
14. RNaseA digestion must be done at 4°C and not on ice. Some manufacturers sell RNaseA products that are already in solution. These products are not recommended for this assay.
15. If RNaseA activity is too strong, it will lyse erythrocytes, which will compromise the accuracy of measurement of micronuclei.
16. Younger animals have more reticulocytes than older ones. RNaseA has to be titrated for different ages.
17. A timecourse experiment is recommended to determine the peak of radiation-induced micronucleus induction under different conditions.
18. Match the tube numbers to the sample numbers in the parameter description so that some confusion by occasional mislabeling can be avoided.
19. Preparing single-colored controls ensures proper staining and efficient trouble-shooting.
20. Wash the sample line for 1 min between samples.
21. This flow cytometric assay is highly sensitive for identifying mouse mutants with higher spontaneous micronucleus frequencies. However, since micronuclei are also caused by chromosome loss events, radiation exposure is recommended to determine whether higher micronucleus frequencies are caused by failure to repair double-strand breaks.

References

1. Shedlovsky, A., Guenet, J.-L., Johnson, L., and Dove, W. (1986) Induction of recessive lethal mutations in the *T/t-H-2* region of the mouse genome by a point mutagen. *Genet. Res. Camb.* **47,** 135–142.
2. Shedlovsky, A., King, T. R., and Dove, W. F. (1988) Saturation germ line mutagenesis of the murine *t* region including a lethal allele at the quaking locus. *Proc. Natl. Acad. Sci. USA* **85,** 180–184.
3. Hitotsumachi, S., Carpenter, D. A., and Russell, W. L. (1985) Dose-repetition increases the mutagenic effectiveness of N-ethyl-N-nitrosourea in mouse spermatogonia. *Proc. Natl. Acad. Sci. USA* **82,** 6619–6621.
4. Hrabe de Angelis, M. H., Flaswinkel, H., Fuchs, H., et al. (2000) Genome-wide, large-scale production of mutant mice by ENU mutagenesis. *Nat. Genet.* **25,** 444–447.

5. Nolan, P. M., Peters, J., Strivens, M., et al. (2000) A systematic, genome-wide, phenotype-driven mutagenesis programme for gene function studies in the mouse. *Nat. Genet.* **25,** 440–443.
6. Schimenti, J. and Bucan, M. (1998) Functional genomics in the mouse: phenotype-based mutagenesis screens. *Genome Res.* **8,** 698–710.
7. Justice, M. J., Noveroske, J. K., Weber, J. S., Zheng, B., and Bradley, A. (1999) Mouse ENU mutagenesis. *Hum. Mol. Genet.* **8,** 1955–1963.
8. Nolan, P. M., Kapfhamer, D., and Bucan, M. (1997) Random mutagenesis screen for dominant behavioral mutations in mice. *Methods* **13,** 379–395.
9. Weber, J. S., Salinger, A., and Justice, M. J. (2000) Optimal N-ethyl-N-nitrosourea (ENU) doses for inbred mouse strains. *Genesis* **26,** 230–233.
10. Yoshida, K., Kondoh, G., Matsuda, Y., Habu, T., Nishimune, Y., and Morita, T. (1998) The mouse RecA-like gene *Dmc1* is required for homologous chromosome synapsis during meiosis. *Mol. Cell* **1,** 707–718.
11. Pittman, D., Cobb, J., Schimenti, K., et al. (1998) Meiotic prophase arrest with failure of chromosome pairing and synapsis in mice deficient for *Dmc1*, a germline-specific RecA homolog. *Mol. Cell* **1,** 697–705.
12. Edelmann, W., Cohen, P. E., Kneitz, B., et al. (1999) Mammalian MutS homologue 5 is required for chromosome pairing in meiosis. *Nat. Genet.* **21,** 123–127.
13. Kneitz, B., Cohen, P. E., Avdievich, E., et al. (2000) MutS homolog 4 localization to meiotic chromosomes is required for chromosome pairing during meiosis in male and female mice. *Genes Dev.* **14,** 1085–1097.
14. Romanienko, P. J. and Camerini-Otero, R. D. (2000) The mouse spo11 gene is required for meiotic chromosome synapsis. *Mol. Cell* **6,** 975–987.
15. Ward, J. O., Reinholdt, L. G., Hartford, S. A., et al. (2003) Toward the genetics of mammalian reproduction: induction and mapping of gametogenesis mutants in mice. *Biol. Reprod.* **69,** 1615–1625.
16. Baudat, F., Manova, K., Yuen, J. P., Jasin, M., and Keeney, S. (2000) Chromosome synapsis defects and sexually dimorphic meiotic progression in mice lacking spo11. *Mol. Cell* **6,** 989–998.
17. Pittman, D., Cobb, J., Schimenti, K., et al. (1998) Meiotic prophase arrest with failure of chromosome synapsis in mice deficient for Dmc1, a germline-specific RecA homolog. *Mol. Cell* **1,** 697–705.
18. Yoshida, K., Kondoh, G., Matsuda, Y., Habu, T., Nishimune, Y., and Morita, T. (1998) The mouse RecA-like gene Dmc1 is required for homologous chromosome synapsis during meiosis. *Mol. Cell* **1,** 707–718.
19. Xu, Y., Ashley, T., Brainerd, E., Bronson, R., Meyn, M., and Baltimore, D. (1996) Targeted disruption of ATM leads to growth retardation, chromosomal fragmentation during meiosis, immune defects, and thymic lymphoma. *Genes Dev.* **10,** 2411–2422.
20. Baker, S., Plug, A., Prolla, T., et al. (1996) Involvement of mouse Mlh1 in DNA mismatch repair and meiotic crossing over. *Nat. Genet.* **13,** 336–342.
21. Woods, L. M., Hodges, C. A., Baart, E., Baker, S. M., Liskay, M., and Hunt, P. A. (1999) Chromosomal influence on meiotic spindle assembly: abnormal meiosis I in female Mlh1 mutant mice. *J. Cell. Biol.* **145,** 1395–1406.

22. Yuan, L., Liu, J. G., Zhao, J., Brundell, J., Daneholt, B., and Hoog, C. (2000) The murine SCP3 gene is required for synaptonemal complex assembly, chromosome synapsis, and male fertility. *Mol. Cell* **5,** 73–83.
23. Celeste, A., Petersen, S., Romanienko, P. J., et al. (2002) Genomic instability in mice lacking histone H2AX. *Science* **296,** 922–927.
24. Yuan, L., Liu, J. G., Hoja, M. R., Wilbertz, J., Nordqvist, K., and Hoog, C. (2002) Female germ cell aneuploidy and embryo death in mice lacking themeiosis-specific protein SCP3. *Science* **296,** 1115–1118.
25. Dobson, M. J., Pearlman, R. E., Karaiskakis, A., Spyropoulos, B., and Moens, P. B. (1994) Synaptonemal complex proteins: occurrence, epitope mapping and chromosome disjunction. *J. Cell Sci.* **107,** 2749–2760.
26. Meuwissen, R. L., Offenberg, H. H., Dietrich, A. J., Riesewijk, A., van Iersel, M., and Heyting, C. (1992) A coiled-coil related protein specific for synapsed regions of meiotic prophase chromosomes. *EMBO J.* **11,** 5091–5100.
27. Ashley, T. and Plug, A. (1998) Caught in the act: deducing meiotic function from protein immunolocalization. In: *Meiosis and Gametogenesis* (Handel, M., ed.), vol. 37. Academic, San Diego, pp. 202–240.
28. Tarsounas, M. and Moens, P. B. (2002) Checkpoint and DNA-repair proteins are associated with the cores of mammalian meiotic chromosomes. *Curr. Top. Dev. Biol.* **51,** 109–134.
29. Heddle, J. A. (1973) A rapid in vivo test for chromosomal damage. *Mutat. Res.* **18,** 187–90.
30. Nusse, M., Miller, B. M., Viaggi, S., and Grawe, J. (1996) Analysis of the DNA content distribution of micronuclei using flow sorting and fluorescent in situ hybridization with a centromeric DNA probe. *Mutagenesis* **11,** 405–413.
31. Morita, T., Asano, N., Awogi, T., et al. (1997) Evaluation of the rodent micronucleus assay in the screening of IARC carcinogens (groups 1, 2A and 2B) the summary report of the 6th collaborative study by CSGMT/JEMS MMS. Collaborative Study of the Micronucleus Group Test. Mammalian Mutagenicity Study Group. *Mutat. Res.* **389,** 3–122.
32. Dertinger, S. D., Torous, D. K., and Tometsko, K. R. (1996) Simple and reliable enumeration of micronucleated reticulocytes with a single-laser flow cytometer. *Mutat. Res.* **371,** 283–292.
33. Dertinger, S. D., Torous, D. K., Hall, N. E., Tometsko, C. R., and Gasiewicz, T. A. (2000) Malaria-infected erythrocytes serve as biological standards to ensure reliable and consistent scoring of micronucleated erythrocytes by flow cytometry. *Mutat. Res.* **464,** 195–200.
34. Shima, N., Hartford, S., Duffy, T., Wilson, L., Schimenti, K., and Schimenti, J. C. (2003) Phenotype based identification of mouse chromosome instability mutants. *Genetics* **163,** 1031–1040.
35. Peters, A. H., Plug, A. W., van Vugt, M. J., and de Boer, P. (1997) A drying-down technique for the spreading of mammalian meiocytes from the male and female germline. *Chromosome Res.* **5,** 66–68.

36. Javois, L. C. (1999) Direct immunofluorescent labeling of cells. *Methods Mol. Biol.* **115,** 107–111.
37. Thompson, L. H., Brookman, K. W., Dillehay, L. E., et al. (1982) A CHO-cell strain having hypersensitivity to mutagens, a defect in DNA strand-break repair, and an extraordinary baseline frequency of sister-chromatid exchange. *Mutat. Res.* **95,** 427–440.
38. Jones, N. J., Cox, R., and Thacker, J. (1987) Isolation and cross-sensitivity of X-ray-sensitive mutants of V79-4 hamster cells. *Mutat. Res.* **183,** 279–286.
39. Jones, N. J., Cox, R., and Thacker, J. (1988) Six complementation groups for ionising-radiation sensitivity in Chinese hamster cells. *Mutat. Res.* **193,** 139–144.
40. Friedberg, E., Walker, G., and Siede, W. (1995) *DNA repair and mutagenesis.* American Society for Microbiology, Washington, DC.

II

Recombination as a Reporter of Genomic Instability

7

Detecting Carcinogens With the Yeast DEL Assay

Richard J. Brennan and Robert H. Schiestl

Summary

The yeast DEL assay is a simple, rapid method for measuring the frequency of reversion of a disrupted *his3* gene by homologous intrachromosomal recombination. Reversion to histidine prototrophy results in deletion (DEL) of the disrupting sequence. The DEL assay has been used to study the effects of various DNA-damaging treatments on the frequency of deletion-recombination and has been shown to have a high level of sensitivity and specificity toward carcinogens, many of which are poorly detected by bacterial mutagenicity and other short-term genotoxicity assays. The DEL assay therefore is a useful addition to the arsenal of predictive tests for genotoxicity and carcinogenicity. This chapter provides an in-depth description of materials and methods for the yeast DEL assay from a user's prospective and should allow the assay to be successfully deployed in any laboratory with basic microbiological capability and minimal user training.

Key Words: DNA deletions, intrachromosomal recombination, carcinogens, xenobiotics, genetic control, genetic toxicology assay, predictive carcinogenesis

1. Introduction

Genomic rearrangements such as inversions, translocations, and deletions are involved in carcinogenesis ***(1–4)***. The *p53* gene is the most commonly altered gene yet identified in human cancers, occurring in a large fraction (perhaps even half) of the total cancers in the United States ***(5)***. Cancer-prone families with Li-Fraumeni syndrome show germline transmission of a mutated *p53* gene ***(6)***. Many human tumors, however, show deletions of a small region of chromosome 17p including the *p53* gene, and where a *p53* point mutation has been identified, the second wild-type allele is lost before tumor development ***(7)***. This process, known as loss of heterozygosity, is frequently the result of genome rearrangements such as deletions. Several genetic diseases cause genomic instability and at the same time a high frequency of carcinogenesis.

From: *Methods in Molecular Biology, vol. 262, Genetic Recombination: Reviews and Protocols*
Edited by: A. S. Waldman © Humana Press Inc., Totowa, NJ

An elevated frequency of recombination and genome rearrangements is found in cells from human patients suffering from the cancer-prone diseases ataxia telangiectasia ***(8)***, Li-Fraumeni syndrome ***(9)***, Bloom's syndrome ***(10)***, Werner's syndrome ***(11)*** Cockayne's syndrome, Fanconi's anemia, Lynch syndromes I and II, Wiskott-Aldrich syndrome, and xeroderma pigmentosum ***(12)***. Defects in the genes responsible for Bloom's syndrome and for Werner's syndrome encode proteins highly homologous to RecQ helicases, which are involved in recombination and are central in the maintenance of genomic stability ***(12)***.

Because of the association among recombination, deletion, and carcinogenesis, systems that measure the frequency of intrachromosomal homologous recombination, resulting in genomic deletions (DEL), have been developed in organisms as diverse as yeast ***(13)***, human ***(14)***, mouse ***(15)***, hamster ***(16)***, tobacco ***(17)***, and arabidopsis ***(18)***. In all cases these recombination systems have been shown to be responsive to genotoxic stresses.

The simplest and most user-friendly of the intrachromosomal recombination systems is the yeast DEL recombination assay. DEL recombination in yeast is induced by a broad range of mutagenic insults including ionizing and nonionizing radiation, alkylating agents, oxidative stress, DNA strand breaks and heat shock ***(19–22)***. The assay responds to many compounds, which, although they cause cancer in long-term rodent bioassays, are poorly detected by most other short-term mutagenicity assays ***(20,23,24)***. For instance, the *Salmonella* reverse mutation (Ames) test, which is based on the reversion of point mutations in *Salmonella typhimurium*, and which is the most widely used short-term mutagenicity test, detects only about 50% of carcinogens ***(25)***.

In a broad interlaboratory study, 60 laboratories investigated the ability of commonly used short-term genotoxicity assays to classify correctly eight carcinogens and two noncarcinogens ***(26)***. Bacterial mutation assays were consistently negative with all chemicals, with the Ames assay correctly classifying only 2 of the 10 (20%). The overall performance (accuracy) of the other assays was between 40 and 60%. In a later study, 60% of these same chemicals were correctly identified in the yeast DEL assay ***(23)***.

Table 1 lists chemicals of known carcinogenicity that have so far been tested in the yeast DEL assay. With this wider set of chemicals, the DEL assay shows a sensitivity of 90% (44/49 carcinogens detected), a specificity of 75% (5/20 noncarcinogens detected as false-positives), and an overall accuracy of 86% (59/69 compounds correctly classified). The Ames assay and other short-term assays incorrectly classify many of these compounds. The yeast assay also discriminates between carcinogen/noncarcinogen pairs of closely related structure ***(20,24,27,28)***.

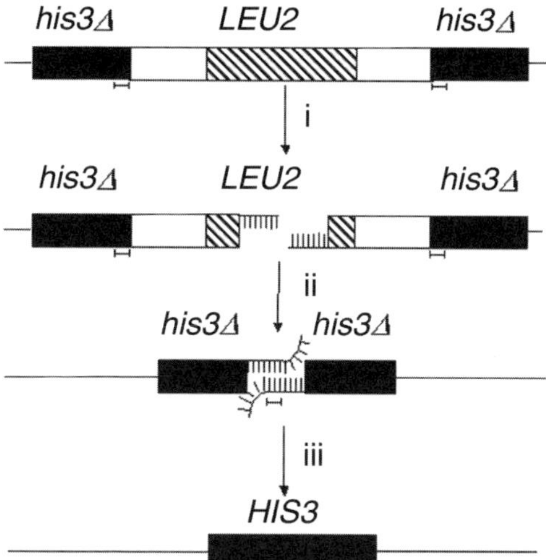

Fig. 1. Schematic representation of the yeast DEL recombination construct and its reversion by the most likely mechanism (single-strand annealing) to regenerate a functional *HIS3* gene. Horizontal bars indicate regions of homology. DNA damage directly or indirectly generates a double-strand break (i); the break is extended by exonucleases to a gap (ii) until a region of homology is reached where strand annealing occurs (iii); the nonhomologous single-stranded overhangs are cleaved by the RAD1/RAD10 endonuclease and the ends ligated.

Because of its promising performance as a sensitive and specific assay for carcinogens, the yeast DEL assay has elicited much interest as a predictive screen for carcinogenic activity and has been used to investigate the genotoxicity of environmental contaminants ***(23)***, agricultural chemicals ***(29)***, and pharmaceuticals ***(30)***. The DEL assay also has great utility as a tool for investigating the mechanism and genetic control of homologous recombination and genomic instability ***(22,31–34)***.

The yeast DEL assay works by measuring the recombination frequency between two tandem *his3*Δ alleles (*his3*Δ3', *his3*Δ5'), each containing approx 400 bp of homology to the other allele. The DEL assay was constructed in the yeast *S. cerevisiae* by disruptive integration of the plasmid pRS6, containing an internal fragment of the HIS3 gene and a *LEU2* marker gene, into the *HIS3* locus ***(32)***, which encodes the enzyme imidazole-glycerolphosphate dehydratase. The resulting strain is unable to synthesize histidine. Homologous recombination between the two *his3*Δ alleles results in the deletion (DEL) of intervening sequences to give a functional *HIS3* gene (**Fig. 1**) and a reversion to histidine prototrophy. Practically, the assay is similar to the liquid preincubation version of the Ames *Salmonella* mutagenicity assay ***(35)***. Cells are

Table 1
Response of Carcinogens and Noncarcinogens in the Yeast DEL Recombination Assay[a]

Compound	Carcinogenicity	Response in DEL	Compound	Carcinogenicity	Response in DEL
Safrole	+	+	EMS	+	+
Ethionine	+	+	Nitrogen mustard	+	+
Urethane	+	+	Epichlorohydrin	+	+
Auramine O	+	+	Aflatoxin B1	+	+
Methylene chloride	+	+	Ethylene dibromide	+	+
Carbon tetrachloride	+	+	Dimethyl-hydrazine	+	+
Cadmium chloride	+	+	Cyclophosphamide	+	+
Cadmium sulfate	+	+	Formaldehyde	+	+
3-Amino-1,2,4-triazole	+	+	Ethylene oxide	+	+
Acetamide	+	+	Propylene oxide	+	+
Thioacetamide	+	+	2,4-Diaminotoluene	+	+
Thiourea	+	+	TPA	+	–
DDE	+	+	Diethyl-stilbestrol	+	–
Ethylene-thiourea	+	+	Peroxysome proliferators	+	–
Aniline	+	+	Diethylhexyl-phthalate	+	–
o-Toluidine	+	+	Phenobarbital	+	–
o-Anisidine	+	+	2,6-Diaminotoluene	–	–
Hexamethyl-phosphoramide	+	+	Hydroxylamine HCl	–	–
Acrylonitrile	+	+	Sodium azide	–	–
Benzene	+	+	5-Bromouracil	–	–
Arsenate	+	+	2-Aminopurine	–	+
Catechol	+	+	Ethidium bromide	–	+

Aroclor 1221 (PCB)	+	+	Benzoin	–	–
UV irradiation	+	+	Methionine	–	–
γ-Ray exposure	+	+	Ethanol	–	–
4-NQO	+	+	Caprolactam	–	+
MMS	+	+	2-Acetoamido-fluorene	+	+
2-Amino anthracene	+	+	Hydroxyurea	+	+
2-Nitrofluorene	+	+	Hydroquinone	+	+
Formaldehyde	+	+	Acetone	–	–
1-Butanol	–	–	Methanol	–	–
Phenol	–	+	1-Propanol	–	–
2-Propanol	–	–	L-Tryptophan	–	–
Benzoin	–	–	Buctril	–	–
o-Xylene	–	–			

[a]Information about the carcinogenicity of the agents was obtained from ref. ***38*** and IARC reports.

exposed to test agents in liquid culture and subsequently washed, counted, diluted, plated, and scored for survival and *HIS* reversion on selective media plates. The assay can be performed in the presence of an external metabolic system such as rat S9 liver microsomes, allowing the effect of mammalian metabolic enzymes on genotoxicity to be determined.

The most widely used yeast DEL strain is the diploid strain RS112, which also carries heteroallelic mutations in both *ADE2* loci, resulting in adenine auxotrophy. RS112 therefore can be used to score the frequency of *inter*chromosomal gene conversion in addition to *intra*chromosomal DEL recombination. The two types of recombination are mechanistically distinct, under different genetic control, and respond differently to various types of genotoxic insult ***(20)***.

Like the Ames assay, the DEL assay can be performed by any operator reasonably skilled in microbiological techniques and with access to a basic microbiological laboratory. Detailed procedures for performing the assay are given here, including a number of precautions to ensure that the highest sensitivity and reproducibility are maintained.

2. Materials

2.1. Yeast Strains

The diploid *S. cerevisiae* strain RS112 is the most widely used yeast DEL assay strain. RS112 contains the DEL system on one homolog (*HIS3*::pRS6) and a deletion of the entire open reading frame of *HIS3* on the other homolog (his3-Δ200). RS112 has the genotype *MATa*/α *ura3-52*/*ura3-52 leu2-3,112*/*leu2-Δ98 trp5-27*/*TRP5 arg4-3*/*ARG4 ade2-40*/*ade2-101 ilv1-92*/*ILV1 HIS3*::pRS6/*his3-Δ200 LYS2*/*lys2-801*.

A haploid strain is available containing the same DEL construct but without the ability to score for gene conversion between *ade2* heteroalleles. The haploid strain RSY6 has the genotype *MATa ura3-52 leu2-3,112 trp5-27 arg4-3 ade2-40 ilv1-92 HIS3*::pRS6 (*see* **Note 1**).

2.2. Media (see Notes 2–6)

1. YPAD plates: 1% yeast extract, 2% peptone, 2% dextrose, 48 mg/mL adenine sulfate, 2% bacto agar. Dissolve in water and autoclave before pouring.
2. Amino acid mix: 1.8 g L-adenine hemisulphate (Sigma, cat. no. A9126), 1.2 g L-arginine-HCl (Sigma, cat. no. A5131), 1.2 g L-histidine-HCl (Sigma, cat. no. H8125), 6.0 g homoserine (Sigma, cat. no. H1001), 1.8 g L-isoleucine (Sigma, cat. no. I2752), 1.8 g L-leucine (Sigma, cat. no. L8000), 1.8 g L-lysine (Sigma, cat. no. L5626), 1.2 g L-methionine (Sigma, cat. no. M9625), 3.0 g L-phenylalanine (Sigma, cat. no. P2126), 2.4 g L-tryptophan (Sigma, cat. no. T0254) 1.8 g L-tyrosine (Sigma, cat. no. T3754), 1.2 g uracil (Sigma, cat. no. U0750), 9.0 g L-valine (Sigma, cat. no. V0500).

   Synthetic complete mix: Consists of all 13 amino acids (and bases).

3. Synthetic complete agar plates (SC): 4.0 g yeast nitrogen base (YNB) without amino acids, 0.4 g amino acid mixture, 12 g dextrose, 10 g agar. Dissolve in 600 mL of water and autoclave.
4. Synthetic leucine dropout agar plates: (SC-Leu): 4.0 g YNB without amino acids, 0.4 g amino acid mixture (minus leucine), 12 g dextrose, 10 g agar. Dissolve in 600 mL of water and autoclave.
5. Histidine dropout agar plates (SC-His): 4.0 g YNB without amino acids, 0.4 g amino acid mixture (minus histidine), 12 g dextrose, 10 g agar. Dissolve in 600 mL of water and autoclave.
6. Adenine dropout agar plates (SC-Ade): 4.0 g YNB without amino acids, 0.4 g amino acid mixture (minus adenine), 12 g dextrose, 10 g agar. Dissolve in 600 mL of water and autoclave.
7. Liquid leucine dropout medium: 4.0 g YNB without amino acids, 12 g dextrose, 0.4 g amino acid mixture (minus leucine). Dissolve in water and autoclave.
8. 30% aqueous glycerol solution. Autoclave, store at room temperature.

2.3. Mammalian Liver Microsome (S9) Mix Components

1. Aroclor-1254-induced S9 fraction (*see* **Note 7**). Store 1-mL aliquots at –70°C.
2. 0.04 *M* Nicotinamide, adenine, dinucleotide phosphate (NADP). Filter-sterilize. Store 1-mL aliquots at –20°C.
3. 1 *M* KCl. Autoclave. Store at 4°C.
4. 0.25 *M* $MgCl_2$. Autoclave. Store at 4°C.
5. 0.2 *M* Glucose-6-PO_4. Filter-sterilize. Store 0.5-mL aliquots at –20°C.
6. 0.2 *M* $NaHPO_4$ buffer, pH 7.4. Store at 4°C.
7. S9 buffer: 0.1 *M* $NaPO_4$, pH 7.8, 0.033 *M* KCl, 8 m*M* $MgCl_2$. Autoclave. Store at 4°C.

2.4. Supplies

1. 15-cm Sterile Petri dishes.
2. Disposable sterile microbiological loops.
3. 15-mL Sterile disposable centrifuge tubes.
4. 50-mL Sterile disposable centrifuge tubes.
5. Hemocytometer.
6. Sterile cryogenic storage vials.
7. Sterile 1.8-mL microcentrifuge tubes.
8. Glass or disposable plastic spreaders.

2.5. Control Compounds

1. Direct-acting positive control for DEL and gene conversion, methyl methane sulfonate.
2. Metabolic activation control, cyclophosphamide.

3. Methods

3.1. Culture and Storage of Yeast Strains

1. Streak out yeast strains received as frozen cultures or agar slants onto fresh YPAD agar plates using a sterile disposable loop. Incubate plates at 30°C for 2–3 d until individual colonies appear. Inoculated plates may be sealed with Parafilm and stored at 4°C for several weeks.
2. Pick half of a well-isolated colony from a YPAD plate into 15-mL leucine dropout liquid medium (SC-Leu; *see* **Note 8**) in a 50-mL sterile disposable centrifuge tube. Incubate overnight at 30°C, shaking at 250 rpm (*see* **Note 9**). Streak the other half of the colony onto histidine and adenine dropout plates and incubate at 30°C for 3 d to ensure the correct unstable His^- and Ade^- phenotypes.
3. For long-term storage, add 30% glycerol in equal volume to a fresh overnight culture of the yeast strain in SC-Leu medium. Pipette 1-mL aliquots of the mixture into sterile cryogenic storage vials and place directly into storage at –70°C (*see* **Note 10**).
4. For ongoing laboratory culture, a small amount of frozen glycerol stock can be picked directly from the storage vial using a sterile plastic micropipet tip and the vial returned to –70°C without thawing. Streak the aliquot of the stock culture onto a YPAD agar plate and incubate at 30°C for 2–3 d to allow growth of individual colonies before short-term storage of the plate at 4°C.

3.2. The DEL Assay

3.2.1. Preparation of 10-mL S9 Microsomal Metabolism Mix (see **Note 11**)

1. Chill a 15-mL sterile disposable centrifuge tube on ice.
2. Add 0.33 mL 1 *M* KCl, 0.32 mL 0.25 *M* $MgCl_2$, 0.25 mL 0.2 *M* glucose-6-PO_4, 1.00 mL 0.04 *M* NADP, 5.00 mL 0.2 *M* $NaHPO_4$, 2.10 mL sterile deionized H_2O to a chilled tube.
3. Mix and allow to chill to 4°C.
4. Thaw a 1.00 mL aliquot of S9 fraction on ice.
5. Add S9 fraction to buffer. Mix gently and store on ice until use.

3.2.2. Exposure of Yeast Cultures to Test Compound

1. Pipet 10 µL of a fresh overnight culture of the yeast strain in SC-Leu into the counting chamber of a hemocytometer underneath a coverslip. Check for even cell distribution and measure the cell density by counting 100–200 cells (*see* **Note 9**).
2. Dilute the culture to 2×10^7 cells/mL in fresh SC-Leu medium.
3. Prepare stock solutions of test agent in SC-Leu medium. Aliquot required amounts (*see* **Notes 12–14**) into a separate sterile 15-mL disposable centrifuge tube for each assay and add additional SC-Leu to a final volume of 4.5 mL. For agents with low aqueous solubility, initial stock solutions in dimethylsulfoxide (DMSO), EtOH, or acetone may be used and diluted in SC-Leu to a maxi-

mum final solvent concentration of 3% in the assay. Equal amounts of solvent must be added to control cultures.

4. When S9 microsomal metabolizing system is to be used, substitute 1.4 mL SC-Leu with 1.4 mL S9 mix. Use 1.4 mL of S9 buffer for –S9 controls.
5. For positive control cultures, add methyl methane-sulfonate (MMS; final concentration 100 µg/mL) or cyclophosphamide (25 mg/mL).
6. Add 0.5 mL of resuspended yeast culture from **step 2**.
7. Cap tubes tightly and incubate for 16–18 h at 30°C with shaking at 250 rpm. Tubes should be shaken at a 45° angle from the horizontal to ensure adequate agitation.
8. Precipitate cells by spinning in a benchtop centrifuge at 3200*g* for 3 min. Wash cells three times with 5 mL sterile deionized water (*see* **Note 15**). Resuspend cells in 5 mL sterile water.

3.2.3. Dilution and Plating of Treated Yeast Cultures

1. Visually inspect the resuspended culture for growth. Cultures that have undergone several generations of cell division during exposure will appear cloudy, opaque, and white in color. Cultures that have achieved few or no cell divisions during exposure will be only slightly or not at all cloudy, and transparent.
2. Recentrifuge cultures that have achieved few or no cell divisions and resuspend in 1 mL sterile deionized H_2O (*see* **Note 16**).
3. Count resuspended cells in hemocytometer (*see* **Subheading 3.3.2., step 1**).
4. Prepare four dilution tubes per treated and control culture by dispensing 900 µL sterile deionized H_2O into sterile 1.8-mL microcentrifuge tubes.
5. Serially dilute each treated and control culture in four 10-fold steps by removing 100 µL of resuspended culture into a dilution tube, vortexing in three short bursts (1–2 s) to mix.
6. Pipette 100-µL aliquots of the appropriate dilutions onto SC plates to assess survival, SC-His plates to assess DEL recombination frequency, and (when desired) SC-Ade plates to assess gene conversion frequency (*see* **Note 17**).
7. Spread the cell suspension evenly over the plate using a sterile spreader, keep plates upright for 30 min to allow absorption of the liquid into the agar, and then incubate inverted at 30°C.
8. Count the number of colonies appearing on the plates after 3 d.
9. Percent survival is calculated by dividing the viable colony count (from SC plates) by the number of cells plated following exposure (from the cell count on the hemocytometer). DEL recombination and gene conversion frequencies are calculated by dividing the number of His^+ cells (from colony counts on SC-His plates) or Ade^+ cells respectively (from colony counts on SC-Ade plates) by the number of viable cells (*see* **Note 18**).

4. Notes

1. The haploid strain has a spontaneous DEL recombination frequency approx fivefold higher than the diploid strain. This may result in a slightly lower sensitivity

to weakly recombinagenic treatments and must be taken into account when diluting cells prior to plating for scoring recombination rates.

2. Preformulated mixes for complete (YPAD), synthetic complete, and dropout (lacking specific nutrients) yeast media are conveniently available commercially (e.g., from Qbiogene/Bio101). Alternatively, the media can be made using the recipes available in Burke et al. ***(36)***.
3. For all the growth media described (with the exception of adenine dropout medium) it is vital to maintain a concentration of adenine sufficient to prevent adenine starvation. Certain adenine-deficient mutants of *S. cerevisiae* grown under adenine-limiting conditions accumulate an adenine precursor, which gives them a noticeable pink or red coloration. The pigment responsible for the coloration impairs the growth of *ade2* mutants and leads to the enrichment of the culture with *ADE2* revertants ***(37)***. The addition of an extra 50–100 mg/L adenine sulfate to each recipe is recommended. This can safely be done prior to sterilization. Colonies or cultures of recombination tester strains that have accumulated a visible pink or red tint should not be used in recombination assays.
4. Glucose in the medium is prone to caramelize if autoclaving is excessive. Media caramelization reduces the growth rate and maximum cell density of the yeast cultures. Synthetic medium should be clear and a very pale yellow color following autoclaving. YPAD should be a copper color. Dark yellow synthetic or dark copper to brown YPAD medium may be caramelized. In this case, reduce the autoclaving time by a few minutes. Ensure, however, that all the agar in medium for plates is dissolved following sterilization. The presence of small, clear agar crystals following autoclaving indicates insufficient sterilization time.
5. It is not generally necessary to take excessive precautions to ensure sterility. In a clean, closed laboratory, with good aseptic practice, media and plate preparation can safely be done on the open benchtop.
6. Liquid media, and plates sealed in plastic bags, may be stored for several weeks at 4°C. Freshly poured agar plates should be allowed to dry for several days prior to storage to avoid contamination. Stored sleeves of plates containing one or more contaminated plates or showing visible water droplets inside the sleeve should be discarded. Similarly, liquid medium should be checked for cloudy precipitates or visible mold growth before use.
7. Available from Microbiological Associates Inc. 9900 Blackwell Rd., Rockville, MD 20850.
8. The DEL recombination construct contains a *LEU2* marker gene in the sequence between the two *his3* alleles. The *LEU2* marker is lost during the intrachromosomal recombination and reversion to histidine prototrophy (**Fig. 1**). The presence of the marker allows positive selection for the recombination system and negative selection against the *HIS* revertants. The growth of liquid cultures in SC-Leu blocks the growth of early "jackpot" recombination events and ensures a low, reproducible spontaneous recombination frequency. The spontaneous His$^+$ frequency for RS112 in SC-Leu is highly reproducible at $0.5–2 \times 10^{-4}$ per viable cell. Experiments in which the spontaneous rate significantly exceeds these parameters should be considered suspect.

9. RS112 will reach a final density of approx 5×10^7 cells/mL in SC-Leu medium.
10. Yeast cultures for long-term storage should not be snap-frozen in dry ice/ethanol or liquid N_2, as rapid freezing significantly reduces viability.
11. S9 mix must be prepared fresh and used immediately. Ten milliliters of mix is sufficient for seven individual assays. Commercial 1-mL aliquots of S9 fraction, which are expensive, often contain slightly more than 1 mL, and it is often possible to make sufficient S9 mix for eight assays (11.2 mL) from each aliquot.
12. Some agents positive in the yeast DEL assay—typically those agents negative in other short-term genotoxicity assays—induce DEL recombination only at doses that are also cytotoxic. It is therefore necessary to test agents over a dose range that covers the killing curve from 0 to 90% cell death. When an agent of unknown toxicity is to be tested, an initial dose-finding experiment with a broad dose range is recommended.
13. Institutional guidelines for the safe storage, handling, and disposal of hazardous materials must be followed. Appropriate personal protective equipment must be used at all times and training in hazardous material handling obtained where necessary. Particular care should be exercised when agents of unknown toxicity are to be tested.
14. When the effects on DEL recombination of exposure to nonchemical agents are to be assessed (e.g., ionizing or nonionizing radiation), cells can be exposed directly to the agent in culture medium. Exposure can be performed to cultures in 15-mL tubes or (e.g., for UV irradiation when the medium may attenuate the dose to the cell) in a thin layer on the bottom of a sterile Petri dish.
15. When S9 mix is used, a fibrous protein deposit may form in the culture. This precipitate complicates dilution and counting of the exposed cells. The cells can be separated from the precipitate prior to washing by filtration of the culture through a sterilized tissue (Kimwipe) using copious backwashing with sterile deionized H_2O.
16. Certain treatments may cause the cells to clump together, complicating counting, dilution, and plating. Vigorous vortexing and light sonication in a sonicating water bath can often disrupt clumps. The use of 1 m*M* EGTA to resuspend the cells may also help.
17. The aim ideally is to plate 100–500 viable cells and generate 100–500 countable colonies. In a control culture that has grown to 5×10^7 cells/mL, 100 µL of a 10^4 dilution will contain 500 cells (viability of control cultures should be 90–100%), and 100 µL of the undiluted resuspension will contain (assuming a recombination frequency of 1×10^{-4}) 500 *HIS*$^+$ cells. The spontaneous frequency of gene conversion in RS112 is more variable than that of DEL recombination but is generally in the range of $1–5 \times 10^{-6}$. In treated cultures, in which viability is reduced, it may be necessary to plate higher cell concentrations (lower dilutions) to achieve a sufficient number of colonies. When viability and/or recombination frequencies are particularly low, it is possible to plate up to 300 µL of cell suspension per plate. Too high cell concentrations, however, can lead to cryptic growth of nonrecombinant cells on dropout plates as viable cells scavenge essential nutrients from dead cells. Cryptic growth is generally identifiable as numer-

ous microcolonies (< 1-mm diameter) or a thin background lawn of cells. Plates exhibiting suspected cryptic growth should not be used in calculating recombination frequencies.

18. Our quality control criteria for a successful experiment are as follows: negative control culture viability ≥ 90%, spontaneous DEL recombination frequency 0.5–2×10^{-4}, positive control DEL recombination frequency with 100 µg/mL MMS approx 25-fold induced over negative control, positive control frequencies with 25 mg/mL cyclophosphamide similar to negative control (–S9) or approx 5-fold induced (+S9). We generally consider a treatment to be a positive inducer of recombination when a twofold or higher increase in recombination frequency is observed, with a statistical significance of ≤0.05 (Student's *t*-test on independent replicate determinations) and a positive dose response.

Acknowledgments

This work was supported by a grant from the National Cancer Institute, NIH RO1 CA82473, as well as funding from the UCLA Center for Occupational and Environmental Health (to R.H.S.).

We thank Zhanna Kirpnick for useful suggestions during the preparation of this manuscript.

References

1. Sandberg, A. A. (1991) Chromosome abnormalities in human cancer and leukemia. *Mutat. Res.* **247,** 231–240.
2. Tlsty, T. D., Briot, A., Gualberto, A., et al. (1995) Genomic instability and cancer. *Mutat. Res.* **337,** 1–7.
3. Bishop, A. J. and Schiestl, R. H. (2002) Homologous recombination and its role in carcinogenesis. *J. Biomed. Biotechnol.* **2,** 75–85.
4. Bishop, J. M. (1987) The molecular genetics of cancer. *Science* **235,** 305–311.
5. Hollstein, M., Sidransky, D., Vogelstein, B., and Harris, C. C. (1991) p53 mutations in human cancers. *Science* **253,** 49–53.
6. Malkin, D., Li, F. P., Strong, L. C., et al. (1990) Germ line p53 mutations in a familial syndrome of breast cancer, sarcomas, and other neoplasms. *Science* **250,** 1233–1238.
7. Vogelstein, B. (1990) Cancer. A deadly inheritance. *Nature* **348,** 681–682.
8. Meyn, M. S. (1993) High spontaneous intrachromosomal recombination rates in ataxia-telangiectasia. *Science* **260,** 1327–1330.
9. Livingstone, L. R., White, A., Sprouse, J., Livanos, E., Jacks, T., and Tlsty, T. D. (1992) Altered cell cycle arrest and gene amplification potential accompany loss of wild-type p53. *Cell* **70,** 923–935.
10. German, J. (1993) Bloom syndrome: a mendelian prototype of somatic mutational disease. *Medicine (Baltimore)* **72,** 393–406.
11. Fukuchi, K., Martin, G. M., and Monnat, R. J., Jr. (1989) Mutator phenotype of Werner syndrome is characterized by extensive deletions. *Proc. Natl. Acad. Sci. USA* **86,** 5893–5897.

12. Ellis, N. A. (1996) Mutation-causing mutations. *Nature* **381,** 110–111.
13. Schiestl, R. H., Gietz, R. D., Mehta, R. D., and Hastings, P. J. (1989) Carcinogens induce intrachromosomal recombination in yeast. *Carcinogenesis* **10,** 1445–1455.
14. Aubrecht, J., Rugo, R., and Schiestl, R. H. (1995) Carcinogens induce intrachromosomal recombination in human cells. *Carcinogenesis* **16,** 2841–2846.
15. Schiestl, R. H., Aubrecht, J., Khogali, F., and Carls, N. (1997) Carcinogens induce reversion of the mouse pink-eyed unstable mutation. *Proc. Natl. Acad. Sci. USA* **94,** 4576–4581.
16. Zhang, L. H. and Jenssen, D. (1994) Studies on intrachromosomal recombination in SP5/V79 Chinese hamster cells upon exposure to different agents related to carcinogenesis. *Carcinogenesis* **15,** 2303–2310.
17. Lebel, E. G., Masson, J., Bogucki, A., and Paszkowski, J. (1993) Stress-induced intrachromosomal recombination in plant somatic cells. *Proc. Natl. Acad. Sci. USA* **90,** 422–426.
18. Kovalchuk, O., Titov, V., Hohn, B., and Kovalchuk, I. (2001) A sensitive transgenic plant system to detect toxic inorganic compounds in the environment. *Nat. Biotechnol.* **19,** 568–572.
19. Brennan, R. J., Swoboda, B. E., and Schiestl, R. H. (1994) Oxidative mutagens induce intrachromosomal recombination in yeast. *Mutat. Res.* **308,** 159–167.
20. Schiestl, R. H. (1989) Nonmutagenic carcinogens induce intrachromosomal recombination in yeast. *Nature* **337,** 285–288.
21. Davidson, J. F., Whyte, B., Bissinger, P. H., and Schiestl, R. H. (1996) Oxidative stress is involved in heat-induced cell death in *Saccharomyces cerevisiae. Proc. Natl. Acad. Sci. USA* **93,** 5116–5121.
22. Galli, A. and Schiestl, R. H. (1998) Effects of DNA double-strand and single-strand breaks on intrachromosomal recombination events in cell-cycle-arrested yeast cells. *Genetics* **149,** 1235–1250.
23. Carls, N. and Schiestl, R. H. (1994) Evaluation of the yeast DEL assay with 10 compounds selected by the International Program on Chemical Safety for the evaluation of short-term tests for carcinogens. *Mutat. Res.* **320,** 293–303.
24. Brennan, R. J. and Schiestl, R. H. (1999) The aromatic amine carcinogens o-toluidine and o-anisidine induce free radicals and intrachromosomal recombination in *Saccharomyces cerevisiae. Mutat. Res.* **430,** 37–45.
25. Ashby, J. and Tennant, R. W. (1991) Definitive relationships among chemical structure, carcinogenicity and mutagenicity for 301 chemicals tested by the U.S. NTP. *Mutat. Res.* **257,** 229–306.
26. Ashby, J., Draper, M., Matter, B. E., et al., eds. (1985) *Evaluation of Short Term Tests for Carcinogens.* Progress in Mutation Research, vol. 5, Elsevier, New York.
27. Brennan, R. J. and Schiestl, R. H. (1998) Chloroform and carbon tetrachloride induce intrachromosomal recombination and oxidative free radicals in *Saccharomyces cerevisiae. Mutat. Res.* **397,** 271–278.
28. Brennan, R. J. and Schiestl, R. H. (1997) Diaminotoluenes induce intrachromosomal recombination and free radicals in *Saccharomyces cerevisiae. Mutat. Res.* **381,** 251–258.

29. Brennan, R. J., Kandikonda, S., Khrimian, A. P., DeMilo, A. B., Liquido, N. J., and Schiestl, R. H. (1996) Saturated and monofluoro analogs of the oriental fruit fly attractant methyl eugenol show reduced genotoxic activities in yeast. *Mutat. Res.* **369,** 175–181.
30. Aubrecht, J., Narla, R. K., Ghosh, P., Stanek, J., and Uckun, F. M. (1999) Molecular genotoxicity profiles of apoptosis-inducing vanadocene complexes. *Toxicol. Appl. Pharmacol.* **154,** 228–235.
31. Schiestl, R. H. and Prakash, S. (1990) RAD10, an excision repair gene of *Saccharomyces cerevisiae*, is involved in the RAD1 pathway of mitotic recombination. *Mol. Cell. Biol.* **10,** 2485–2491.
32. Schiestl, R. H., Igarashi, S., and Hastings, P. J. (1988) Analysis of the mechanism for reversion of a disrupted gene. *Genetics* **119,** 237–247.
33. Schiestl, R. H. and Prakash, S. (1988) RAD1, an excision repair gene of *Saccharomyces cerevisiae*, is also involved in recombination. *Mol. Cell. Biol.* **8,** 3619–3626.
34. Brennan, R. J. and Schiestl, R. H. (2001) Persistent genomic instability in the yeast *Saccharomyces cerevisiae* induced by ionizing radiation and DNA-damaging agents. *Radiat. Res.* **155,** 768–777.
35. Ames, B. N., McCann, J., and Yamasaki, E. (1975) Methods for detecting carcinogens and mutagens with the *Salmonella*/mammalian-microsome mutagenicity test. *Mutat. Res.* **31,** 347–364.
36. Burke, D., Dawson, D., and Stearns, T. (2000) *Methods in Yeast Genetics.* Cold Spring Harbor Laboratory, Cold Spring Harbor, NY.
37. Ugolini, S. and Bruschi, C. V. (1996) The red/white colony color assay in the yeast *Saccharomyces cerevisiae*: epistatic growth advantage of white ade8-18, ade2 cells over red ade2 cells. *Curr. Genet.* **30,** 485–492.
38. McCann, J., Choi, E., Yamasaki, E., and Ames, B. N. (1975) Detection of carcinogens as mutagens in the *Salmonella*/microsome test: assay of 300 chemicals. *Proc. Natl. Acad. Sci. USA* **72,** 5135–5139.

8

In Vivo DNA Deletion Assay to Detect Environmental and Genetic Predisposition to Cancer

Ramune Reliene, Alexander J. R. Bishop, Jiri Aubrecht, and Robert H. Schiestl

Summary

Large-scale genomic rearrangements such as DNA deletions play a role in the etiology of cancer. The frequency of DNA deletions can be elevated by exposure to carcinogens or by mutations in genes involved in the maintenance of genomic integrity. The in vivo DNA deletion assay allows a visual detection of deletion events within the pink-eyed unstable (p^{un}) locus in developing mouse embryos. A deletion of one copy of a duplicated 70-kb DNA fragment within the p^{un} locus restores the pink-eyed dilute (*p*) gene, which encodes a protein responsible for the assembly of a black color melanin complex. Deletion events occurring in premelanocytes cause visible black patches (fur-spots) on the light gray fur of offspring and black pigmented cells (eye-spots) on the unpigmented retinal pigment epithelium (RPE). In the fur-spot assay, 10-d-old pups are observed for black spots on the fur. In the eye-spot assay, mice are sacrificed at d 20, eyes are removed, and the wholemount RPE slides are prepared for eye-spot analysis. The frequency, size, and position relative to the optic nerve of the eye-spots are determined. This assay can be used to study the effect of environmental chemicals and physical agents as well as the genetic control of DNA deletions in vivo.

Key Words: p^{un} reversion, homologous recombination, DNA deletion, mice, in vivo, fur-spot assay, eye-spot assay

1. Introduction

Genetic instability (such as chromosomal translocations, deletions, and duplications) is associated with carcinogenesis ***(1–3)***. Determining the frequency of DNA deletions in vivo provides a powerful tool to assess the carcinogenicity of environmental chemicals and/or the cancer-predisposing mutations of an organism. The in vivo DNA deletion assay utilizes the

From: *Methods in Molecular Biology, vol. 262, Genetic Recombination: Reviews and Protocols*
Edited by: A. S. Waldman © Humana Press Inc., Totowa, NJ

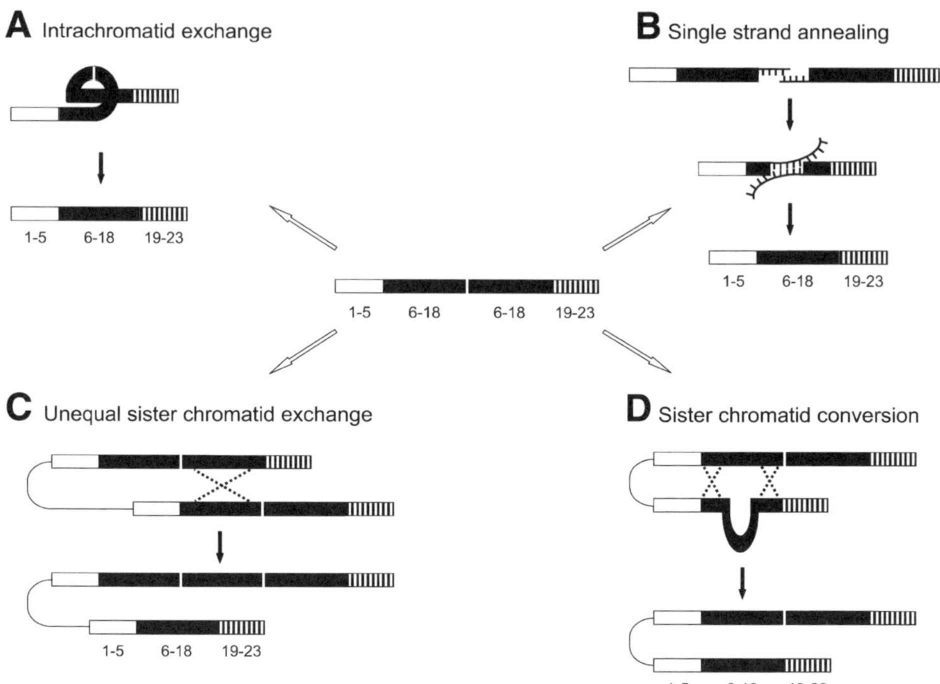

Fig. 1. The p^{un} mutation and possible mechanisms of intrachromosomal recombination resulting in deletion of the repeated exons 6–18. The deletion can occur by several homologous recombination pathways, such as **(A)** intrachromosomal crossing-over, **(B)** single-strand annealing, **(C)** unequal sister chromatid exchange, and **(D)** sister chromatid conversion *(8)*.

C57BL/6J p^{un}/p^{un} mouse strain (Jackson Laboratory, Bar Harbor, ME), which has a 70-kb tandem duplication at the pink-eyed dilution (*p*) locus *(4,5)* (**Fig. 1**), termed the pink-eyed unstable (p^{un}) mutation. The assay is based on the determination of the frequency of deletion events of one copy of the 70-kb DNA duplication within the p^{un} locus in developing embryos. The *p* gene is transcribed in melanocytes and retinal pigment epithelium (RPE) cells and encodes a melanosomal integral membrane protein responsible for the assembly of a high-molecular-weight melanin complex that produces a black pigment in the hair and retinal epithelium of wild-type mice *(6)*.

The p^{un} mutation is autosomal recessive and results in a dilute, light gray coat color and pink eyes *(7)*. p^{un} reversion events occurring in premelanocytes in the p^{un}/p^{un} embryo reconstitute the *p* gene and cause visible black spots on the gray fur (fur-spot) of offspring after the amplification of premelanocytes

Fig. 2. A fur-spot on the p^{un} mouse. The upper p^{un} mouse is a control, and the lower mouse shows a black spot on its back and side.

(8,9) (**Fig. 2**). The p^{un} reversion in a precursor cell of the RPE results in pigmented cells (eye-spots) on the unpigmented background *(10)* (**Fig. 3**). In the fur-spot assay, the black spots are determined on the fur of 10-d-old offspring, and the DNA deletion frequency is expressed in percent spotted animals. In the eye-spot assay, mice are sacrificed at d 20, and the number of pigmented spots per RPE is determined by microscopic examination. An eye-spot may consist of one pigmented cell or a clone of pigmented cells.

Spontaneous p^{un} reversions occur at a relatively high frequency, at least three to five orders of magnitude greater than other recessive mutations at other coat-color loci *(11)*. The p^{un} reversion frequency has been calculated to be $1.8–5.8 \times 10^{-4}$ at d 10 of embryonic life in various types of matings *(12)*. On the C57BL/6J inbred strain, four to six eye-spots are observed in the RPE that consists of approx 54,000 cells *(13,14)*, and 5–10% of mice display fur-spots *(4,8)*.

The eye-spot assay also allows determination of the approximate time during embryo development at which a DNA deletion event occurred by mapping an eye-spot relative position within the RPE *(13)*. The RPE development begins at E8 *(15)*, when the RPE precursors invade the embryonic eye-cup around the

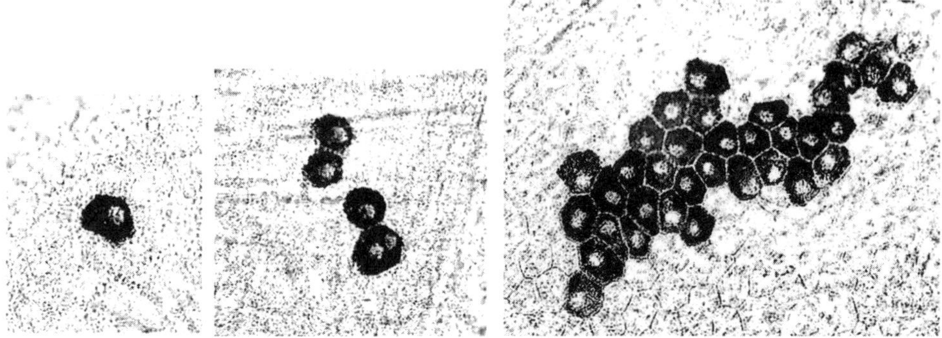

Fig. 3. Close-up of eye-spots on the RPE of a p^{un} mouse.

optic nerve head and proliferate radially into a near spherical eye-cup. During development, the RPE develops radially away from the optic nerve, with most proliferation occurring in the most peripheral cells. A DNA deletion event occurring at the p^{un} locus is displayed by the observed pigmentation of the RPE cell. Examining the location, or position, of any pigmented cell, relative to the distance from the optic nerve head, gives an indication as to when during development that reversion took place. The eye-spot position is measured by dividing the radial distance from the centrally located optical nerve head to the most proximal edge of the eye-spot over the radial distance from the optical nerve head through the eye-spot to the circumferential edge of the RPE. Following an acute exposure to benzo(a)pyrene or X-rays on the 10th day of gestation, an increased frequency of reversion events was detected in a distinct region of the adult RPE ***(16)***. Examination of exposure at different times of eye development (E8, E10, E12, and E17) reveals distinct regions where the induced events lie, with the position of each region found at an increasing distance from the optic nerve the later the time of exposure ***(17)***. This pattern of induction fits nicely with the developmental and cellular proliferation pattern of the RPE ***(13)***.

The mechanism of p^{un} reversion involves intrachromosomal homologous recombination between two identical tandemly repeated 70-kb DNA fragments spanning exons 6–18 (**Fig. 1**), resulting in deletion of one of the fragments and thereby reconstituting the *p* gene. The deletion can occur by several homologous recombination pathways, such as intrachromosomal crossing-over, single-strand annealing, unequal sister chromatid exchange, and sister chromatid conversion ***(8)*** (**Fig. 1**). Some of these events are initiated by a double-strand break (DSB) between the duplicated fragments. As an example, the frequency of p^{un} reversion is elevated in the offspring of dams treated with X-rays ***(18)***, which are known to be potent inducers of DSBs. Furthermore, a number of carcinogens were tested using the fur-spot assay in which X-rays ***(18)***, ethylmethane sulfonate (EMS), methylmethane sulfonate (MMS), ethyl nitrosourea (ENU), benzo(*a*)pyrene (B*a*P), trichloroethylene (TCE), sodium arsenate (SOA), benzene (BEN) ***(8)***, the polychlorinated biphenyls Aroclor 1221 and Aroclor 1260, 2,3,7,8-tetrachlorodibenzo-*p*-dioxin (TCDD) ***(19)***, and cigarette smoke ***(9)*** elevated the fur-spot frequency significantly compared with untreated controls (**Table 1**). The elevated frequency of DNA deletions was also observed in mice with inherent genomic instability owing to mutations in *Trp53*, *Atm*, *Gadd45*, and *Wrn* genes by using the fur-spot and/or eye-spot assay ***(20,21)***.

There are several advantages of the eye-spot assay over the fur-spot assay. The eye-spot assay requires fewer animals to obtain significant results and provides extremely high sensitivity, with resolution to the single cell level. In the fur-spot assay, fur-spot production is only detected when the revertant cells have undergone a multiple number of divisions, enough to form a visible fur-spot. On the other hand, the two assays complement each other, since they allow study of deletion events in tissues of different origin. The hair follicle melanocytes are derived from the neural crest, where RPE cells are derived from neural epithelium. The development of both tissues begins at approximately the same time. The hair follicle melanocytes are first detected at 10 dpc dividing slowly and beginning to migrate by 11 dpc ***(22)***. This proliferation pattern continues throughout embryo development, and then shortly after birth they begin to proliferate rapidly before migrating into hair follicles. The neuroepithelial melanocytes are first detected in the developing eye-cup at 8 dpc ***(15)***; the RPE reaches its highest rate of mitotic division between 11.5 and 15.5 dpc ***(23)*** and ceases in the first postnatal weeks ***(13)***.

The in vivo DNA deletion assay allows assessment of an intermediate biological marker in carcinogenicity, e.g., DNA deletions, and takes relatively little time to perform. It requires 2–3 mo compared with the several years needed for long-term carcinogenicity studies. A possible shortcoming of the assay might be its limitation to proliferating RPE and/or skin premelanocyte cells, when proliferation starts at about embryonic d 9 and ceases after birth. Thus, this assay is used to determine the frequency of DNA deletions occurring in embryos, but it cannot be applied for adult animals. The assay, however, seems to be an excellent tool for studying the effects of acute exposure by carcinogens and other environmental chemicals ***(8,9,18,19)*** as well as transgenerational effects ***(24,25)***. The assay also allows the determination of genetic control for the maintenance of genome integrity in vivo ***(14,20)***.

Table 1
Carcinogens Induce Deletions In Vivo in Mice

Chemical	Dose	No. of offspring	No. of spotted offspring	Spotting frequency (%)	*p* value	Ref.
Control		498	28	5.6		*18*
X-rays	100 cGy	172	40	23	$<10^{-6}$	*18*
Control		585	62	11	<<0.0005	*8*
EMS	100 mg/kg	94	27	29	<<0.0005	*8*
MMS	100 mg/kg	83	21	25	<<0.0005	*8*
ENU	25 mg/kg	57	30	53	<<0.0005	*8*
Control		337	18	5.3		*8*
SOA	20 mg/kg	56	16	29	<<0.0005	*8*
Vehicle control (corn oil)	0.2 mL	51	2	3.9		*8*
B*a*P	150 mg/kg	32	20	63	<<0.0005	*8*
TCE	200 mg/kg	41	13	32	<0.005	*8*
BEN	200 mg/kg	48	13	27	<0.01	*8*
Vehicle control (corn oil)	0.2 mL	51	2	3.9		*19*
Aroclor 1221	300 mg/kg	71	10	14	NS	*19*
	500 mg/kg	40	8	20	<0.025	*19*
	1000 mg/kg	40	10	25	<0.005	*19*
Aroclor 1260	75 mg/kg	54	6	11	NS	*19*
	500 mg/kg	140	36	26	<0.005	*19*
TCDD	2.25 μg/kg	69	17	25	<0.005	*19*

Control			475	60	12.6		*9*
Cigaret smoke	Smoke concentration						*9*
	CO (ppm)	TPM (mg/m^3)					
Unfiltered cigarets	32 ± 2.5	30.9 ± 2.0	43	10	23.3	<0.05	*9*
	51.5 ± 8.5	57.0 ± 7.0	30	7	23.3	NS	*9*
	99 ± 4	96 ± 15.1	36	10	27.8	<0.05	*9*
Filtered cigarets	40 ± 14.0	12 ± 0.0	35	11	31.4	<0.05	*9*
Vehicle control (DMSO)	0.2 mL		24	4	16.7		*9*
CSC	15 mg/kg		52	15	28.8	<0.05	*9*

Abbreviations: EMS, ethylmethane sulfonate; MMS, methylmethane sulfonate; ENU, ethyl nitrosourea; B*a*P, benzo(*a*)pyrene; TCE, trichloroethylene; SOA, sodium arsenate; BEN, benzene; TCDD, 2,3,7,8-tetrachlorodibenzo-*p*-dioxin; CO, carbon monoxide; TPM, total particulate mass; DMSO, dimethyl sulfoxide; CSC, cigarette smoke condensate; NS, not significant.

2. Materials

1. C57BL/6J p^{un}/p^{un} mouse strain (Jackson Laboratory).
2. Fixative: 4% paraformaldehyde in 0.1 *M* phosphate buffer, pH 7.4 (Sigma).
3. Phosphate-buffered saline (PBS): 8 g/L NaCl, 0.2 g/L KCl, 0.2 g/L KH_2PO_4, 2.31 g/L $Na_2HPO_4 \cdot 7H_2O$, pH 7.2.
4. 90% Glycerol (Sigma).
5. Clear nail polish.
6. Dissecting microscope.
7. Micro Dissecting Tweezers (length 110 mm, tip 0.05 × 0.01 mm, material stainless steel; Roboz Surgical, Rockville, MD).
8. Micro Dissecting Spring Scissors (Vannas, straight cutting edge 3 mm, comb tip width 0.2 mm, overall length 7.5 cm; Roboz Surgical).
9. Microscope slides (75 × 25 mm) and cover glasses (22 × 22 mm) (Fisher Scientific).
10. DC120 digital camera (Eastman Kodak) or similar microscope-mounted digital imaging setup.
11. DMLB microscope (Leica Microsystems) or similar.
12. Computer software: Adobe Photoshop 5.0, Microsoft Excel, or similar image manipulation and data-file programs.

3. Methods

3.1. Mice Breeding and Care

1. Breed C57BL/6J-p^{un}/p^{un} mice in the animal facility under standard conditions with a 12-h light/dark cycle and feed standard diet and water *ad libitum* (*see* **Note 1**).
2. Time pregnancy by checking for vaginal plugs, with noon of the day of discovery counted as 0.5 d *post coitum* (dpc).
3. The time of birth of a litter is timed, with the noon of discovery counted as 0.5 d *post partum* (dpp).

3.2. Mouse Exposure

1. To determine the effect of a certain chemical or physical agent on the frequency of DNA deletions on the fur or RPE of developing embryos, treat dams in the second half of gestation, preferably at about 10.5 dpc ***(17,26)***. During this time the effect is most notable, since the hair follicle melanocyte proliferation begins at approx 10 dpc ***(22)*** and the RPE formation begins in the developing eye-cup at approx 8 dpc ***(15)***.
2. Treat dams with a reagent of interest under desired scenario. A single dose or multiple doses of intraperitoneal (ip) injection ***(8)***, gavage ***(17)***, and/or inhalation exposure ***(9)*** can be used.

3.3. Fur-Spot Assay

1. Examine pups for the presence of the black fur-spots at the 10th dpp.
2. Record number, location, and shape of spots.

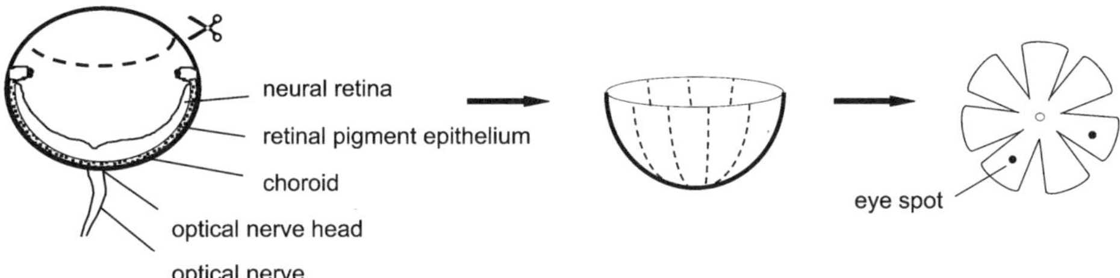

Fig. 4. Preparation of an RPE. A round incision is made at the upper corneo-scleral border of the eye to allow removal of cornea and lens. Six to eight incisions are made in the eye-cup from the corneo-scleral margin toward the centrally positioned optic nerve head, so that the eye-cup is laid flat. The neural retina is then gently removed to reveal the RPE.

3. Count the mice with one or more spots on the fur in relation to those without any spot.
4. Score the frequency of p^{un} reversion/deletion events as percentage of spotted animals (*see* **Note 2**).

3.4. Eye-Spot Assay

3.4.1. Fixation of Eyes

1. Sacrifice mice at the 20th dpp.
2. Remove eyes from the orbit and immerse in fixative.
3. Incubate eyes in the fixative for 1 h at room temperature, and then store in PBS at 4°C until dissection (*see* **Note 3**).

3.4.2. RPE Dissection and Wholemount Slide Preparation

1. Place the eye under the dissecting scope. Use 4× or (if required) higher magnification.
2. Make a radial incision at the upper corneo-scleral border (**Fig. 4**).
3. Remove cornea and lens.
4. Make six to eight radial incisions from the corneo-scleral margin toward the centrally positioned optic nerve head (**Fig. 4**).
5. Flatten the eye-cup, put a drop of PBS on the flattened eye-cup, and gently remove the neural retina.
6. Place the remaining specimen consisting of RPE attached to the underlying choroid into a drop of 90% glycerol (immerse in glycerol) on a glass slide with the RPE facing up.
7. Carefully flatten the eye-cup.
8. Put a small drop of 90% glycerol on a cover glass, and gently overlay the RPE specimen on the slide.
9. Overlay the sides of the cover glass, i.e., the junction between the slide and cover glass, with the nail polish to prevent drying out of the RPE mount.

See **Notes 4–6**.

3.4.3. Scoring of DNA Deletion Events on the RPE

1. Count eye-spots under a microscope under 10× or (if required) higher magnification. An eye-spot is defined as a single pigmented cell or groups of pigmented cells separated from each other by no more than one unpigmented cell ***(13)***. Such an eye-spot corresponds to one p^{un} reversion/deletion event.
2. Count spots in each of six or eight "petals" of the RPE (**Fig. 4**) and the number of cells that comprise each eye-spot.
3. Score the frequency of DNA deletion events in the RPE by number of eye-spots per RPE.

3.4.4. Distance Analysis of Eye Spots From the Optic Nerve

1. Photograph images of the RPEs with a 2.5× N-plan objective or similar magnification on a DMLB microscope attached to a computer DC120 digital camera. With a 2.5× objective, it is not possible to obtain a single complete image of the RPE within a single field of view. Therefore, take multiple, overlapping images of the entire RPE until full coverage is achieved. It is important to keep the resolution settings of the imaging system consistent throughout these examinations to facilitate assembling the images into one complete image.
2. Assemble images into a complete RPE image. Take the individual, overlapping images of an RPE, and (using the natural landmarks that occur—tissue debris or other shapes and patterns), align images to obtain an accurate reconstruction of the entire RPE. Examination of the digital image usually clearly reveals any remains of the cornea-sclera divide as a slightly darker/browner region. Removing all regions beyond the start of this divide will aid in the position analysis.
3. Examine eye-spot position in Adobe Photoshop 5.0 using a measuring tool.
   a. Identify pigmented eye-spots by microscopic examination of the RPE at a higher magnification (10, 20, and 40×). Count and record the number of pigmented cells that comprise an eye-spot. We have defined a single eye-spot as being a group of immediately adjacent pigmented cells, or a group that is separated by no more than one unpigmented cell. By comparing the reconstructed digital image with the actual RPE at this higher magnification, identify each eye-spot on the digital image.
   b. Use any ruler, or measuring tool within the imaging program, to measure two distances for each eye spot: the *eye-spot distance* is the distance from the center of the optic nerve head to the most proximal edge of the eye spot, and the *RPE distance* of an eye-spot is the distance from the optic nerve through the eye-spot to the outer edge of the RPE. These data are recorded in a spreadsheet program such as Microsoft Excel.
   c. Determine eye-spot position by dividing the eye-spot distance by the RPE distance, to compensate for minor differences in the size of the eyes.
4. Analysis of eye-spot positional distribution.
   a. *Eye-spot frequency*. The number of eye-spots for each RPE can be determined, as can the size (number of cells per eye-spot). This overall frequency between

groups of RPE cells can be compared in a statistical analysis such as the highly sensitive Mann-Whitney test. The frequency of each size eye-spot can offer also some insight into any developmental effect.

b. *Eye-spot position frequency and distribution.* Analyzing the frequency of each eye-spot by position and size is often the most informative test that can be performed. The size and position of each eye-spot should already be recorded. For convenience, we divide the distance from the optic nerve head to the edge of the RPE into 10 equal regions and plot the position of each eye-spot within these regions. By examining the relative position distribution of eye-spots per all eye-spots identified as a group, the timing of events can be determined. Differences in these distribution patterns can be examined with a Mann-Whitney test. In addition, a straight examination of the frequency of eye-spots in a particular region compared with the frequency in the rest of the RPE also provides a means of defining a region where a difference lies. Statistically this can be done with a chi-square contingency test. Finally, we have noted that there is a difference in single-cell eye-spot events and multiple-cell eye-spots. Multiple-cell eye-spots that occur at the same time as single-cell eye-spots tend to lie more distal to the optic nerve. So performing all the examinations described in this subheading with single-cell or multiple-cell eye-spot populations independently yields valuable information.

3.4.5. Statistical Analysis

Perform a statistical evaluation of the data using an appropriate test (Student's *t*-test, Mann-Whitney, or a contingency chi-square test).

4. Notes

1. In case mutant mice of a strain background different from C57BL/6J-p^{un}/p^{un} are desired for study, the mutation desired has to be crossed into the C57BL/6J-p^{un}/p^{un} genetic background by at least six backcrosses. The desired genotype of the mice is determined by the presence of diluted gray coat color morphologically similar to the parental C57BL/6J-p^{un}/p^{un} strain. The presence of the mutation of interest is determined by genotyping as recommended by the provider of the mutant mice.
2. Normally one or (less commonly) two fur-spots per mouse are observed. Molecular analysis of some fur-spots can be performed by reverse transcriptase-polymerase chain reaction (RT-PCR) as described previously *(8)*.
   a. Briefly, sacrifice 3–4-d-old mice, excise black and gray pieces of skin, and isolate total RNA using guanidinium thiocyanate-phenol extraction *(27)*. As a positive control, use pieces of black skin of wild-type C57BL/6J mice.
   b. Synthesize first-strand cDNA using the SuperScript II reverse transcription preamplification system and oligo(dT)$_{12–18}$ (Gibco/BRL, Gaithersburg, MD). The first-strand cDNA reaction contains 5 µg total RNA, 1× PCR buffer (Gibco/BRL), 2.5 m*M* $MgCl_2$, 0.5 m*M* dNTP mix, and 10 n*M* dithiothreitol (DTT).
   c. Perform a PCR reaction of cDNA by using *Taq* DNA polymerase (Gibco/BRL) and the following primers: 3' primer, CAA CCA GAT GGC ACC CAG AAT

AGC; 5' primer, CTG TGT CAC CGC TGG AAA ACT ACT. These primers are homologous to sequences outside the duplicated region and yield a PCR product 1.3 kb long for the wild-type *p* gene cDNA and 2.6 kb long for p^{un} cDNA. The PCR reaction contains $\frac{1}{10}$ of the cDNA reaction mixture, 1× PCR buffer, 1.5 m*M* $MgCl_2$, 200 µ*M* dNTP mix, 100 n*M* of each primer, and 2 units of *Taq* DNA polymerase.

d. After initial denaturation for 3 min, perform 35 PCR cycles under the following conditions: annealing at 55°C for 1 min, synthesis at 72°C for 1.5 min, and denaturation at 95°C for 1 min.

e. Run PCR products on 1% agarose gel electrophoresis and stain with ethidium bromide.

f. Photograph stained gels and analyze the relative intensity of the wild-type *p* mRNA as the ratio of the bands' intensities to *p* and p^{un} by scanning densitometry quantification.

g. The ratio between the 1.3-kb *p* fragment and the 2.6-kb p^{un} PCR fragment is from 4:1 to more than 10:1, as found in PCRs from cDNA from the black spots from X-ray-, ethylmethane sulfonate-, sodium arsenate-, or benzene-treated mice ***(8)***. The presence of the 2.6-kb p^{un} PCR fragment is expected in the spots for the following two reasons. First, contaminating surrounding tissue may be excised together with the spots. Second, most likely only one of the two alleles of the homozygous p^{un} alleles has recombined to the *p* gene, leaving the other allele as a p^{un} allele. PCR product from cDNA from gray fur contains a ratio of 1:1 of the two species. The presence of *p* transcript might be owing to an expected frequency of reversion events of about 10^{-4}. This reversion frequency may yield a ratio of 1:1, since the *Taq*-based PCR underrepresents large fragments, such as the p^{un} cDNA PCR fragment.

3. Fixation of eyes for an extended period (for example, overnight) leads to crosslinking of the neural retina to the underlying RPE, which makes it difficult to remove it from the RPE during dissection of the eye. This may lead to disruption and loss of parts of the RPE. The fixed eye can be stored in PBS but not for more than several years, because the eyes may dry out and tissue layers in the RPE crosslink irreversibly.
4. The RPE slide preparation is less satisfactory when the eyes of older animals (more than 3 mo old) are used. The layer of neural retina becomes progressively more attached to the underlying RPE, which may lead to disruption and a removal of fragments of RPE while removing the neural retina. The RPE and/or the RPE underlying choroid become increasingly pigmented so as to obscure the pigmented cells with p^{un} reversions.
5. The pigmentation of cells with p^{un} reversions may fade on the RPE slides over several years; therefore the counting and measurements must be performed soon upon successful mounting of the RPE.
6. The number of eye-spots varies from 1 to more than 10 but rarely reaches 15 spots. Most eyes display four to six eye-spots ***(13,14)***. A p^{un} reversion that occurs in a precursor cell early in development may produce mosaic mice. This is observed

as multiple black patches on the fur or multiple pigmented cells or clones of cells dispersed throughout the transparent RPE. Early p^{un} reversion events result in mosaic fur in 1% or fewer mice and occur in approx 1–2% of RPEs. Such events are not to be scored as single p^{un} reversions and are to be excluded from analysis. If the p^{un} reversion occurs in the germline, the resulting offspring will have a black coat and pigmented (black) RPE. Such germline reversions occur in less than 1% of mice.

Acknowledgments

Thanks to Nicholas Carls from the Schiestl lab for insightful help in the development of this assay. Our work was supported by grants from the National Institute of Environmental Health Sciences, NIH, RO1 grant ES09519, NIEHS KO2 award ES00299, funding from the UCLA Center for Occupational and Environmental Health (to R.H.S.), NIH RCDA Award F32GM19147 (to A.J.R.B.), and a postgraduate research fellowship from the UC Toxic Substances Research and Teaching Program (to R.R.).

References

1. Mitelman, F., Johansson, B., and Mertens, F. (2003) Mitelman Database of Chromosomal Abberations in Cancer. *http://cgap.nci.nih.gov/Chromosomes/Mitelman.*
2. Khanna, K. K. and Jackson, S. P. (2001) DNA double-strand breaks: signaling, repair and the cancer connection. *Nat. Genet.* **27,** 247–254.
3. Knudson, A. G. (2001) Two genetic hits (more or less) to cancer. *Nat. Rev. Cancer* **1,** 157–162.
4. Brilliant, M. H., Gondo, Y., and Eicher, E. M. (1991) Direct molecular identification of the mouse pink-eyed unstable mutation by genome scanning. *Science* **252,** 566–569.
5. Gondo, Y., Gardner, J. M., Nakatsu, Y., et al. (1993) High-frequency genetic reversion mediated by a DNA duplication: the mouse pink-eyed unstable mutation. *Proc. Natl. Acad. Sci. USA* **90,** 297–301.
6. Rosemblat, S., Durham-Pierre, D., Gardner, J. M., Nakatsu, Y., Brilliant, M. H., and Orlow, S. J. (1994) Identification of a melanosomal membrane protein encoded by the pink-eyed dilution (type II oculocutaneous albinism) gene. *Proc. Natl. Acad. Sci. USA* **91,** 12,071–12,075.
7. Lyon, M. F., King, T. R., Gondo, Y., et al. (1992) Genetic and molecular analysis of recessive alleles at the pink-eyed dilution (*p*) locus of the mouse. *Proc. Natl. Acad. Sci. USA* **89,** 6968–6972.
8. Schiestl, R. H., Aubrecht, J., Khogali, F., and Carls, N. (1997) Carcinogens induce reversion of the mouse pink-eyed unstable mutation. *Proc. Natl. Acad. Sci. USA* **94,** 4576–4581.
9. Jalili, T., Murthy, G. G., and Schiestl, R. H. (1998) Cigarette smoke induces DNA deletions in the mouse embryo. *Cancer Res.* **58,** 2633–2638.

10. Searle, A. G. (1977) The use of pigment loci for detecting reverse mutations in somatic cells of mice. *Arch. Toxicol.* **38,** 105–108.
11. Schlager, G. and Dickie, M. M. (1967) Spontaneous mutations and mutation rates in the house mouse. *Genetics* **57,** 319–330.
12. Melvold, R. W. (1971) Spontaneous somatic reversion in mice. Effects of parental genotype on stability at the *p*-locus. *Mutat. Res.* **12,** 171–174.
13. Bodenstein, L. and Sidman, R. L. (1987) Growth and development of the mouse retinal pigment epithelium. I. Cell and tissue morphometrics and topography of mitotic activity. *Dev. Biol.* **121,** 192–204.
14. Bishop, A. J. R., Barlow, C., Wynshaw-Boris, A. J., and Schiestl, R. H. (2000) Atm deficiency causes an increased frequency of intrachromosomal homologous recombination in mice. *Cancer Res.* **60,** 395–399.
15. Nakayama, A., Nguyen, M. T., Chen, C. C., Opdecamp, K., Hodgkinson, C. A., and Arnheiter, H. (1998) Mutations in microphthalmia, the mouse homolog of the human deafness gene MITF, affect neuroepithelial and neural crest-derived melanocytes differently. *Mech. Dev.* **70,** 155–166.
16. Bishop, A. J. R., Kosaras, B., Sidman, R. L., and Schiestl, R. H. (2000) Benzo(a)pyrene and X-rays induce reversions of the pink-eye unstable mutation in the retinal pigment epithelium of mice. *Mutat. Res.* **457,** 31–40.
17. Bishop, A. J., Kosaras, B., Carls, N., Sidman, R. L., and Schiestl, R. H. (2001) Susceptibility of proliferating cells to benzo[a]pyrene-induced homologous recombination in mice. *Carcinogenesis* **22,** 641–649.
18. Schiestl, R. H., Khogali, F., and Carls, N. (1994) Reversion of the mouse pink-eyed unstable mutation induced by low doses of x-rays. *Science* **266,** 1573–1576.
19. Schiestl, R. H., Aubrecht, J., Yap, W. Y., Kandikonda, S., and Sidhom, S. (1997) Polychlorinated biphenyls and 2,3,7,8-tetrachlorodibenzo-p-dioxin induce intrachromosomal recombination in vitro and in vivo. *Cancer Res.* **57,** 4378–4383.
20. Bishop, A. J. R., Hollander, M. C., Kosaras, B., Sidman, R. L., Fornace, J., and Schiestl, R. H. (2003) Atm, p53 and Gadd45 deficient mice have an increased frequency of homologous recombination at different times during development. *Cancer Res.* **63,** 5335–5539.
21. Lebel, M. (2002) Increased frequency of DNA deletions in pink-eyed unstable mice carrying a mutation in the Werner syndrome gene homologue. *Carcinogenesis* **2,** 213–216.
22. Cable, J., Jackson, I. J., and Steel, K. P. (1995) Mutations at the W locus affect survival of neural crest-derived melanocytes in the mouse. *Mech. Dev.* **50,** 139–150.
23. Kong, Y., Usuda, N., and Nagata, T. (1992) Radioautographic study on DNA synthesis of the retina and retinal pigment epithelium of developing mouse embryos. *Cell. Mol. Biol.* **38,** 263–272.
24. Carls, N. and Schiestl, R. H. (1999) Effect of ionizing radiation on transgenerational appearance of *p*(un) reversions in mice. *Carcinogenesis* **20,** 2351–2354.

25. Shiraishi, K., Shimura, T., Taga, M., et al. (2002) Persistent induction of somatic reversions of the pink-eyed unstable mutation in F1 mice born to fathers irradiated at the spermatozoa stage. *Radiat. Res.* **157,** 661–667.
26. Aubrecht, J., Secretan, M. B., Bishop, A. J., and Schiestl, R. H. (1999) Involvement of p53 in X-ray induced intrachromosomal recombination in mice. *Carcinogenesis* **20,** 2229–2236.
27. Chomczynski, P. and Sacchi, N. (1987) Single-step method of RNA isolation by acid guanidinium thiocyanate- phenol-chloroform extraction. *Anal. Biochem.* **162,** 156–159.

III

Recombination as a Tool for Producing Targeted Genetic Modification

9

Gene Targeting at the Chromosomal Immunoglobulin Locus

A Model System for the Study of Mammalian Homologous Recombination Mechanisms

Mark D. Baker

Summary

Plasmid DNA transfected into mammalian cells can integrate into mammalian chromosomes by homologous recombination. This phenomenon, known as gene targeting, can be used as a tool to investigate mammalian homologous recombination mechanisms. The chromosomal immunoglobulin (Ig) genes as they are presented in mouse hybridoma cells have several advantages that render them particularly amenable to gene targeting. Here, we present the basic methods currently in use in our laboratory that exploit this system to investigate mammalian gene targeting mechanisms.

Key Words: gene targeting, immunoglobulin genes, hybridoma, myeloma, mouse, mammalian, sequence replacement vector, sequence insertion vector, enhancer-trap vector, homologous recombination, methods

1. Introduction

Gene targeting is the process by which transfected DNA undergoes homologous recombination with the resident genome; it permits predetermined changes to be introduced directly into chromosomal genes *(1,2)*. Consequently, gene targeting has wide-ranging applications in the study of gene structure and function, in the creation of animal models of human genetic diseases *(3)*, and perhaps, ultimately, in areas of human gene therapy *(4)*. It also serves as a model system in the study of homologous recombination mechanisms *(5)*.

Gene targeting relies on the use of either sequence insertion (O-type) or sequence replacement (Ω-type) vectors. Sequence insertion vectors contain a continuous segment of genomic DNA that is homologous to the desired chromosomal target locus inserted into a recipient vector bearing a selectable

From: *Methods in Molecular Biology, vol. 262, Genetic Recombination: Reviews and Protocols*
Edited by: A. S. Waldman © Humana Press Inc., Totowa, NJ

genetic marker. In contrast, most common replacement vectors consist of a selectable marker flanked by two genomic DNA segments bearing homology to the target locus. For most mammalian chromosomal loci, the frequency of gene targeting with either vector type is rare (approx 10^{-6} recombinants/cell in an unselected population, or approx 10^{-3} recombinants/cell in a population selected for drug resistance), making the detection and recovery of recombinants difficult ***(1)***. However, several features of the chromosomal immunoglobulin (Ig) heavy and light chain genes as they are presented in mouse hybridoma cells render them particularly advantageous as targets for homologous recombination:

1. The Ig genes are well mapped and extensively sequenced.
2. They are usually present in only one copy per cell.
3. The Ig genes are expressed at a high level, such that Ig constitutes approx 1% of the total cell protein.
4. Powerful recombinant isolation procedures are available.
5. Mouse hybridoma cells have a doubling time of approx 18 h, a period short enough to be of practical importance.

These unique properties facilitate the study of gene targeting and homologous recombination mechanisms (for example, see refs. ***6–19***), and the engineering of specifically altered Ig through gene targeting approaches ***(6,8,20–22)***. Our gene targeting system is based on the mouse hybridoma cell line Sp6 ***(23,24)***, which contains a single copy of the chromosomal Ig μ heavy chain gene and synthesizes IgM(κ). Here, we present the basic methods in use in our laboratory that exploit this system to investigate gene targeting mechanisms.

2. Materials

1. Mouse hybridoma cells (*see* **Note 1**).
2. Supplemented DMEM (Dulbecco's modified Eagle's medium [DMEM; Gibco, cat. no. 12100] containing 13% bovine calf serum [BCS; Hyclone] and 0.0035% 2-mercaptoethanol).
3. *Escherichia coli* (strain DH5α).
4. Oligonucleotide primers.
5. Restriction enzymes, T4 DNA ligase, T4 DNA polymerase, Taq polymerase, proteinase K.
6. Agarose and gel electrophoresis equipment.
7. LB (Luria-Bertani) medium.
8. Ampicillin (AMP).
9. 1× Phosphate-buffered saline (PBS): 137 m*M* NaCl, 2.7 m*M* KCl, 10 m*M* HPO_4, 1.5 m*M* KH_2PO_4. Usually made up in a 1000 mL volume.
10. Permeabilization buffer: 140 m*M* KCl, 1 m*M* $MgCl_2$, 1 m*M* ATP, 10 m*M* glucose, 1 m*M* EGTA, 0.2 m*M* $CaCl_2$, 10 m*M* HEPES (adjusted to pH 7.3 with

N-methyl-D-glucammonium). Usually made up in a 100 mL volume, and then frozen in 10-mL aliquots at –20°C.

11. Lysis buffer (Applied Biosystems, cat. no. 400676).
12. G418 (Invitrogen, cat. no. 11811-031).
13. Freezing medium (FM). In a 100 mL volume contains supplemented DMEM, 86.6 mL; BCS, 4.7 mL; DMSO, 8.7 mL).
14. 1× TE: 10 m*M* Tris-HCl, pH 8.0, 1 m*M* EDTA.
15. Expand Long Template polymerase chain reaction (PCR) kit (Roche, cat. no. 1681842).
16. Nunc polypropylene cryotubes (VWR, cat. no. 363401B).

3. Methods

The methods described in this subheading outline (1) the construction of enhancer-trap gene targeting vectors, (2) *E. coli* transformation and plasmid preparation, (3) hybridoma cell transfection, (4) the isolation of independent transformants, (5) the rapid extraction of hybridoma genomic DNA, and (6) the identification and storage of targeted recombinants.

3.1. Construction of Enhancer-Trap Gene Targeting Vectors

The procedures used in the construction of enhancer-trap sequence insertion and sequence replacement vectors are standard in the field of molecular biology ***(25)*** and are not described in detail here. Both vector types share two main features. First, they are based on a derivative of the plasmid pSV2neo ***(26)***, in which the 372-bp *Nsi*I/*Nde*I fragment encompassing the SV40 early region enhancer responsible for *neo* gene expression has been deleted. As explained in **Note 2**, the enhancerless-*neo* gene greatly enriches for gene targeting events at the chromosomal μ locus. Second, the homology segments are well above the minimum size required for efficient gene targeting (approx 1 kb) ***(1)*** and are isogenic with the chromosomal μ locus. Several such segments have been used successfully in our studies, and examples are presented next.

3.1.1. Sequence Insertion Vector

Enhancer-trap sequence insertion vectors are constructed by using oligonucleotide adaptors to insert the 4.3-kb *Xba*I fragment from the wildtype chromosomal μ gene constant (Cμ) region (**Fig. 1A**) into the *Eco*RI site of the enhancerless pSV2neo vector (**Fig. 1B**). Homologous recombination is stimulated by introducing a double-strand break (DSB) into the vector-borne Cμ region; several unique restriction enzyme sites are available for this purpose (for example, *Bam*HI, *Bst*XI, *Kpn*I). When cleaved, optimal Cμ region homology exists on each side of the DSB to stimulate recombination.

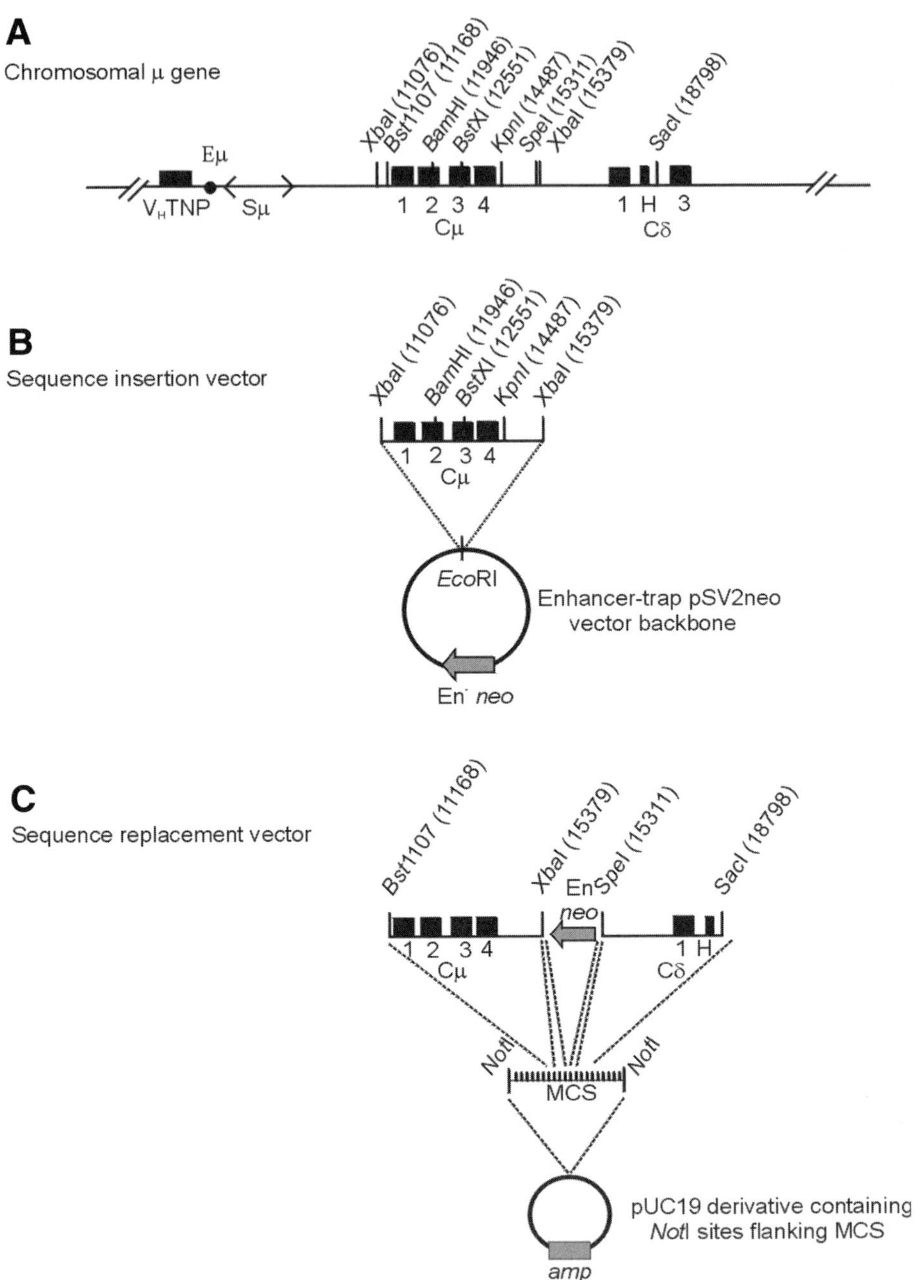

Fig. 1. Schematic drawings of **(A)** the endogenous chromosomal Ig μ heavy chain locus and the enhancer-trap **(B)** sequence insertion and **(C)** sequence replacement vectors. The numbering system used to denote restriction enzyme sites is based on the wildtype μ sequence *(27)*.

3.1.2. Sequence Replacement Vector

For the sequence replacement vector (**Fig. 1C**), it has been convenient to utilize a targeting segment consisting of the 3.3-kb *Nsi*I/*Swa*I fragment encoding the enhancerless *neo* gene flanked by the 4.2-kb *Bst*1107/*Xba*I Cμ region fragment and the 3.5-kb *Spe*I/*Sac*I Cδ fragment. The fragments are appropriately end-modified and inserted into the multiple cloning site (MCS) of a derivative of pUC19 in which unique *Not*I sites have been created at the termini of the polylinker by oligonucleotide mutagenesis. In this way, the entire targeting segment can be recovered following *Not*I digestion and prepared for transfection.

3.2. E. coli *Transformation and Plasmid Preparation*

The following procedures are used to verify the correct plasmid constructs:

1. Transform *E. coli* (a standard strain such as DH5α) with the ligation mix according to published procedures *(25)*.
2. Plate *E. coli* cells on LB agar containing 100 μg/mL ampicillin (LB-AMP) and incubate overnight at 37°C.
3. Pick single colonies, streak on LB-AMP agar, and incubate overnight at 37°C.
4. Pick a single colony from each streaked culture, inoculate tubes containing 5 mL of LB-AMP, and incubate overnight at 37°C on a rotary shaker.
5. Prepare plasmid minipreps by standard procedures *(25)*.
6. For construct verification, digest plasmid DNA with appropriate restriction enzymes and analyze fragment sizes by gel electrophoresis.
7. Following identification of correct isolates, freeze two independent fresh overnight cultures at –70°C in LB containing 60% sterile glycerol.

For mammalian cell transfection, large scale plasmid preparation is performed as follows:

1. Inoculate 250 mL of LB-AMP with 1 mL of an overnight culture of *E. coli* containing the plasmid of interest, and incubate overnight at 37°C on a rotary shaker.
2. Perform large-scale isolation of plasmid DNA by alkaline lysis according to standard procedures *(25)*.
3. Determine plasmid DNA concentration by UV spectrophotometry, and adjust concentration to approx 1 μg/μL with 1× TE.

3.3. Hybridoma Cell Transfection

The next steps involve transfection of the hybridoma cells with the enhancer-trap gene targeting vectors and the determination of cell viability.

3.3.1. Tissue Culture

Maintain suspension cultures of the hybridoma cells in tissue culture flasks in supplemented DMEM in a humidified, 7% CO_2 atmosphere. Discard cultures after approx 1 mo and restart from a frozen stock. For electroporation, grow the cells in 100 mL of supplemented DMEM in 75-cm^2 flasks (Gibco-BRL) to a final density of approx 4×10^5 cells/mL at 37°C. Determine the density of viable cells/mL by trypan blue exclusion using a hemocytometer under low-power magnification.

3.3.2. Preparation of Plasmids for Electroporation

In the case of the enhancer-trap sequence insertion vector, create the recombination-initiating DSB by digesting 50 µg of plasmid DNA with one of the restriction enzymes indicated in **Subheading 3.1.1.**, whose recognition site resides within the vector-borne Cµ region (**Fig. 1B**). In the case of the sequence replacement vector, release the targeting segment containing the enhancerless-*neo* gene from the derivative pUC19 backbone by digesting 50 µg of plasmid with *Not*I. If desired, a gel purification step can be used to separate the targeting segment from the pUC19 backbone; in this case, cut approx 100 µg of plasmid with *Not*I. For both vector types, the cut DNA is treated with phenol/chloroform, ethanol precipitated, and resuspended in 50 µL of 1× PBS and kept on ice. Prior to electroporation, a 0.25-µL aliquot of the plasmid preparation is analyzed by gel electrophoresis to verify the correct concentration.

3.3.3. Electroporation

Throughout the following procedure, everything is kept on ice unless noted otherwise.

1. Centrifuge 2×10^7 hybridoma cells in a 50-mL centrifuge tube for 10 min at 800 rpm (approx 200*g*). Wash the cells with 1× PBS, and resuspend in 0.75 mL of permeabilization buffer.
2. Transfection is performed by electroporation using a Gene Pulser (Bio-Rad). Combine the cut plasmid DNA and hybridoma cells in the electroporation cuvet (0.4-cm electrode gap) and, using the pipetor, gently mix the contents, taking care not to create any air bubbles. Dry the outside of the cuvet with a paper towel. Pulse the contents 2× at 700 V, 25 µF, with gentle mixing between pulses.
3. Place the cuvet on ice for 10 min, and then add 1 mL of supplemented DMEM (warmed to 37°C) with gentle mixing. Transfer the cuvet to a humidified, 7% CO_2 incubator at 37°C for 20 min to aid in cell recovery.
4. Transfer the approx 1 mL of cuvet contents into 96 mL of supplemented DMEM in a tissue culture flask. Carefully wash the internal cavity of the cuvet 3× with 1 mL of supplemented DMEM, combining the 3 mL of washes with the rest of the culture to yield a final volume of 100 mL.
5. Determine hybridoma cell survival by placing 0.5 mL of the well-mixed 100 mL electroporated culture in each of 2 wells in a 24-well tissue culture plate.

Add 0.5 mL of DMEM to each well. Assuming complete cell survival, there should be approx 1×10^5 cells in each well. Repeat the same procedure with a sample of the non-electroporated control culture. Place the plates in a humidified, 7% CO_2 incubator overnight at 37°C.

6. The next day, determine the number of cells/mL in both the electroporated and non-electroporated culture by trypan blue exclusion as explained in **Subheading 3.3.1.**, and calculate % cell survival. Under these electroporation conditions, hybridoma cell survival typically averages approx 50% (range approx 40–70%).

3.4. Isolation of Independent Transformants

1. After the platings for hybridoma cell survival are completed, make up the remaining 99 mL of the electroporated culture to 333 mL with supplemented DMEM and mix well.
2. Dispense a portion of the culture into a sterile, square plastic Petri dish, and, using an ethanol-cleaned, 8-channel pipetor, distribute 0.1-mL aliquots containing approx 6000 hybridomas into the individual wells of 96-well tissue culture plates. Repeat this process until all the electroporated culture has been distributed (in total, approx 34 tissue culture plates encompassing approx 3333 individual culture wells). The plating is usually completed within approx 2 h of the electroporation.
3. Set up several plates containing non-electroporated cells in the same way to serve as controls. Place the plates in a humidified, 7% CO_2 incubator. As indicated in **Subheading 3.3., step 6**, post-electroporation hybridoma cell survival averages approx 50%. Therefore, each culture well will contain approx 3000 viable hybridoma cells.
4. Twenty four hours after plating, place 0.1 mL of supplemented DMEM containing G418 (DMEM-G418) into each culture well to achieve a final concentration of 1.2 mg/mL.
5. Continue incubation at 37°C with periodic examination to monitor the selection process. After a few days, the non-electroporated control cells are completely dead, as are the bulk of electroporated cells. However, in the latter case, small colonies are visible in some culture wells. Continue the selection for approx 14 d, after which the total number of positive culture wells is enumerated. According to the Poisson distribution, the plating procedures make it highly likely that each recombinant represents the progeny of a single $G418^R$ cell (*see* **Note 3**).
6. Recover each transformant from the 96-well plates and save by freezing according to the following procedure. Transfer a portion of each transformant culture into 1.5 mL of supplemented DMEM-G418 in a 24-well tissue culture plate. After 3 or 4 d, cell growth has progressed to the stage at which the culture medium is slightly orange. At this point, mix the culture well with a 1-mL pipet, and transfer 0.75 mL into a sterile 1.5-mL Eppendorf microfuge tube, and spin for 10 min at 800 rpm (approx 200*g*) in a refrigerated microcentrifuge. Aspirate the supernatant and

resuspend the cell pellet in 200 μL of FM. Gently mix the contents by tapping, and store at –70°C. Cultures frozen in this way are viable for approx 6 mo.

7. Add 0.75 mL of DMEM-G418 to each of the culture wells and resume growth at 37°C for several more days to prepare the culture for isolation of genomic DNA.

3.5. Rapid Extraction of Hybridoma Genomic DNA

1. For extraction of genomic DNA from 24-well plate cultures, hybridoma cell density should be approx 5–7 × 10^5 cells/mL. (The culture medium should be slightly yellow.) Gently mix the cells with a 2-mL pipet, and transfer the entire 1.5-mL culture to an Eppendorf microfuge tube. Centrifuge at 800 rpm (approx 200*g*) for 10 min.
2. Wash the cells with 1× PBS.
3. Resuspend in 50 μL of 1× PBS. Add 100 μL of lysis buffer together with 17.5 μL of proteinase K.
4. Incubate overnight in a waterbath at 65°C.
5. To precipitate the DNA, add 450 μL of 95% ethanol and 5 μL of 5 *M* NaCl, and mix by rotation for 20 min. Using a pipet tip, gently drag the clump of genomic DNA up the side of the microfuge tube until it sticks, and then aspirate the 95% ethanol/NaCl mix. Wash the genomic DNA two times with 1 mL of 70% ethanol, and resuspend the DNA in 75 μL of 1× TE. To remove any excess ethanol, place the opened tube in a waterbath at 65°C for 30 min. Replace the cap and store the DNA at 4°C.

3.6. Identification of Targeted Recombinants

A combination of Southern and PCR analysis is used to identify targeted G418^R recombinants from among the random, G418^R transformants. In a sequence insertion event, the target sequence is duplicated and flanks the vector backbone, whereas in a sequence replacement event, target chromosomal sequences are exchanged for those in the transfected vector (**Fig. 2**) (*see* **Note 4**).

1. Initially, screening is performed to eliminate the vast majority of random, G418^R transformants bearing an unmodified chromosomal μ locus. This is accomplished by PCR analysis of hybridoma genomic DNA using primers 9931FCμ (5' GCAAGAGTGAGTAGAGCTGGCTGG 3') and 17390RCμ (5' GGTTCGGTTCTGTCTGCACTACTC 3') to generate a specific 7.5-kb Cμ fragment (**Fig. 3A**). PCR amplification is performed using the Expand Long Template PCR kit (Roche) according to the manufacturer's specifications.
2. Genomic DNA from the remaining cell lines is examined further by Southern analysis for the unique chromosome:vector junction fragments characteristic of the sequence insertion (**Fig. 3B**) and sequence replacement (**Fig. 3C**) gene targeting events. Restriction enzyme digestion of genomic DNA, gel electrophoresis and blotting to nitrocellulose, and hybridization with ^{32}P-labeled DNA probes

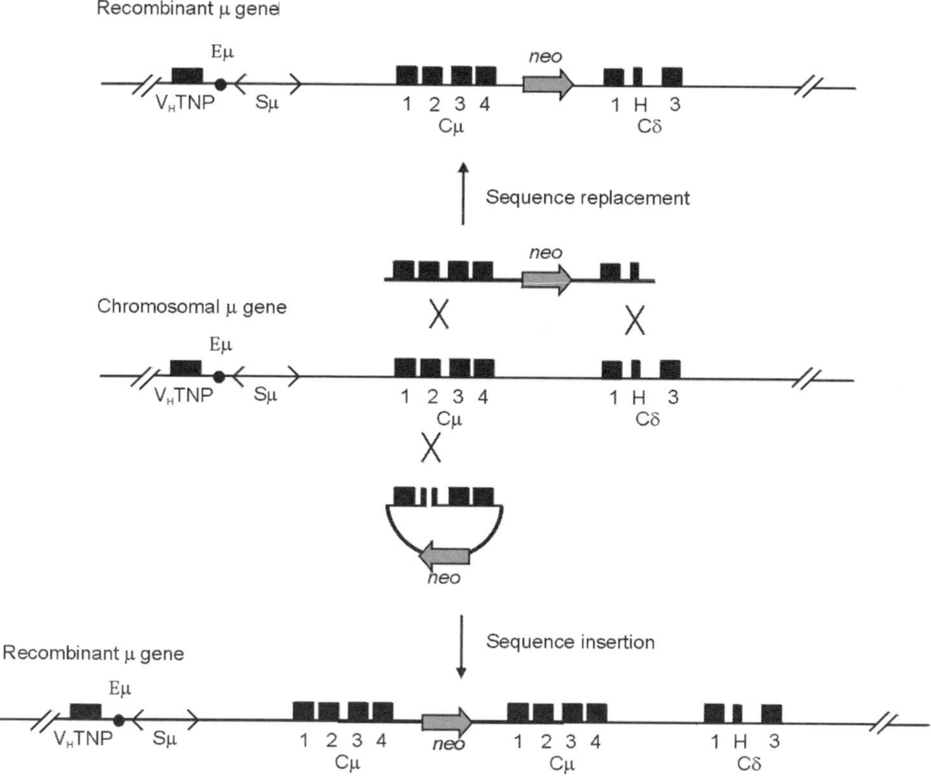

Fig. 2. Recombination between transferred and chromosomal Ig μ genes by sequence insertion and sequence replacement vectors.

are all performed according to standard procedures ***(25)***. The following DNA probe fragments are used: Cμ-specific probe F, an 870-bp *Xba*I/*Bam*HI fragment; probe XR, a 913-bp *Xho*I/*Eco*RI fragment; and probe G, a 762-bp *Pvu*II fragment from the *neo* gene of pSV2neo.

3. Following recombinant identification, the appropriate frozen cells are thawed by immersing the Eppendorf tube in a 45°C waterbath until a small portion of ice remains in the tube. The tube is then placed on ice momentarily, until the contents are completely thawed. The contents are then added to 10 mL of DMEM containing 600 μg/mL of G418 in a 25-cm^2 tissue culture flask, and the flask placed in a humidified, 7% CO_2 incubator. Following growth, the cells are refrozen for long-term storage in Nunc polypropylene cryotubes by resuspending a cell pellet derived following centrifugation of 5 mL of culture (density of 2×10^5 cells/mL) in 1 mL of FM. The cryotubes containing the cell mixtures are first frozen at –70°C overnight, and then transferred to liquid nitrogen for more permanent storage.

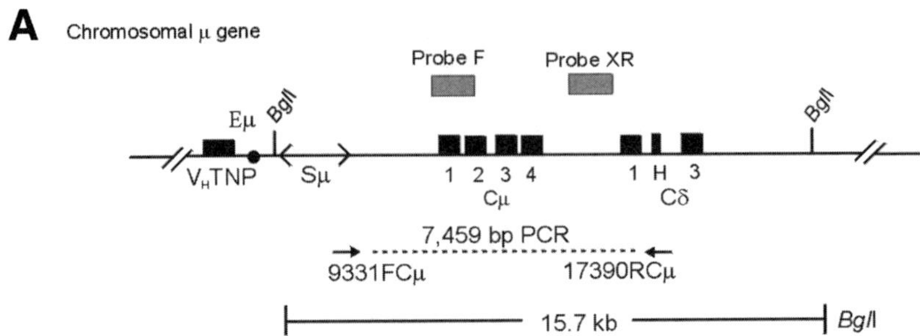

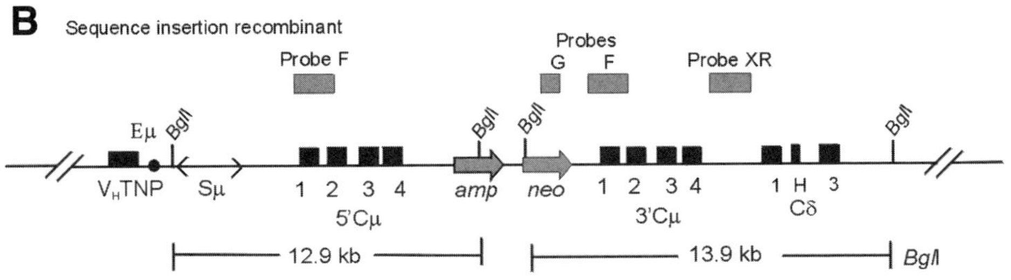

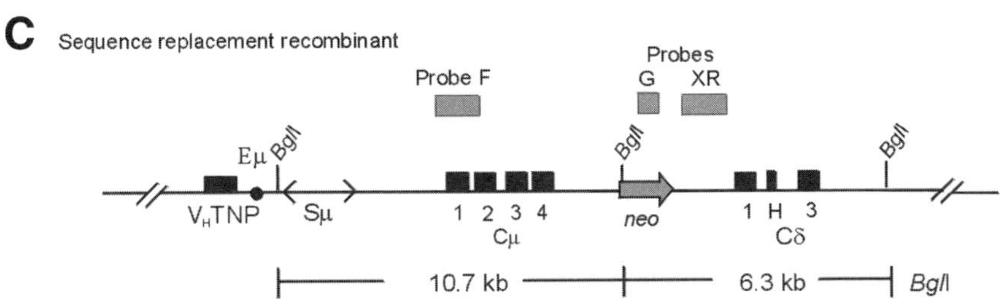

Fig. 3. DNA fragments predicted for endogenous and recombinant μ genes. The expected DNA structures are shown for **(A)** the endogenous chromosomal μ gene, and recombinant μ genes derived following gene targeting with enhancer-trap **(B)** sequence insertion and **(C)** sequence replacement vectors.

4. Notes

1. Our gene targeting system is based on the mouse Sp6 hybridoma cell line ***(23,24)***, which contains a single copy of the chromosomal Ig μ heavy chain gene and synthesizes IgM(κ). We utilize the cloned μ and δ region from the Sp6 hybridoma to prepare gene targeting segments. However, the same DNA can be isolated from genomic libraries prepared from the cell lines of choice. Alternatively, the desired DNA can be obtained by PCR amplification using primers designed

from publicly available DNA sequence information. The methods described here are applicable to any mouse hybridoma (or myeloma) cell line bearing a chromosomal Ig μ gene. Furthermore, the enhancer-trap vector strategy can be adapted for gene targeting in cell lines bearing other rearranged Ig heavy or light chain genes, because the position of chromosomal enhancer elements important in driving expression of the dominant selectable marker in the targeted vector is usually conserved within the genes.

2. In an enhancer-trap vector, the enhancer sequence responsible for expression of the vector-borne dominant selectable marker is removed. This strategy enriches for gene targeting events providing that the target locus supplies the enhancer (or equivalent) activity that is required for expression of the vector-borne dominant selectable marker ***(7,9,28)***. In our system, the Ig μ gene enhancer within the major intron (Eμ) (**Fig. 1A**) is expected to supply the activity required for expression of the enhancerless-*neo* gene in the targeted cell lines (**Fig. 2**). Typically, with an enhancer-trap vector, targeted recombinants comprise approx 10% of the transformant population compared to with approx 0.5% with the corresponding enhancer-positive vector ***(6,9,11,13)***.
3. The likelihood that independent recombinants represent the progeny of individual G418[R] cells is illustrated by the following example. From a total of 6666 culture wells plated, Li et al. ***(13)*** recovered 281 independent G418[R] transformants, of which 26 were correctly targeted recombinants. The recovery of recombinants among the culture wells is expected to follow the Poisson distribution, and, under these circumstances, the probability that the recombinants in a well actually derived from more than one independent recombinant is approx 0.002.
4. In some applications, it might be desirable to utilize a second homologous recombination step to remove the vector-backbone and dominant selectable marker from the targeted Ig locus. Vector excision is facilitated by incorporating a negative-selectable marker into the enhancer-trap gene targeting vector, for example, the herpes simplex-1 thymidine kinase (*tk*) gene, or the *E. coli gpt* gene, which can be selected against using FIAU ***(9,29)*** or MPA ***(7)***.

Acknowledgments

The technical assistance of Erin Birmingham is appreciated. This work was supported by operating grants from the Canadian Institutes of Health Research (CIHR) and the Natural Sciences and Engineering Research Council (NSERC) of Canada.

References

1. Waldman, A. S. (1992) Targeted homologous recombination in mammalian cells. *Crit. Rev. Oncol. Hematol.* **12,** 49–64.
2. Bertling, W. M. (1995) Gene targeting. In: *Gene Targeting* (Vega, M. A., ed.), CRC Press, Boca Raton, FL. pp. 1–44.
3. Guénet, J. (1995) Gene targeting. In: *Gene Targeting* (Vega, M. A., ed.), CRC Press, Boca Raton, FL. pp. 45–64.

4. Vega, M. A. (1995) Gene targeting. In: *Gene Targeting* (Vega, M. A., ed.), CRC Press, Boca Raton, FL. pp. 211–229.
5. Vasquez, K. M., Marburger, K., Intody, Z., and Wilson, J. H. (2001) Manipulating the mammalian genome by homologous recombination. *Proc. Natl. Acad. Sci. USA* **98,** 8403–8410.
6. Baker, M. D., Pennell, N., Bosnoyan, L., and Shulman, M. J. (1988) Homologous recombination can restore normal immunoglobulin production in a mutant hybridoma cell line. *Proc. Natl. Acad. Sci. USA* **85,** 6432–6436.
7. Bautista, D. and Shulman, M. J. (1993) A hit-and-run system for introducing mutations into the immunoglobulin heavy chain locus of hybridoma cells by homologous recombination. *J. Immunol.* **151,** 1950–1958.
8. Baker, M. D. and Shulman, M. J. (1988) Homologous recombination between transferred and chromosomal immunoglobulin κ genes. *Mol. Cell. Biol.* **8,** 4041–4047.
9. Ng, P. and Baker, M. D. (1998) High efficiency, site-specific modification of the chromosomal immunoglobulin locus by gene targeting. *J. Immunol. Methods* **214,** 81–96.
10. Ng, P. and Baker, M. D. (1999) Mechanisms of double-strand-break repair during gene targeting in mammalian cells. *Genetics* **151,** 1127–1141.
11. Li, J. and Baker, M. D. (2000) Use of a small palindrome genetic marker to investigate mechanisms of double-strand-break repair in mammalian cells. *Genetics* **154,** 1281–1289.
12. Li, J. and Baker, M. D. (2000) Formation and repair of heteroduplex DNA on both sides of the double-strand break during mammalian gene targeting. *J. Mol. Biol.* **295,** 505–516.
13. Li, J., Read, L. R., and Baker, M. D. (2001) The mechanism of mammalian gene replacement is consistent with the formation of long regions of heteroduplex DNA associated with two crossing-over events. *Mol. Cell. Biol.* **21,** 501–510.
14. Baker, M. D. and Birmingham, E. C. (2001) Evidence for biased Holliday junction cleavage and mismatch repair directed by junction cuts during double-strand-break repair in mammalian cells. *Mol. Cell. Biol.* **21,** 3425–3435.
15. Raynard, S. L. and Baker, M. D. (2002) Incorporation of large heterologies into heteroduplex DNA during double-strand-break repair in mouse cells. *Genetics* **162,** 977–985.
16. Baker, M. D. and Read, L. R. (1992) Ectopic recombination within homologous immunoglobulin μ gene constant regions in a mouse hybridoma cell line. *Mol. Cell. Biol.* **12,** 4422–4432.
17. Baker, M. D. and Read, L. R. (1995) High-frequency gene conversion between repeated Cμ sequences integrated at the chromosomal immunoglobulin μ locus in mouse hybridoma cells. *Mol. Cell. Biol.* **15,** 766–771.
18. Shulman, M. J., Collins, C., Connor, A., Read, L. R., and Baker, M. D. (1995) Interchromosomal recombination is suppressed in mammalian somatic cells. *EMBO J.* **14,** 4102–4107.

19. Baker, M. D., Read, L. R., Beatty, B. G., and Ng, P. (1996) Requirements for ectopic homologous recombination in mammalian somatic cells. *Mol. Cell. Biol.* **16,** 7122–7132.
20. Fell, H. P., Yarnold, S., Hellström, I., Hellström, K. E., and Folger, K. R. (1989) Homologous recombination in hybridoma cells: heavy chain chimeric antibody produced by gene targeting. *Proc. Natl. Acad. Sci. USA* **86,** 8507–8511.
21. Wood, C. R., Morris, G. E., Alderman, E. M., Fouser, L., and Kaufman, R. J. (1991) An internal ribosome binding site can be used to select for homologous recombinants at an immunoglobulin heavy-chain locus. *Proc. Natl. Acad. Sci. USA* **88,** 8006–8010.
22. Sun, W., Xiong, J., and Shulman, M. J. (1994) Production of mouse V/human C chimeric κ genes by homologous recombination in hybridoma cells. *J. Immunol.* **152,** 695–704.
23. Köhler, G. and Shulman, M. J. (1980) Immunoglobulin M mutants. *Eur. J. Immunol.* **10,** 467–476.
24. Köhler, G., Potash, M. J., Lehrach, H., and Shulman, M. J. (1982) Deletions in immunoglobulin mu chains. *EMBO J.* **1,** 555–563.
25. Sambrook, J., Fritsch, E. F., and Maniatis, T. (1989) *Molecular Cloning, A Laboratory Manual*, 2nd ed. Cold Spring Harbor Laboratory Press, Cold Spring Harbor, NY.
26. Southern, P. J. and Berg, P. (1981) Transformation of mammalian cells to antibiotic resistance with a bacterial gene under control of the SV40 early region promoter. *J. Mol. Appl. Genet.* **1,** 327–341.
27. Bilofsky, H. F., Burks, C., Fickett, J. W., et al. (1986) The GenBank genetic sequence databank. *Nucleic Acids Res.* **14,** 1–4.
28. Jasin, M. and Berg, P. (1988) Homologous integration in mammalian cells without target gene selection. *Genes Dev.* **2,** 1353–1363.
29. Hasty, P., Ramirez-Sollis, R., Krumlauf, R., and Bradley, A. (1991). Introduction of a subtle mutation into the *Hox*-2.6 locus in embryonic stem cells. *Nature* **350,** 243–246.

10

DNA Fragment Transplacement in *Saccharomyces cerevisiae*

Some Genetic Considerations

Glenn M. Manthey, Michelle S. Navarro, and Adam M. Bailis

Summary

The ability to make specific genomic alterations is an invaluable tool to researchers who use genetics and biochemistry to study problems in biology. We have investigated some of the parameters governing DNA fragment transplacement in two commonly used strains of *Saccharomyces cerevisiae*, S288C and W303-1A. These strains exhibited a marked difference in their capacity to take up plasmid DNA and utilize linear DNA fragments as substrates for transplacement. The contributions of transformation efficiency, length of homology, and alternative target site configuration were assessed. This analysis indicates that several genetic parameters are important for optimizing the efficiency of gene transplacement.

Key Words: transplacement, lithium acetate transformation, electroporation, short-sequence recombination, yeast, *Saccharomyces cerevisiae*

1. Introduction

The ability to generate specific changes in a genomic sequence is of great utility to researchers studying the genetic and biochemical basis of biological phenomena. The yeast *Saccharomyces cerevisiae* is an especially facile model system with which to carry out these sorts of studies. A variety of techniques have been developed to make specific changes to genomic sequences. These include the use of single-stranded oligonucleotides, yeast-integrating plasmids, and marked DNA fragments ***(1–6)***. Various techniques for introducing these DNA sequences into the cell have also been developed, including spheroplast transformation, lithium acetate transformation, and electroporation ***(7–9)***. Despite their wide use, little is known about the factors that contribute to the success or failure of these approaches. In this report we provide an overview of

From: *Methods in Molecular Biology, vol. 262, Genetic Recombination: Reviews and Protocols*
Edited by: A. S. Waldman © Humana Press Inc., Totowa, NJ

the techniques for transplacing DNA fragments into the genome of *S. cerevisiae* and also examine some of the important parameters that may contribute to the success of the procedure.

2. Materials

1. Yeast strains: W303-1A, S288C, and isogenic derivatives.
2. Ethidium bromide, 10 mg/mL.
3. Plasmids: pRS413, pRS416, pLAY196.
4. Ampicillin, 25 mg/mL.
5. Restriction enzymes, Taq DNA polymerase.
6. Reaction buffers:
   a. 10× Restriction endonuclease buffer 2: 500 m*M* NaCl, 100 m*M* Tris-HCl, 100 m*M* $MgCl_2$, 10 m*M* dithiothreitol, pH 7.9 at 25°C.
   b. 10× PCR buffer: 200 m*M* Tris-HCl, pH 8.4, 500 m*M* KCl.
7. Klett-Summerson colorimeter.
8. Gene Pulser II (Bio-Rad), 0.2-cm cuvets.
9. Side-arm flasks.
10. Oligonucleotide primers.
11. High-speed centrifuge.
12. Transformation reagents.
    a. 1 *M* LiAc.
    b. 50% Polyethylene glycol (PEG) MW3350.
    c. Single-stranded DNA 10 mg/mL.
    d. Lithium acetate solution: 0.1 *M* LiAC, 10 m*M* Tris-HCl, pH 8.0, 1 m*M* ethylenediamine tetraacetic acid (EDTA).
    e. Sterile PEG solution: 40% PEG 3350, 10 m*M* Tris-HCl, pH 8.0, 1 m*M* EDTA, 0.1 *M* LiAc.
    f. Tris-EDTA (TE): 10 m*M* Tris-base, pH 8.0, 1 m*M* EDTA.
    g. 1 *M* Sorbitol.
13. CsCl preparation reagents.
    a. Solution 1: 25 m*M* Tris-base, pH 8.0, 10 m*M* EDTA. Mix 2.5 mL 1 *M* Tris-base, pH 8.0, 2 mL 0.5 *M* EDTA, and 95.5 mL H_2O.
    b. Solution 2: 1% sodium dodecyl sulfate (SDS), 0.2 *N* NaOH. Mix 1 mL 10% SDS, 0.4 mL 5 *N* NaOH, and 8.6 mL H_2O.
    c. Solution 3: 5 *M* acetate, 3 *M* potassium. Mix 60 mL 5 *M* potassium acetate, 11.5 mL glacial acetic acid 17.4 *M*, and 28.5 mL H_2O.
    d. Phenol/chloroform, 1:1 mixture.
    e. 100% Isopropanol.
    f. 100% Ethanol.
    g. 70% Ethanol.
    h. RNase 10 mg/mL.
    i. CsCl.
    j. Water-saturated butanol.

14. NVT100 rotor (Beckman).
15. Beckman Optima-XL ultracentrifuge.
16. Agarose gel electrophoresis.
17. Plate turntable and spreader.
18. Darkfield counter.
19. Media (for plates add 20 g agar/L):
    a. Luria-Bertani (LB): Dissolve 10 g tryptone, 10 g NaCl, and 5 g yeast extract in 1 L H_2O.
    b. YPD: Dissolve 20 g peptone, 20 g dextrose, and 10 g yeast extract in a final volume of 1 L H_2O.
    c. Synthetic complete (SC) $^-$Ura sorbitol: Dissolve 6.7 g yeast nitrogen base without amino acids, 20 g dextrose, 182.2 g sorbitol, 20 mg each of L-adenine sulfate, L-arginine sulfate, L-histidine-HCl, L-methionine, and L-tryptophan, 30 mg of L-isoleucine and L-lysine-HCl, 50 mg L-phenylalanine, 60 mg L-leucine, and 150 mg L-valine in a final volume of 1 L H_2O.
    d. SC $^-$His sorbitol: Dissolve 6.7 g yeast nitrogen base without amino acids, 20 g dextrose, 182.2 g sorbitol, 20 mg each of L-adenine sulfate, L-arginine sulfate, L-methionine, and L-tryptophan, uracil, 30 mg of L-isoleucine and L-lysine-HCl, 50 mg L-phenylalanine, 60 mg L-leucine, and 150 mg L-valine in a final volume of 1 L H_2O.
    e. SC $^-$His $^-$Ura: Dissolve 6.7 g yeast nitrogen base without amino acids, 20 g dextrose, 20 mg each of L-adenine sulfate, L-arginine sulfate, L-methionine, and L-tryptophan, 30 mg of isoleucine and L-lysine-HCl, 50 mg L-phenylalanine, 60 mg L-leucine, and 150 mg L-valine in a final volume of 1 L H_2O.
    f. SC sorbitol: Dissolve 6.7 g yeast nitrogen base without amino acids, 20 g dextrose, 182.2 g sorbitol, 20 mg each of L-adenine sulfate, L-arginine sulfate, L-histidine-HCl, L-methionine, and L-tryptophan, uracil, 30 mg of L-isoleucine and L-lysine-HCl, 50 mg L-phenylalanine, 60 mg L-leucine, and 150 mg L-valine in a final volume of 1 L H_2O.

3. Methods

The methods described here are the preparation of plasmid DNA, growth curve determination, transformation with lithium acetate and electroporation, and the design of a DNA fragment transplacement strategy.

3.1. Plasmid Transformation Analysis

We assessed the relative ability of two typical laboratory strains of yeast, S288C and W303-1A, to be transformed with plasmid DNA using two popular methods, lithium acetate treatment and electroporation. The first method uses an alkali cation to induce changes in cell membrane permeability, and the second relies on a quick pulse of electrical current to facilitate uptake of DNA. Both strains were transformed with the yeast centromere plasmids, pRS413 and pRS416, that are marked with *HIS3* and *URA3*, respectively ***(13)***.

3.1.1. Plasmid Preparation

The plasmids used in these studies were amplified and purified using a cesium chloride gradient sedimentation protocol derived from a previously described protocol ***(14)***.

1. *E. coli* bearing the plasmid of interest was grown in a 400-mL LB culture overnight at 37°C (*see* **Note 1**).
2. The cells were harvested with a Sorvall SLA-3000 rotor at 6000 rpm (6000*g*) for 5 min at 4°C.
3. The supernatant was discarded and the cells resuspended in 16 mL of solution 1.
4. Four-milliliter aliquots were transferred to 40-mL Oakridge tubes and placed on ice.
5. Eleven milliliters of solution 2 were added to each tube and mixed by inversion.
6. Seven milliliters of solution 3 were added to each tube and mixed by inversion.
7. The tubes were incubated on ice for 15–30 min.
8. The tubes were spun at 10,000 rpm (14,500*g*) for 10 min in a Sorvall SA-600 rotor to pellet the precipitate.
9. Supernatants were transferred to fresh Oakridge tubes, and 4 mL phenol/chloroform (1:1 solution) were added to each tube.
10. The tubes were mixed well by inversion and centrifuged at 10,000 rpm (14,500*g*) for 10 min.
11. The aqueous upper phase was transferred to a new tube, and 20 mL of isopropanol were added to each tube.
12. The tubes were mixed by inversion and centrifuged at 12,000 rpm (21,000*g*) for 15 min to precipitate the nucleic acid.
13. The pellets were washed with 2 mL of 70% ethanol and dried at room temperature for 30 min.
14. 500 µL of TE/RNase 10 µg/mL were added to each tube to resuspend the nucleic acid pellet and digest the RNA component. Samples were digested for 2 h at 25°C.
15. Samples were pooled into 1-mL aliquots, and exactly 1 g of CsCl was added to each tube and mixed.
16. 60 µL of ethidium bromide (10 mg/mL) were added to each sample and mixed.
17. The CsCl/DNA/ethidium bromide solution was transferred into a Beckman quick seal centrifuge tube (0.5 in. × 2 in.). A stock solution of CsCl/TE (approx 4 mL for each sample) was made by adding exactly 1 g of CsCl to each mL of TE.
18. The tubes were then filled to the neck of the tube with the stock solution. Each tube was balanced against a partner to within 1–2 mg and sealed with a Beckman heat sealer.
19. The balanced tubes were loaded into a Beckman NVT100 rotor and spun for at least 4 h at 80,000 rpm (480,000*g*), at 25°C.
20. The DNA was collected, extracted, precipitated, and resuspended in an appropriate volume of TE. The concentration and purity were determined as described previously ***(14)***.

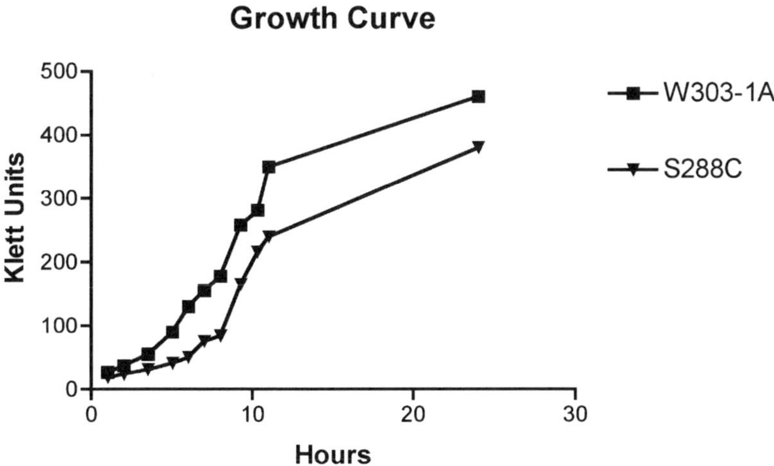

Fig. 1. Growth curve analysis of W303-1A and S288C grown in YPD at 30°C.

3.1.2. Growth Curve Determination

The following experiments rely on the transformation of yeast strains with plasmids or DNA fragments. To maximize the efficiency of transformation, it is important to harvest the cells while they are in the early to mid-exponential phase of the growth curve. Because growth characteristics vary between strains and growth media, the determination of a growth curve is important for optimizing the transformation. To maintain sterile cultures while measuring the growth of yeast, side-arm flasks were used in conjunction with a Klett-Summerson colorimeter (*see* **Note 2**).

1. Twenty-five milliliters of YPD, in a side-arm flask, were inoculated with sufficient saturated culture to result in a Klett reading of approx 25 (1×10^7 cells/mL).
2. Cultures were grown at 30°C in an environmental shaker at 300 rpm.
3. The growth of the culture was monitored hourly.
4. Cell density was plotted vs time.

In the growth curve displayed in **Fig. 1**, the cells should be harvested at 80–150 Klett units ($5–8 \times 10^7$ cell/mL).

3.1.3. Lithium Acetate Transformation

Since the introduction of the lithium acetate transformation method ***(15)***, several modifications have been made to increase efficiency ***(8,15–17)***. The protocol used in this report is based on that of Schiestl and Gietz ***(8)***.

1. In each of two 500-mL side-arm flasks, 300 mL of YPD medium were inoculated with a single colony of either S288C or W303-1A (**Table 1**).

Table 1
Strains Used in This Study

Strain[a]	Genotype	Ref.
W303-1A	*MAT***a** *ade2-1 can1-100 his3-11, 15 leu2-3, 112 trp1-1 ura3-1*	***10***
S288C	*MAT***a** *ade2-101 his3-Δ200 lys2-801 ura3-52*	***11***
W1011-3B	*MATα ade2-1 can1-100 leu2-3, 112 trp1-1 ura3-1*	***10***
ABT407	*MAT***a** *ade2-101 lys2-801 ura3-52*	This study
ABM44	*MAT***a** *ade2-1 can1-100 leu2-3, 112 trp1-1 ura3-1 rad50::hisG*	This study
ABX129-2B	*MATα ade2-1 can1-100 leu2-3, 112 trp1-1 ura3-1 rad1::LEU2*	This study
ABX81-4B	*MATα ade2-1 can1-100 leu2-3, 112 trp1-1 ura3-1 rad3-G595R*	***12***

[a]ABT407 is isogenic with S288C. ABM44, ABX129-2B, and ABX81-4B are isogenic with W303-1A.

2. The cultures were incubated overnight at 30°C, and shaken at 300 rpm.
3. Growth was monitored until the cultures reached a density of approx 80–90 Klett units (5×10^7 cells/mL). Based on the growth curve analysis described earlier (**Fig. 1**), this value corresponds to cultures that are in early to mid-log phase.
4. Cells were pelleted in 40-mL Oakridge tubes and spun at 5000 rpm (3600*g*) for 5 min using a Sorvall SA-600 rotor.
5. The pellets were washed in 10 mL of sterile water and pelleted again at 5000 rpm (3600*g*) for 5 min.
6. The pellets were resuspended in 1.5 mL sterile lithium acetate solution and incubated for 1 h in a 30°C waterbath, with constant agitation.
7. For each transformation, the following were combined in an 1.7-mL Eppendorf tube: 10 μL of 10 μg/μL of salmon sperm DNA, 1 μL of 1 μg/μL pRS416 or pRS413 plasmid DNA, and 200 μL of yeast cell suspension (*see* **Note 3**).
8. The contents of the tube were mixed and incubated at 30°C with constant agitation for 30 min.
9. 1.2 mL of fresh, sterile PEG solution, were then added to each tube, mixed, and incubated for 30 min at 30°C, with constant agitation.
10. The cells were pelleted for 2 min at 6000 rpm (3000*g*) in a microcentrifuge, washed twice with 500 μL of sterile TE, and resuspended in 400 μL of TE.
11. 100 μL of cell suspension were plated on each of four plates containing the appropriate selective medium (SC $^-$His or SC $^-$Ura) (*see* **Note 4**).
12. Appropriate dilutions of the final cell suspension were plated onto nonselective medium to determine the number of viable cells in the suspension.
13. The plates were incubated at 30°C for 3–4 d.
14. The transformant colonies were counted using a Leica darkfield colony counter.

This protocol typically yielded 1000–5000 transformants per μg of DNA. The efficiency of transformation per viable cell was determined by dividing the number of transformant colonies by the number of viable cells that were plated.

3.1.4. Electroporation

1. Twenty-five-milliliter YPD cultures, in 250-mL side-arm flasks, were inoculated with either W303-1A or S288C to a density of about 25 Klett units (1×10^7 cells/mL) from overnight cultures (*see* **Note 5**).
2. Cultures were grown at 30°C with shaking at 300 rpm to a density of about 90 Klett units (5×10^7 cells/mL). This density represents the early to mid-log phase in the growth curve as described in **Subheading 3.1.2., step 4** (**Fig. 1**).
3. The cultures were transferred to 40-mL Oakridge tubes and the cells pelleted by centrifugation at 5000 rpm (3600*g*) for 5 min.
4. The pelleted cells were washed once in 20 mL of ice-cold sterile water and repelleted at 2500 rpm (1000*g*) for 5 min. This step was repeated with 10 mL of ice-cold water.
5. Finally, the cells were washed with 2.5 mL of ice-cold 1 *M* sorbitol and pelleted at 2000 rpm (600*g*) for 5 min and resuspended in 75 µL ice-cold 1 *M* sorbitol. The cells were then ready for electroporation.

Using the Bio-Rad Gene Pulser II, the settings for electroporating yeast are 1.5 kV, 200 Ω, and 25 µF. The electroporations were carried out in electroporation cuvets, with a 0.2-cm electrode gap, that were prechilled on ice while the cells were being prepared.

6. For each electroporation, 1 µL of the pRS413 or pRS416 DNA (100 ng/µL) was placed into a cuvet.
7. 40 µL of cells were added to the cuvet and mixed with the DNA such that the mixture moved to the bottom of the cuvet. The cuvet was placed into the electroporation chamber and pulsed.
8. Following the pulse, the cells were resuspended in 1 mL of ice-cold 1 *M* sorbitol, transferred to a prechilled Eppendorf tube, and placed on ice. It is important that this step be accomplished as rapidly as possible. After the pulse, the time constant that was displayed typically fell between 5.02 and 4.8 (*see* **Note 6**).
9. Following electroporation, four 50-µL aliquots of the cells, transformed with pRS413 or pRS416, were plated onto SC $^-$His or SC $^-$Ura media containing 1 *M* sorbitol, respectively, and incubated for 3–4 d at 30°C.
10. To determine the number of survivors following electroporation, a serial dilution of the cells was performed using 1 *M* sorbitol as diluent.
11. The dilutions were plated onto SC medium containing 1 *M* sorbitol and incubated at 30°C for 3–4 d.

Significant differences in transformation efficiency were observed with the different plasmids, strains, and methods of transformation (**Table 2**). Transformation efficiency varied from two- to ninefold with pRS413 and pRS416, reflecting small but significant differences in the ability of the cells to take up the two plasmids. In general, electroporation was the more efficient transformation method; however, the differences varied from none at all to

Table 2
Transformation Efficiency (Transformants/Viable cell)[a]

	Lithium acetate		Electroporation	
Strain	pRS413	pRS416	pRS413	pRS416
W303-1A	$78.0 \times 10^{-7} \pm 7.1$	$9.1 \times 10^{-7} \pm 1.1$	$57.0 \times 10^{-7} \pm 11$	$160.0 \times 10^{-7} \pm 78$
S288C	$280 \times 10^{-7} \pm 32$	$120 \times 10^{-7} \pm 7.3$	$260 \times 10^{-7} \pm 85$	$1000 \times 10^{-7} \pm 320$

[a]Transformation efficiency from a minimum of five separate trials is reported and expressed as the ratio of the total number of Ura^+ or His^+ transformants obtained to the total number of survivors on nonselective medium.

approx 18-fold. The S288C strain was better able to take up DNA, with transformation efficiencies that ranged from 4- to 13-fold higher than those of W303-1A. To investigate the potential genetic basis for this difference, we crossed two S288C- and W303-1A-derived strains, resulting in the diploid strain ABX730. ABX730 was sporulated, several tetrads were dissected, and the efficiency of transformation by electroporation with pRS416 was assessed among the resulting spore colonies. No consistent pattern of segregation was observed for transformation efficiency, with levels that varied from 16-fold below that of W303-1A to 10-fold above that of S288C (data not shown). This indicates that transformation is under the influence of multiple genes and that a strain with substantially higher transformation efficiencies than either strain may be obtainable.

3.2. DNA Fragment Transplacement

Single-step gene disruption ***(3)*** has become the standard technique, in yeast, for generating null alleles and genetically marking loci of interest. The technique involves transforming yeast cells with DNA fragments containing a dominant selectable marker flanked by sequences that are homologous to the locus to be disrupted. The fragments are inserted into the locus by homologous recombination, resulting in the permanent association of the marker with the locus. More recently, polymerase chain reaction (PCR) technology has been used to generate DNA fragments that can precisely replace the coding sequence of a gene with a selectable marker ***(4,5,17)***. This method generally makes use of 45 bp of the DNA sequence immediately flanking the 5' and 3' ends of the coding sequence to direct the recombination event. Previous studies have shown that the length of the flanking homology and the genotype of the yeast cells both profoundly affect the efficiency of this process ***(5,12,19,20)***.

We have assessed the relative ability to generate transplacements with defined recombination substrates and isogenic derivatives of W303-1A and

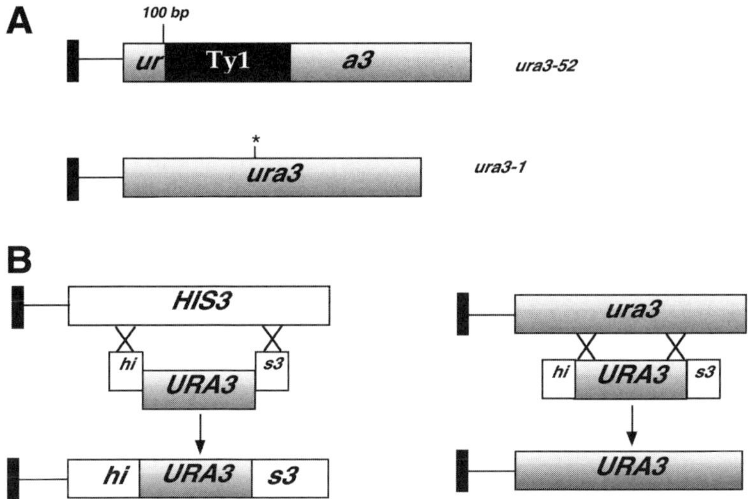

Fig. 2. (A) Structure of the *ura3-52* and *ura3-1* alleles. **(B)** Possible recombination events leading to a Ura⁺ transformant following electroporation of DNA fragments bearing the *URA3* gene flanked by sequences homologous to the *HIS3* locus.

S288C. This allowed us to assess the importance of the genetic background, the length of homology shared by the DNA fragment and the genomic target, and the configuration of alternative targets in determining transplacement efficiency. We performed transplacement experiments in strains derived from W303-1A and S288C, W1011-3B and ABT407, which are identical to the parent strains except that they carry a wild-type *HIS3* allele. We also examined W303-1A derivatives bearing the *rad3-G595R*, *rad1::LEU2*, or *rad50::hisG* alleles to study the affects of these mutations on transplacement. These alleles have previously been shown to affect DNA repair and recombination events that are similar to those described here ***(12,19,20)***.

To address the role of alternative target configuration as an effector of transplacement efficiency, we have taken advantage of the different *ura3* alleles found in W303-1A and S288C strains. W303-1A bears the *ura3-1* allele, which has a missense mutation in the middle of the locus. S288C carries the *ura3-52* allele, in which a 4.5-kb Ty1 retrotransposable element is inserted 100 bp downstream of the start codon (**Fig. 2A**) ***(21)***. We predicted that these strains might respond differently when presented with long and short transplacement substrates, as described in **Subheading 3.2.3.**

Two substrates were prepared, in vitro, for use in generating the gene transplacement. The first substrate, a 2.0-kb *Eco*RI/*Xho*I restriction fragment from pLAY196, bears the *URA3* gene flanked by 900 bp of *HIS3* sequence ***(19)***. The second substrate was generated by PCR, using pLAY196 as a template, and produced a fragment bearing the *URA3* gene flanked by 90 bp of *HIS3* sequence. These PNA fragments represent long and short *HIS3* transplacement substrates. Following transformation into the desired cells, the fragments can use the flanking *HIS3* homology for recombination at the *HIS3* locus generating a *his3::URA3* allele by transplacement, or use the *URA3* sequence and recombine at the mutant *ura3* locus, restoring this allele to the wild-type configuration (**Fig. 2B**).

3.2.1. Preparation of a Substrate Fragment by Restriction Endonuclease Digestion

The long *HIS3* transplacement substrate was generated by restriction endonuclease digestion of the plasmid pLAY196.

1. Five reactions tubes containing 30 μL of CsCl-purified pLAY196 (5 μg/μL), 30 μL 10× Buffer 2, 4 μL (10 U/μL) *Eco*RI, 4 μL (10 U/μL) *Xho*I, and 232 μL H_2O were incubated at 37°C overnight. The enzymes and buffer used in this reaction were supplied by New England Biolabs.
2. Following an overnight digestion, 2-μL aliquots of the reactions were analyzed on a 0.6% agarose gel to assess the efficiency of cutting. If digestion was not complete, a second aliquot of enzyme, 10–20 U, was added.
3. Following digestion, the reactions were extracted twice with phenol/chloroform (1:1) with phase separation at 10,000 rpm (14,500*g*) for 10 min in a microcentrifuge.
4. The aqueous phase was transferred to a new Eppendorf tube and precipitated with 1/10 volume of 3 *M* sodium acetate, pH 5.2, and 0.7 vol of ice-cold isopropanol followed by centrifugation at 13,000 rpm (21,000*g*) for 20 min.
5. The precipitated DNA was washed twice with 500 μL of ice-cold 70% ethanol.
6. The DNA samples were dried, resuspended in 25 μL TE, and pooled.
7. The concentration of the DNA solution was determined, as discussed previously, and the final concentration adjusted to 2 μg/μL with TE.

3.2.2. Preparation of a Substrate Fragment by PCR

The short *HIS3* transplacement substrate was generated by PCR (*see* **Note 7**). The primers HIS3-829F

(5'GATTTGCGCCTTTGGATGAGGCACTTTCCAGAGCGGTGGTAGATC-3')

and HIS3-974R

(5'TGCAAAGCTTTCAAGAAAATGCGGGATCATCTCGCAAGAGAGATC-3')

were used to amplify the desired fragment using pLAY196 as a template.

1. The following reaction conditions were used to produce the desired 1.2-kb fragment: 1 μL (2 ng/μL) pLAY196, 5 μL 10× buffer, 4 μL 2.5 m*M* dNTP, 1.5 μL 50 m*M* $MgCl_2$, 1 μL (10 pmol) HIS3-829F, 1 μL (10 pmol) HIS3-974R, 0.5 μL (5 U/μL) Platinum Taq (Invitrogen), and 36.5 μL H_2O.

2. The reaction was carried out as follows: one cycle of denaturation at 94°C for 10 min, followed by 38 cycles of denaturation at 94°C for 30 s, annealing at 65°C for 30 s, and extension at 72°C for 1.5 min, followed by one cycle of extension at 72°C for 7 min.
3. The products from 20 such reactions were pooled, extracted with phenol/chloroform (1:1), and precipitated with 1/10 vol of 3 *M* sodium acetate, and 0.7 vol of isopropanol.
4. The resulting pellets were washed extensively with 70% ethanol to remove any excess salt, dried, and resuspended in TE at a concentration of 1 μg/μL.

3.2.3. Generation and Analysis of Transformants

The strains ABT407, W1011-3B, ABX129-2B, ABM44, and ABX81-4B (**Table 1**) were used to make electrocompetent cells and were electroporated as described in **Subheading 3.1.4.** For each electroporation 1 μL of digested pLAY196 (2 μg/μL) or 1 μL of the PCR fragment (1 μg/μL) were added to the cuvet. Following electroporation, the cells were plated onto SC $^{-}$Ura medium containing 1 *M* sorbitol and incubated for 3–4 d at 30°C. Since Ura^{+} transformants can be generated by several different recombination events, including fragment transplacement at the *HIS3* locus, gene conversion or transplacement at the *URA3* locus, or random insertion, the fraction of events at the *HIS3* locus was determined by replica plating the transformants to medium lacking uracil or both uracil and histidine. It is important to replicate plates to medium lacking uracil to distinguish between legitimate transformants and background colonies that may appear on the original selection plates. These plates were incubated overnight at 30°C. The total number of Ura^{+} transformants was determined by counting the colonies on the SC $^{-}$Ura medium. The number of *his3::URA*3 recombinants was determined by counting the number of colonies that were unable to grow on the SC $^{-}$Ura $^{-}$His medium. The transplacement efficiency is the ratio of transformant colonies that failed to grow in the absence of histidine vs the total number of Ura^{+} transformants. Using the electroporation conditions described in **Subheading 3.1.4.**, a typical electroporation trial generated 200–2500 transformants.

These results indicate that there are statistically significant differences between W1011-3B and ABT407 for the efficiency of transplacement with both fragments (**Table 3**). In the case of W1011-3B, the efficiency of transplacement with the long substrate is 11-fold higher than the transplacement efficiency with the short substrate. This increased efficiency was predicted based on the amount of *HIS3* homology associated with the individual substrate fragments. Interestingly, in ABT407, the transplacement efficiency with the long substrate is only twofold higher than the efficiency observed with the short transplacement substrate. When comparing the two

Table 3
Efficiency of Transplacement at the *HIS3* Locus in W303-1A- and S288C-Derived Strains with Substrates Bearing 90 bp and 900 bp of *HIS3* Homology[a]

		Transplacement efficiency[b]	
Strain	Relevant genotype	90 bp *HIS3*	900 bp *HIS3*
W1011-3B	Wild type	.084 ± .018	.92 ± .01
ABT407	Wild type	.46 ± .02	.98 ± .002
ABX81-4B	*rad3-G595R*	.35 ± .03	.99 ± .001
ABX129-2B	*rad1::LEU2*	.0019 ± .0004	.73 ± .01
ABM44	*rad50::hisG*	.027 ± .002	.76 ± .006

[a]The spontaneous reversion frequency for *ura3-1* and *ura3-52* was determined to be $8 \times 10^{-9} \pm 8.2$ and $<3.5 \times 10^{-9} \pm 1.8$, respectively. These values are substantially lower than the frequencies of transformation observed in our strains, suggesting that spontaneous reversion of the *ura3* alleles is unlikely to have contributed significantly to the results.

[b]The transplacement efficiency from a minimum of five separate trials is reported and expressed as a ratio of His$^-$ transformants versus total Ura$^+$ transformants.

strains, the transplacement efficiency of the long substrate was slightly higher in ABT407 than W1011-3B, whereas the efficiency with the short transplacement substrate was fivefold higher. These results may indicate differences in the genetic control of transplacement in these strains. Alternatively, it may be explained by the differences between the *ura3* alleles borne by these strains, which can serve as alternative targets for the transplacement substrates.

As described in **Subheading 3.2.**, W1011-3B and ABT407 have different kinds of mutations at their *URA3* loci. W1011-3B carries the *ura3-1* allele, which has a single mis-sense mutation. This allele may be readily restored to the wild type by gene conversion or transplacement using the *URA3* sequence on the substrate fragment. By comparison, ABT407 carries the *ura3-52* allele in which a 4.5-kb Ty1 element is inserted 100 bp downstream of the start codon. Since removal of this insertion by recombination with the substrate fragment requires interaction with homologous genomic sequences on both sides of the Ty element ***(22)***, and there is very little homology 5' to the insertion sequence, any loss of these sequences from the substrate fragment will dramatically reduce the ability of this fragment to generate a wild-type *URA3* locus. Because of these constraints, the ABT407 cell can only acquire a wild-type *URA3* locus by recombining with fragments that have undergone very limited processing. Thus, a lower proportion of the total population of Ura$^+$ transformants may result from recombination events at the *URA3* locus. This will be reflected as a change in the His$^-$/Ura$^+$ ratio describing transplacement efficiency (*see* **Note 8**).

Mutations in the *rad3*, *rad1*, and *rad50* genes have previously been shown to alter transplacement efficiency ***(12,19,20)***. ABX81-4B, a W303-1A derivative bearing the *rad3-G595R* allele, exhibited an increase in efficiency of transplacement at *HIS3* with both the long and short *HIS3* transplacement substrates (**Table 3**). The more profound effect, a fourfold increase, is observed with the short substrate. This is in agreement with previous results that have shown that *rad3-G595R* increases the stability of broken DNA ends and is hyper-rec when presented with fragments bearing short lengths of homology ***(12,20)***. This is consistent with the idea that factors governing the processing of the ends of DNA molecules are particularly important determinants of the efficiency of transplacement when using short homologous sequences.

By comparison, ABX129-2B and ABM44, W303-1A derivatives bearing the *rad1* and *rad50* mutant alleles, respectively, showed a decrease in their ability to utilize these fragments to recombine at the *HIS3* locus (**Table 3**). Once again, the most dramatic affect is observed with the short substrate fragment. The *rad1* mutant strain exhibits a 44-fold drop in transplacement efficiency compared with wild type, whereas the *rad50* mutant strain exhibits a 3-fold reduction. These results are consistent with previous work that has shown these mutants to be hypo-rec when presented with fragments with limited homology ***(19,20)***. Taken together, these results demonstrate the importance of the length of homology shared by the fragment and genomic targets and the genotype of the recipient strain in the design of DNA fragment transplacement strategies.

4. Notes

1. This protocol is a modified form of the protocol outlined in Birnboim and Doly ***(23)***. Several additional modifications are possible to increase the yield of plasmid DNA. These include initially growing the cells in a richer medium such as TY or inoculating a larger culture. In the absence of the Beckman centrifuge and NVT100 rotor used here, a variety of other ultracentrifuges and rotors may be substituted; however, the protocol may require modification.
2. If a Klett-Summerson colorimeter is not available, a standard Erlhenmeyer flask in conjunction with a visible wavelength spectrophotometer set to 600 nm may be used. The advantage of the side-arm flask is that the volume of the culture does not change throughout the assay, and because the flask remains closed there is minimal chance for contamination of the culture.
3. Transformation efficiency increases when salmon sperm DNA is sheared or sonicated to an average size of 7 kb and boiled for 5 min prior to use ***(8,16)***. A salmon sperm DNA concentration of 100–200 μg produces optimal results ***(8)***. When adding large amounts of DNA to the transformation mix, make sure to maintain a final concentration of 0.1 *M* LiAC.

4. Cells must be washed adequately before plating to remove any PEG that can reduce transformation efficiency. Adjustments to the final resuspension volume may be made to prevent overcrowding the plates with transformants.
5. A 25-mL culture will provide enough cells to carry out five to seven electroporations. If more electroporations are planned, the culture volume can be scaled up accordingly. Also, if it is advantageous to harvest the culture at a low cell density after a minimal number of generations (i.e., because the strain has a high mutation rate), scaling up the initial volume will provide a sufficient number of cells for a successful experiment.
6. We have noted empirically, with the conditions described in **Subheading 3.1.4.**, that time constants below 4.8 are consistent with a reduced electroporation efficiency. If the time constant is lower than 4.8 or arcing is observed, this may indicate that there is too much salt in the system. The salt may be in the DNA sample or the competent cells. Excess salt can be remove from the DNA sample by re-precipitation and multiple 70% ethanol washes. Excess salt can be removed from the cells with additional water washes during preparation.
7. The generation of DNA substrates for transplacements by PCR has become very popular. The most popular strategy involves designing primers that start with 40–50 nucleotides of sequence complementary to the desired genomic target linked to 19–20 nucleotides that are complementary to a genetic marker of choice. Plasmids containing these markers can be used as template ***(13)***. This approach was described previously ***(17)***.
8. If one were to extrapolate this finding to a strain that lacks all *URA3* sequences, one might hypothesize a further increase in the efficiency of transplacement to the *HIS3* locus. The use of markers that do not have homologous sequences in yeast can potentially alleviate this problem. An example of such a marker is the *KAN/MX* cassette ***(24)***, which has been used successfully in this lab.

Acknowledgments

This work was supported by US Public Health Service grant GM57484 to A. M. B., as well as funds from the Beckman Research institute of the City of Hope, and the City of Hope National Medical Center.

The authors thank J. McDonald, E. Alani, N. Kleckner, H. Ronne, R. S. Sikorski, B. Thomas Rathstein, P. Garfinkel, and P. Hieter for yeast strains and plasmids. We also thank the members of the Bailis laboratory for stimulating discussions.

References

1. Moerschell, R. P., Tsunawawa, S., and Sherman, F. (1988) Transformation of yeast with synthetic oligonucleotides. *Proc. Natl. Acad. Sci. USA* **85,** 524–528.
2. Scherer, S. and Davis, R. W. (1979) Replacement of chromosome segments with altered DNA sequences constructed *in vitro*. *Proc. Natl. Acad. Sci. USA* **76,** 4951–4955.
3. Orr-Weaver, T. L., Szostak, J. W., and Rothstein, R. J. (1983) Genetic applications of yeast transformation with linear and gapped plasmids. *Methods Enzymol.* **101,** 228–245.

4. Erdeniz, N., Mortensen, U. H., and Rothstein, R. (1997) Cloning-free PCR-based allele replacement methods. *Genomic Res.* **12,** 1174–1183.
5. Manivasakam, P. Weber, S. C., McElver, J., and Schiestl, R. H. (1995) Micro-homology mediated PCR targeting in *Saccharomyces cerevisiae. Nucleic Acid Res.* **23,** 2799–2800.
6. Rothstein, R. (1991) Targeting, transplacement, replacement and allele rescue: integrative DNA transformation in yeast. *Methods Enzymol.* **194,** 281–301.
7. Hinnen, A., Hicks, J. B., and Fink, G. R. (1978) Transformation of yeast. *Proc. Natl. Acad. Sci. USA* **75,** 1929–1933.
8. Schiestl, R. H. and Gietz, R. D. (1989) High efficiency transformation of intact yeast cells using single stranded nucleic acid an a carrier. *Curr. Genet.* **16,** 339–346.
9. Becker, D. M. and Guarente, L. (1991) High-efficiency transformation of yeast by electroporation. *Methods Enzymol.* **194,** 182–187.
10. Thomas, B. J. and Rothstein, R. (1989) The genetic control of direct-repeat recombination in *Saccharomyces*: the effect of *rad52* and *rad1* on mitotic recombination at *GAL10*, a transcriptionally regulated gene. *Genetics* **123,** 725–728.
11. Fink, G. R. (1970) The biochemical genetics of yeast. *Methods Enzymol.* **7,** 59–78.
12. Bailis, A. M., Maines, S., and Negritto, M. T. (1995) The essential helicase gene *RAD3* suppresses short-sequence recombination in *Saccharomyces cerevisiae. Mol. Cell. Biol.* **15,** 3998–4008.
13. Sikorski, R. S. and Hieter, P. (1989) A system of shuttle vectors and yeast host strains designed for efficient manipulation of DNA in *Saccharomyces cerevisiae. Genetics* **122,** 19–27.
14. Sambrook, J., Fritsch, E. F., and Maniatis, T. (1989) *Molecular Cloning, A Laboratory Manual*, 2nd ed. Cold Spring Harbor Laboratory Press, Cold Spring Harbor, NY.
15. Ito, H., Fukuda, Y., Murata, K., and Kimura, A. (1983) Transformation of intact yeast cells treated with alkali cations. *J. Bacteriol.* **153,** 163–168.
16. Geitz, R. D. and Woods, R. A. (2002) Transformation of yeast by lithium acetate/single-stranded carrier DNA/polyethylene glycol method. *Methods Enzymol.* **350,** 87–96.
17. Burke, D., Dawson, D., and Stearns, T. (2000) *Methods in Yeast Genetics, A Cold Spring Harbor Laboratory Course Manual*, 2000 ed. Cold Spring Harbor Laboratory Press, Cold Spring Harbor, NY.
18. Sherman, F. and Hicks J. (1991) Micromanipulation and dissection of asci. *Methods Enzymol.* **194,** 21–37.
19. Manthey, G. M. and Bailis, A. M. (2002) Multiple pathways promote short-sequence recombination in *Saccharomyces cerevisiae. Mol. Cell. Biol.* **22,** 5347–5356.
20. Bailis, A. M. and Maines, S. (1996) Nucleotide excision repair gene function in short-sequence recombination. *J. Bact.* **178,** 2136–2140.
21. Rose, M. and Winston, F. (1984) Identification of a Ty insertion within the coding sequence of the *S. cerevisiae URA3* gene. *Mol. Gen. Genet.* **193,** 557–560.

22. Negritto, M. T., Wu, X., Kuo, T., Chu, S., and Bailis, A. M. (1997) Influence of DNA sequence identity on efficiency of targeted gene replacement. *Mol. Cell. Biol.* **17,** 278–286.
23. Birnboim, H. C. and Doly, J. (1979) A rapid alkaline extraction procedure for screening recombinant plasmid DNA. *Nucleic Acids Res.* **7,** 1513–1523.
24. Wach, A., Brachat, A., Pohlmann, R., and Philippsen, P. (1994) New heterologous modules for classical or PCR-based gene transplacements in *Saccharomyces cerevisiae. Yeast* **10,** 1793–1798.

11

Targeted Gene Modification Using Triplex-Forming Oligonucleotides

Jean Y. Kuan and Peter M. Glazer

Summary

In recent years, triplex-forming oligonucleotides (TFOs) have emerged as powerful tools for site-specific gene modification. Their sequence specificity, binding affinity, and ability to provoke repair and recombination make them promising reagents for altering gene expression. This chapter highlights the binding requirements for triplex formation, identifies a number of chemical modifications that have been used with some success, and discusses studies using TFOs for inhibiting transcription. It also reviews work done using TFOs and related molecules to direct site-specific DNA damage, inducing mutagenesis or sensitizing a site to recombination. TFOs were initially used as positioning devices for nonspecific mutagens but were later discovered to have mutagenic properties of their own in cells with functional nucleotide excision repair (NER) and transcription-coupled repair (TCR) pathways. In subsequent studies triplex formation was able to induce both intramolecular and intermolecular homologous recombination, revealing its potential application for gene therapy. Recent reports demonstrate the ability of these molecules to locate and modify their cognate sites in chromosomal DNA in both cell culture and live animals, laying the foundation for triplex technology in vivo.

Key Words: TFO, triplex, PNA, DNA repair, nucleotide excision repair, antigene, gene therapy, recombination, gene-targeting, mutagenesis

1. Introduction

Triplexes were first observed in 1957 when Felsenfeld et al. *(1)* demonstrated that a single-stranded polyuridine oligonucleotide could bind stably and specifically to a polyuridine/polyadenosine duplex. Although third-strand binding was weak compared with Watson-Crick hydrogen bonds, the interactions were stabilized in the presence of divalent cations such as Mg^{2+}. Later work by other groups discovered that poly (dT-dC) and poly (dG-dA) also have the ability to form triple helices. Subsequently, a number of groups demonstrated

From: *Methods in Molecular Biology, vol. 262, Genetic Recombination: Reviews and Protocols*
Edited by: A. S. Waldman © Humana Press Inc., Totowa, NJ

triplex formation by synthetic oligonucleotides at polypurine/polypyrimidine sites in duplex DNA, without the need for repeated sequence motifs and helped establish the third strand binding code *(2–4)*. As the binding was so sequence-specific, these studies raised the possibility of using triplex-forming oligos (TFOs) as gene targeting reagents.

2. The Binding Code and Limitations of Triplex Binding

Oligonucleotides can bind as a third strand in polypurine:polypyrimidine tracts via Hoogsteen or reverse Hoogsteen hydrogen bonds, fitting into the major groove of double-stranded DNA. Homopyrimidine oligos bind in a parallel orientation (in terms of 5'-3' direction) to the purine strand in the DNA, and in this motif, T forms hydrogen bonds to A:T base pairs and C to G:C base pairs *(5)*. Protonation at N3 of cytosine is required for optimal Hoogsteen bonding with N7 of guanine (**Fig. 1A**), and a nonmodified pyrimidine TFO will not bind DNA at a pH above 6 *(6)*. The pH dependence of pyrimidine oligos limits their usefulness as intracellular gene targeting reagents. Homopurine oligos, on the other hand, can form triplexes readily at physiological pH binding in an antiparallel manner with A hydrogen bonding to A:T and G to G:C base pairs *(2,4,7)*. T may also be included in the purine motif, as it is capable of forming hydrogen bonds with A:T base pairs (**Fig. 1B**) *(2,8)*. Triplex formation by G-rich oligos in the purine motif, however, is inhibited by physiological levels of K^+ *(9,10)*. K^+ favors aggregation of TFOs by stabilizing secondary structures such as G quartets *(11,12)*, thus diminishing the number of oligos capable of reaching their target sites. Both motifs also require divalent metal cations, particularly Mg^{2+}, to help neutralize charge repulsion between the three negatively charged phosphodiester backbones.

TFOs are very specific for their target sites in duplex DNA, and in addition to some of the physiologic limitations, interruptions in the polypurine tract or mismatches in the TFO destabilize the binding of the third strand. Some oligonucleotide length and sequence studies in the purine motif have suggested that an uninterrupted stretch of at least 12–14 purines is necessary for stable triplex formation *(13)*. Single base pair inversions interrupting a homopurine target site (e.g., AT or GC inverted to TA or CG, respectively) can significantly destabilize TFO binding.

3. Chemistries and Analogs

To overcome some of the limitations of triplex binding under physiological conditions, a number of changes in TFO chemistry (base, sugar, and backbone modifications) have been developed and tested. TFOs in the pyrimidine motif require protonation of the cytosines to form both Hoogsteen bonds with G:C base pairs. Replacement of cytosine on the third strand with 5-methylcytosine

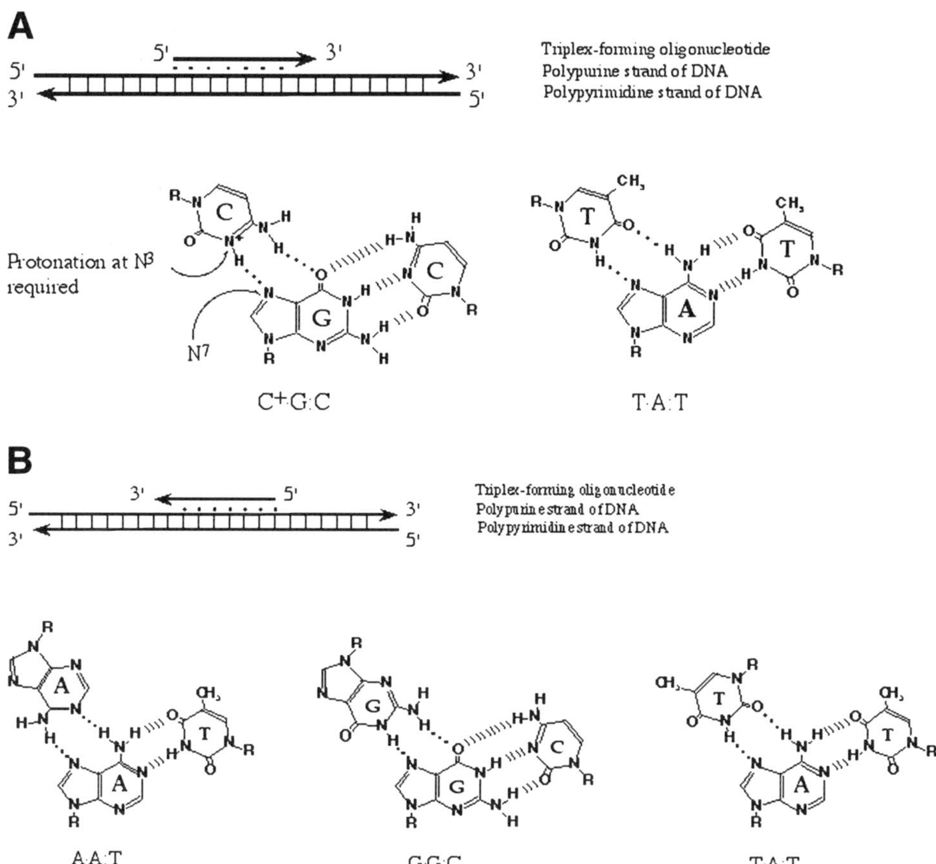

Fig. 1. The binding code for triple helix formation. **(A)** The pyrimidine binding motif. **(B)** The purine binding motif. In the pyrimidine motif, binding of the TFO occurs in a parallel orientation with respect to the polypurine strand of duplex DNA. In the purine motif, the TFO binds in an antiparallel orientation. The canonical base triplets in each motif are shown.

stabilizes the triplex at neutral pH. Although protonation is still lacking at N3, the additional methyl group appears to facilitate hydrophobic interactions with adjacent thymidines and to enhance base stacking ***(14)***. Other analogs, such as 2'-*O*-methylpseudoisocytidine, deal with the same problem by having a constitutive N3 protonation ***(15)***. Other cytosine substitutions that have shown some success include 8-oxoadenine ***(16)***, 7,8-dihydro-8-oxoadenine ***(17)***, 6-oxocytidine ***(18)***, 8-oxo-2'-deoxyadenosine ***(19,20)***, and 8-aminoguanine ***(21)***. In the purine motif, replacement of natural bases with modified bases such as 6-thioguanine ***(9,22,23)*** counteracted some of the potassium inhibitions by preventing undesirable hydrogen bonding between same-strand guanines. 7-Deazaxanthine, a purine analog of thymine, also showed affinity for triplex binding at physiological pH and salt conditions ***(24,25)***.

Ribose modifications such as 2'-*O*-methyl, 2'-*O*-methoxyethyl, and 2'-*O*-aminoethyl are chemically more stable than DNA or RNA and are resistant to both DNA and RNA nucleases ***(26)***. The 2'-*O*-aminoethyl substitution also provides a positive charge to help neutralize charge repulsion ***(27)***. Charge repulsions of the DNA backbones normally stabilized by Mg^{2+} can also be overcome by positively charged or neutral backbone modifications. Replacement of the TFOs phosphodiester backbone with cationic *N,N*-diethylethylenediamine (DEED) phosphoramidate linkages increase triplex binding affinity in vitro and reduce both K^+ inhibition and Mg^{2+} dependence ***(28,29)***. Peptide nucleic acids (PNAs) are a DNA mimic in which the DNA bases are attached to a neutral polyamide backbone and form very stable triplexes at low salt conditions ***(30,31)***. However, these are not DNA:DNA:PNA triplexes but complexes containing two strands of PNA and one strand of DNA. PNA oligomers strand-invade the DNA duplex and form Watson-Crick base pairs with one of the complementary strands. The other PNA oligo then forms Hoogsteen hydrogen bonds with the DNA:PNA duplex, and a stable D-loop is formed ***(32–35)***. Continued improvements in TFO chemistry and analog development will help to increase binding affinity and extend the triplex binding code.

4. Applications of Triplex Binding

Triplex formation is several orders of magnitude slower than duplex formation, but, once formed, it has been shown to compete successfully with other DNA binding molecules ***(36–40)***. The aforementioned sequence specificity combined with high-affinity binding gives TFOs the potential to modify gene expression in several different ways. Targeting a site in the c-myc promoter in HeLa cells, Postel et al. ***(41)*** first showed that TFOs could hinder the transcription of c-myc mRNA, presumably by competing with transcription factors or preventing elongation by RNA polymerase (**Fig. 2**). Since then, several other groups have used TFOs in this manner to inhibit regulatory protein binding, replication, or transcription elongation both in vitro and in vivo ***(42–48)*** (for reviews, *see* refs. ***49*** and ***50***). Inhibition of transcription by TFOs offers an advantage compared with antisense strategies, which rely on the ability of the oligonucleotide to eliminate many mRNA transcripts before an effect is seen. Triplex-mediated antigene strategies, however, control expression by manipulating the source of these transcripts and may be more effective owing to the smaller number of targets per cell. In order for TFOs to maintain long-term transcription inhibition though, constant administration of oligos or a great increase in intracellular stability would be required owing to their short half-

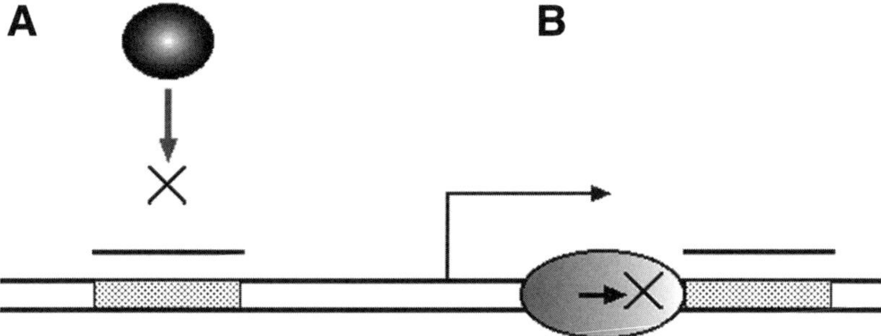

Fig. 2. Blocking of transcription by TFOs. **(A)** TFO competition with transcription factor binding. **(B)** Obstruction of RNA polymerase during elongation.

lives in cells. An alternative antigene approach is the use of TFOs as gene-targeting reagents for mutagenesis and recombination since the specificity of their binding allows for site-directed DNA modification. Triplex-induced site-specific mutation or recombination results in changes that are permanent and heritable.

5. Targeted Gene Modification

5.1. TFO-Induced Mutagenesis

The sequence-specific binding abilities of TFOs have been used to direct nonspecific mutagens to specific locations. Our laboratory and others have successfully used TFOs tethered to psoralen, a photoreactive mutagen, to induce DNA damage at an expected region near a TFO binding site ***(51–54)***. Psoralen, after being correctly positioned, can intercalate into DNA and covalently crosslink thymines on both strands at 5' TpA 3' or 5' ApT 3' sites after activation by long-wave UVA irradiation (**Fig. 3A**) ***(55)***. Subsequent DNA repair and/or replication of the lesion in mammalian cells lead to characteristic T:A to A:T transversions ***(53,56–58)***. Helene and colleagues ***(55)*** first demonstrated triplex-directed targeting using psoralen-linked TFOs designed to photo-induce crosslinks in segments of double-stranded DNA. It was shown that the psoralen crosslinks were highly dependent on triplex formation and were not present at nontargeted sites.

To show site-specific mutations in a gene, Havre et al. ***(53)*** designed plasmids containing the suppressor tRNA reporter gene *supF* and targeted *supF* gene both in vitro and in vivo by psoralen-conjugated TFOs. After photoactivation, increased mutagenesis was detected. A mutation frequency of 0.23% (at least 100-fold higher than an untargeted gene) was seen when *supF* was targeted to phage DNA in vitro. Of these mutations, 96% occurred at the targeted region, and 56% were T:A to A:T transversions at the targeted base

Fig. 3. Site-directed mutagenesis by TFOs (**A**) Triplex formation can induce mutations when tethered to a mutagen such as psoralen. (**B**) TFOs can also induce mutations by themselves, possibly owing to recognition and repair of the three-stranded complex by DNA repair pathways.

pair *(53)*. In mammalian cells, an SV40 shuttle vector pSP189 containing *supF*, which could be replicated in mammalian cells, was targeted with psoralen-TFOs and then transfected into COS cells. Mutation frequencies of up to 7% were seen in this system *(52)*, and later it was shown that oligonucleotides could be simply added to the media of COS cells already containing the vector. After allowing triplexes to form in vivo followed by UVA irradiation, mutation frequencies of about 2.1% (vs a background rate of 0.03%) were observed *(57)*. Furthermore, Wang et al. *(57)* observed that high intracellular targeting frequencies were dependent on both TFO concentration and binding affinity, as the 10-bp target site and complementary TFO had to be lengthened to 30 bp (until its dissociation constant was in the nanomolar range) before effects were seen.

Illustrating the importance of high-affinity triplex binding to targeted mutagenesis in vivo, another attempt to target the 10-bp polypurine site in the chromosomal reporter gene *supF* proved successful when the TFOs were modified by replacing the phosphodiester linkages with cationic DEED linkages *(29)*. The modified psoralen-TFOs both bound the 10-bp target site with high affinity (K_d approx 10^{-8} *M*) and enhanced *supF* mutation frequency by 6–10-fold over background, depending on the type of modification. Not only did these cationic oligos display high-affinity binding in vitro, but also the binding affinity was high even in the presence of low Mg^{2+} or high K^+ concentrations.

5.2. Mutagenesis Using PNAs

PNAs, which were shown to inhibit transcription initiation and elongation in vitro *(59–61)*, also appeared to be attractive molecules for inducing mutagenesis owing to their high-affinity binding capabilities *(62)*. In Faruqi et al. *(32)*, dimeric PNAs were designed to form a clamp via both double- and triple-helix formation within a chromosomally integrated *supFG1* reporter gene in a mouse fibroblast cell line. PNAs specific for short (either 8- or 10-bp) sites were able to induce mutations in the *supFG1* gene with a frequency of 0.1% (10-fold over background). Delivery remains an issue with PNAs since their uncharged backbone makes many traditional cationic lipid transfection protocols inefficient; and in this study, cells were permeabilized with streptolysin O and cellular uptake was confirmed by fluorescein labeling and fluorescent microscopy.

5.3. Mutagenesis at Chromosomal Sites by TFOs

In a natural progression, mutagenesis was later demonstrated at chromosomal sites in mammalian cells. Although it was initially believed that chromatin structure may have an inhibitory effect on triplex recognition and binding of duplex DNA inside cells ***(63–65)***, several groups have since shown the ability of TFOs to mutagenize chromosomal loci effectively in cell culture ***(29,66,67)***. Majumdar et al. ***(66)*** used psoralen-conjugated TFOs (also containing the intercalators acridine or pyrene, which stabilize triplexes in cells) to induce specific mutations at the endogenous *hprt* gene in Chinese hamster ovary (CHO) cells, 85% of which were within the triplex target region. Vasquez et al. ***(67)*** constructed a mouse fibroblast cell line containing multiple copies of a λsupFG1 reporter gene. After treatment of the cells with psoralen-TFOs, they observed a 6–10-fold induction of site-specific *supFG1* mutations over background. Interestingly, mutagenesis was induced by TFOs in the presence and absence of psoralen crosslinking, suggesting that under these experimental conditions, covalent crosslinks had no additional effect on the mutation frequency.

This finding supported a study carried out 3 yr earlier by Wang et al. ***(68)***, which showed that triplex formation itself, without a tethered mutagen, can be mutagenic in cells transfected with the *supFG1* reporter vector (**Fig. 3B**). As with psoralen-conjugated TFOs, mutation frequency using TFOs alone correlates with binding affinity. While investigating a possible role for DNA repair in triplex-induced mutagenesis, they observed that the ability of TFOs to cause mutations was not detected in Xeroderma pigmentosum group A (XPA) cells or Cockayne's syndrome group B (CSB) cells. XPA cell lines are deficient in nucleotide excision repair (NER) since they contain a defect in a DNA damage-recognition protein essential to the repair complex ***(69)***. CSB cell lines are defective in the transcription-coupled repair subset of the NER pathway since they are defective in the protein responsible for recruitment of the repair apparatus to sites of stalled transcription ***(70)***. By cotransfecting a vector expressing either XPA or CSB into the respective deficient cell lines, triplex-induced mutagenesis could be restored. It was hypothesized that formation of the triple helix could block transcription at that site, triggering gratuitous and possibly error-prone repair by the NER pathway ***(71)*** or that the triplex structure itself produces sufficient helical distortion to provoke DNA repair, even in the absence of chemical damage to the DNA.

More recently, it has been demonstrated that TFOs can induce mutagenesis at sites within a chromosomally integrated reporter gene in transgenic mice

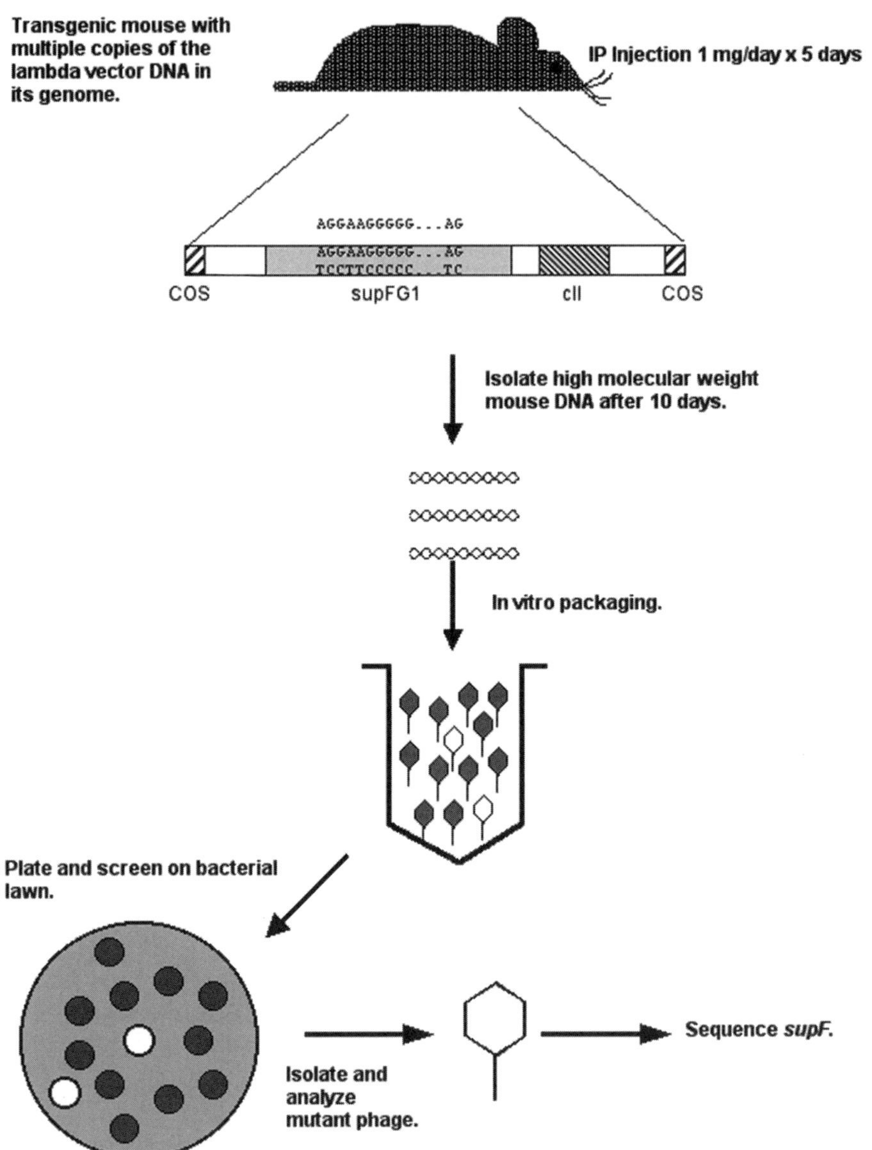

Fig. 4. Protocol to detect TFO-directed genomic mutations by intraperitoneal injection of oligonucleotides. Transgenic mice carrying integrated copies of the λsupFG1 shuttle vector (*supFG1* encodes a tRNA suppressor gene) were injected with oligos for 5 d. Ten days later, various tissues were collected, and genomic DNA was isolated for in vitro packaging into viable phage particles by λ packaging extracts. Phage were analyzed in a lacZ(am) strain of *Escherichia coli*. In the presence of a wild-type *supFG1* gene, the amber mutation in the β-galactosidase gene is suppressed, and the resulting plaques are blue on indicator plates. Phage that contain mutated *supFG1* are white.

following intraperitoneal injection of 3'-protected phosphodiester TFOs (**Fig. 4**) *(72)*. A 30-bp polypurine region within the integrated *supF* reporter was targeted with a 30-nucleotide (nt) purine TFO, which was injected at a dose of 50 mg/kg for five consecutive days. Ten days after the last injection, mouse tissues were collected for mutation analysis. The combined mutation frequencies showed an average fivefold enhancement of mutagenesis over background in all the tissues assayed, and 85% of the mutations were concentrated within the 30-bp target site. Brain tissue was unaffected, however, consistent with the notion that oligonucleotides cannot cross the blood-brain barrier *(73)*. Until this study, TFOs had only been examined in cell culture systems in which the oligo was either transfected or microinjected. Although overall efficiency of mutagenesis was relatively low, these results provided evidence that triplex technology is capable of targeted gene modification in live animals.

5.4. Triplex-Induced Recombination

Based on the ability of TFOs to direct site-specific damage, either using a tethered mutagen or in the form of the triple helix itself, it was hypothesized that TFOs could also induce recombination. It is well established that gene targeting via homologous recombination is limited by both the low rate of homologous recombination in mammalian cells and the high rate of random (nonhomologous) integration (reviewed in ref. *74*). However, homologous recombination (HR) can be enhanced by DNA damage near the target site, including UV irradiation, alkylating agents, and the crosslinking agent psoralen *(75–77)*, as well as by double-strand breaks created by endonucleases *(78–82)*. Based on these studies, and also on the evidence that TFOs could either direct DNA damage by a mutagen or create it in the form of a triple helix, TFOs were tested as a tool to provoke both intramolecular and intermolecular recombination.

5.5. Intramolecular Recombination Induced by TFOs

Our lab and others have used several different reporter constructs to detect increases in homologous recombination. Originally, psoralen-TFOs were used to induce recombination between tandem mutant copies of a reporter gene in plasmids *(83)* and later in a similar construct in chromosomal sites of mouse cell lines (**Fig. 5A**). These studies used TFOs both conjugated and unconjugated to psoralen *(84)*. In the latter study *(84)*, TFOs designed to bind between two mutant copies of the herpes simplex virus thymidine kinase (TK) gene, each carrying a different mutation, were microinjected into the nuclei of mouse LTK^- cells carrying this dual TK construct. Recombination in these cells can be detected by restoration of a functional TK gene. Recombination frequencies were observed in the range of 1%, over 2500-fold above background, and in the range of the results in model systems using site-specific endonucleases

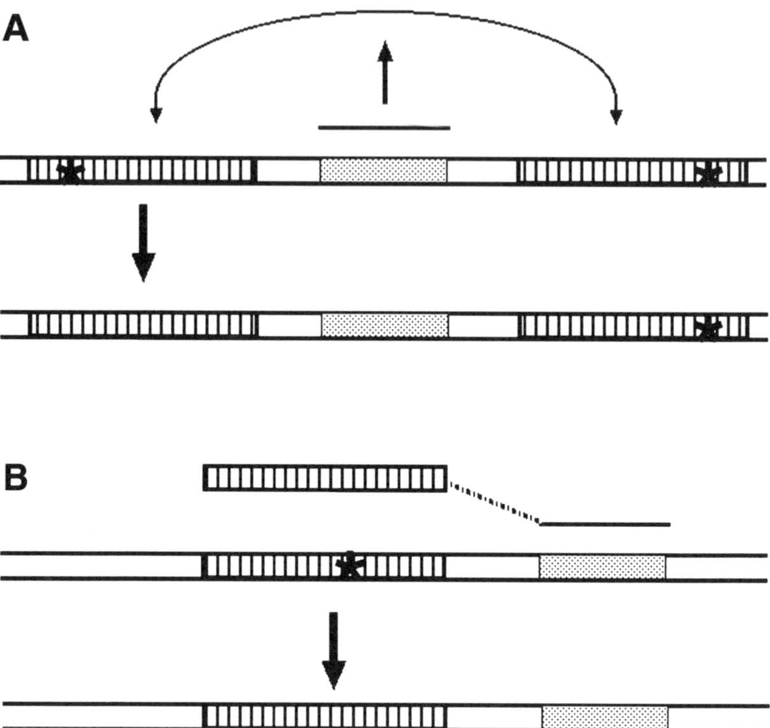

Fig. 5. (A) Intrachromosomal recombination: TFO binding at a site between direct repeats of two mutant genes can induce homologous recombination, resulting in a gene conversion event. **(B)** Interchromosomal recombination: tethered donor TFOs (TD-TFOs), bifunctional oligos with a TFO domain on one end to provoke repair and recombination attached to a donor domain to deliver sequence information. Several in vitro studies indicate that these domains do not need to be attached if local concentrations of both are high enough.

such as *Sce*I to generate double-strand breaks. These results support the utility of TFOs as tools to sensitize genomic sites to recombination but at the same time illustrate the importance of successful intranuclear TFO delivery, as TFOs transfected in this study using cationic lipids generated only a six to sevenfold increase in recombinants.

Additional work was done to elucidate the pathways involved in these intrachromosomal recombination events. Using a dual *supF* system in an SV40-based vector, Faruqi et al. ***(85)*** demonstrated that triplex formation, itself, is capable of promoting recombination and that this effect is dependent on a functional NER pathway. In cell lines deficient in the damage recognition factor XPA, triplex-induced recombination frequencies did not exceed background levels. However, when these cells were complemented by expression of XPA cDNA, recombination was restored to near wild-type levels. Recombination by triplex-directed psoralen adducts was partially but not completely dependent on XPA, consistent with current models of crosslink repair that involve multiple repair pathways ***(86)***.

5.6. Intermolecular Recombination Using TFOs With Donor Molecules

Studies using short DNA fragments as substrates for gene correction ***(87,88)*** have provided some promising results, although a direct method of targeting donor DNA to the gene of interest is lacking in these experiments. The process by which a DNA fragment searches for homology and participates in strand exchange in mammalian cells is not fully understood, but it is thought that Rad51 and related factors catalyze this energy-dependent reaction ***(89)***. In contrast, a TFO can find its cognate site within complex DNA without the need for any associated enzyme activity ***(90,91)***. With the evidence that triplexes could stimulate DNA repair and recombination, Chan et al. ***(92)*** developed a gene-targeting strategy using TFOs tethered to a donor DNA fragment with the purpose of inducing intermolecular recombination (**Fig. 5B**).

On one end of this oligonucleotide, a TFO domain is designed to bind the target gene and stimulate repair and recombination. Attached covalently via a flexible polyethylene glycol linker is a short double- or single-stranded donor fragment, homologous to the gene except for the nucleotide or nucleotides to be altered, which would then be positioned to participate in the information transfer reaction. After showing that the attached donor domain minimally affected binding affinity, the oligos were preincubated with a plasmid containing a G-to-C point mutation in a *supFG1* reporter gene. These plasmid-oligo complexes were transfected into COS cells and reisolated 48 h later for characterization. A single-stranded donor tethered to a 30-nt TFO region gave reversion frequencies of 0.17%, and a double-stranded donor tethered to the same TFO region gave frequencies of 0.68% (57-fold and 227-fold over background, respectively). Reversion frequencies with the single- and double-stranded donor fragments alone were much lower (0.051% sense strand, 0.062% antisense strand, and 0.17% annealed). Transfection of the TFO domain alone showed little effect over background. Although double-stranded donor DNA appeared to be more effective when the triplex and donor complexes were preformed, the single-stranded donor linked to the TFO was better at inducing recombination when cells already containing the vector as an episomal plasmid were targeted. The reversion frequency in this intracellular targeting protocol was lower than when the triplexes were preformed, implying that oligonucleotide delivery may be a limiting factor in this approach. As with the earlier studies ***(83,85)***, the TFO-donor directed recombination was shown to be at least partially dependent on the NER pathway, as XPA-deficient cells show diminished TFO-donor-induced reversion frequency.

Consistent with this work, in vitro experiments by Gamper et al. ***(93)*** provided physical evidence for strand invasion in a supercoiled plasmid (and subsequent D-loop formation) by a similarly designed TFO domain attached to a single-stranded homologous DNA. Results by the Sun group ***(94,95)*** support this strategy and have attached the DNA donor domain noncovalently via an adapter region so that the recombining strand is not limited to areas directly adjacent to the TFO binding region.

To explore further the mechanism of this induced recombination, the roles of the NER damage recognition factor XPA and the human recombinase HsRad51 were directly tested by manipulating the levels of these proteins in human cell-free extracts ***(96)***. Using a plasmid-based reversion assay and bifunctional oligonucleotides containing the TFO and donor domains, triplex-induced recombination was detected in the extracts but was reduced to the levels of donor alone in XPA-immunodepleted extracts and then restored by readdition of the protein. Similarly, recombination was reduced to the same levels in HsRad51-immunodepleted extracts and restored when the extracts were supplemented with resolubilized HsRad51. (Readdition of purified HsRad51 protein did not fully restore activity. It was hypothesized that immunoprecipitation of this protein had also removed other physically associated factors necessary for recombination.) Supplementing whole extracts with additional HsRad51 boosted activity in a dose-dependent manner. Surprisingly, linkage between the TFO and donor domains was not necessary, as linked and unlinked donors generated similar reversion frequencies. This suggested that when local concentrations of donor and TFO are high enough, as in these in vitro experiments, physical linkage is not required, and the ability of the TFO to provoke DNA repair and recombination is more important than its role as a positioning molecule.

5.7. PNAs as Agents to Induce Repair and Recombination

In a recent related study, Rogers et al. ***(97)*** reasoned that because PNAs can bind DNA as PNA clamps with high affinity and specificity ***(98,99)***, they could be substituted for the sequence-specific TFO domain in a tethered donor approach (**Fig. 6**) ***(92,96)***. A PNA clamp designed to bind to a 10-nt stretch within the *supFG1* reporter gene was postsynthetically coupled to a DNA donor molecule. The cytosines in the triplex-forming portion of the PNA molecule were replaced by pseudoisocytosine, an analog that mimics N-3 protonation of cytosine and allows stable triplex formation at neutral pH ***(98)***. In HeLa cell-free extracts, the PNAs (either linked or unlinked to a donor molecule) were able to induce reversion of the supFG1 G-to-C point mutation at frequencies of

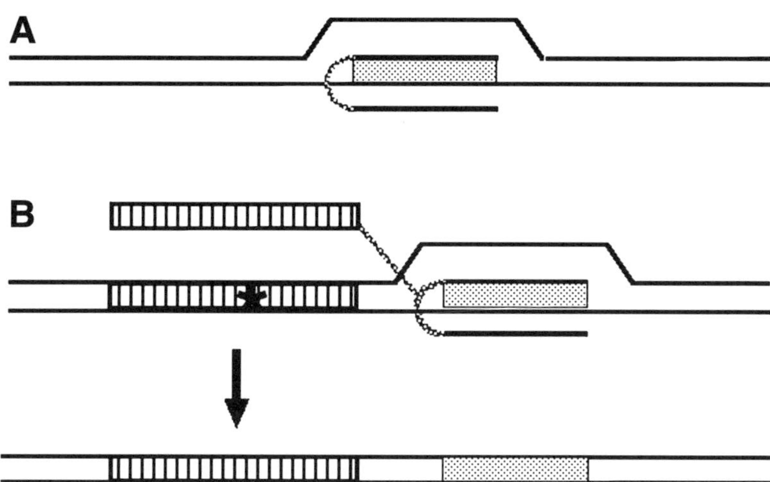

Fig. 6. **(A)** Dimeric bis-PNAs can be designed to strand invade and form a clamp structure. One portion of the PNA displaces one of the duplex strands and forms Watson-Crick bonds with the other strand. The other PNA portion, attached by a flexible linker, can then form Hoogsteen hydrogen bonds with the PNA:DNA duplex, resulting in a PNA:DNA:PNA triplex. **(B)** Recombination using PNAs: PNA clamps can be conjugated to a donor domain postsynthetically. The resulting hybrid molecule can mediate information transfer and homologous recombination.

6.2×10^{-4} and 8.1×10^{-4}, respectively, 3- and 5-fold above the effects of donor alone and more than 50-fold above background.

The PNAs and PNA-DNA conjugates were also assessed for their ability to stimulate DNA repair by an [α-^{32}P]dCTP incorporation assay in the plasmid substrate and were found to induce repair synthesis 16- and 7-fold above background, respectively. In fact, PNA clamps were shown to be a more potent substrate for induction of repair than DNA triplexes. The manner in which PNA clamps strand invade to form a PNA:DNA:PNA triplex displacing the other DNA strand as a D-loop may create a more severe helical distortion of the duplex than TFO binding in which the third strand fits entirely within the major groove, leaving the duplex intact. It is also possible that the non-natural chemical composition of the PNA may be a factor in the induced repair. Both recombination and repair induced by PNA clamps were shown to be at least partially dependent on XPA and the NER pathway, as extracts depleted of the protein demonstrated diminished activity in both assays.

5.8. Factors and Proteins Involved in TFO Recognition and Repair

Additional interest in the pathways involved in repair of triplex structures has led to several in vitro studies investigating the proteins and factors responsible in recognizing triplexes. In an in vitro study using an [α-^{32}P]dCTP incorporation assay to detect DNA repair synthesis in HeLa cell extracts, high-affinity triplex formation was shown to induce repair at a highly transcribed site on a plasmid substrate ***(100)***. Comparisons between regions with a strong SV40 promoter and regions lacking this promoter showed elevated levels of repair in the former resulting from the different levels of transcription. Depletion of individual transcription components such as RNA Pol II, TFIID, TFIIH, and rNTPs resulted in reduced levels of TFO-mediated DNA repair activity, suggesting that transcription itself was necessary. This study also confirmed earlier observations of XPA and CSB protein involvement ***(68,96)***. Vasquez et al. ***(101)*** recently provided the first physical evidence of repair protein binding and recognition of triplexes. In a series of electrophoretic mobility shift assays and antibody supershifting experiments, it was found that RPA and XPA participate in triplex binding.

5.9. Factors Influencing Chromosomal Accessibility

The in vivo effectiveness of TFOs and PNAs for inducing mutagenesis and recombination is dependent on their ability to find and bind to their target sites. Although TFOs have been shown to induce site-specific mutagenesis at chromosomal sites in yeast ***(102)***, mammalian cells ***(66,67)***, and mouse tissues ***(72)***, triplex formation on DNA sequences organized into nucleosomes was inhibited in in vitro studies ***(64,65)***. It has been hypothesized that this inconsistency could be owing to the dynamic nature of chromatin in vivo, as cellular processes such as transcription, recombination, repair, and replication would change chromatin structure. Using a restriction protection assay to physically detect triplex-directed psoralen crosslinks in genomic DNA, it was demonstrated that transcription increases the extent of triplex formation at the TFO target site ***(103)***. A human 293 cell line was created containing a TFO binding site downstream of an ecdysone-inducible promoter so that transcription through the target site could be manipulated. When transcription was induced by adding an ecdysone analog, ponasterone, TFO binding increased threefold over the uninduced controls. About 10% binding was observed in the absence of induced transcription, suggesting that even minimally or untranscribed regions are still accessible to triplex formation. Inhibition of RNA polymerase activity by α-amanitin caused a substantial decrease in chromosomal TFO binding, even in the presence of ponasterone, indicating that active transcription through the site and not just transcription factor binding is responsible for the increased binding.

5.10. Oligonucleotide Delivery into Cells

Intracellular delivery of TFOs into cells may be one of the most significant challenges to this technology. Passive uptake of oligos by endocytosis can occur at a low level in cell culture, and transfection by cationic lipids or electroporation may be more efficient for laboratory studies by increasing the amount of oligo that can successfully reach the nucleus ***(104)***. Permeabilization by streptolysin O or digitonin ***(32,103,105,106)*** also increases uptake but can damage the membrane and in the case of digitonin is irreversible. Microinjection of oligo directly into cells or nuclei has shown some promise. Regardless, with the eventual objective of gene modification in human patients, these delivery methods are unrealistic. Intraperitoneal injections of DNA TFOs have shown some success in mice ***(72)***, although the rates of mutagenesis were low. Other antisense experiments with PNA oligomers have also demonstrated effective, sequence-specific upregulation of gene expression in mice by high-dose ip injection (50 mg/kg) ***(107)***.

6. Future Challenges in Triplex-Mediated Gene Modification

TFOs have potential as sequence-specific tools for gene modification, but the requirements for polypurine binding sites, oligonucleotide stability, and intranuclear delivery all remain obstacles. Synthetic base analogs, as well as backbone and sugar modifications, are being developed to help extend the binding code, increase binding stability, and prevent intracellular degradation. For example, both PNAs and DEED-modified oligonucleotides have mediated gene targeting at relatively short (10-bp) target sites in cells ***(29,32,97)***. Furthermore, although the binding code may be a limitation, many human genes contain polypurine/polypyrimidine regions that may be suitable for targeting with appropriately modified TFOs.

Acknowledgments

We thank M. Knauert and M. Macris for their suggestions as well as the other members of the Glazer laboratory for their help. This work was supported by National Institutes of Health grants CA64186 and GM54731 (to P.M.G). J.Y.K. was supported by the Anna Fuller Fellowship Fund.

References

1. Felsenfeld, G., Davies, D. R., and Rich, A. (1957) Formation of a three-stranded polynucleotide molecule. *J. Am. Chem. Soc.* **79,** 2023–2024.
2. Beal, P. A. and Dervan, P. B. (1991) Second structural motif for recognition of DNA by oligonucleotide-directed triple-helix formation. *Science* **251,** 1360–1363.
3. Francois, J. C., Saison-Behmoaras, T., and Helene, C. (1988) Sequence-specific recognition of the major groove of DNA by oligodeoxynucleotides via triple helix formation. Footprinting studies. *Nucleic Acids Res.* **16,** 11,431–11,440.
4. Letai, A. G., Palladino, M. A., Fromm, E., Rizzo, V., and Fresco, J. R. (1988) Specificity in formation of triple-stranded nucleic acid helical complexes: stud-

ies with agarose-linked polyribonucleotide affinity columns. *Biochemistry* **27,** 9108–9112.
5. Moser, H. E. and Dervan, P. B. (1987) Sequence-specific cleavage of double helical DNA by triple helix formation. *Science* **238,** 645–650.
6. Singleton, S. F. and Dervan, P. B. (1992) Influence of pH on the equilibrium association constants for oligodeoxyribonucleotide-directed triple helix formation at single DNA sites. *Biochemistry* **31,** 10,995–11,003.
7. Broitman, S. L., Im, D. D., and Fresco, J. R. (1987) Formation of the triple-stranded polynucleotide helix, poly(A.A.U). *Proc. Natl. Acad. Sci. USA* **84,** 5120–5124.
8. Cooney, M., Czernuszewicz, G., Postel, E. H., Flint, S. J., and Hogan, M. E. (1988) Site-specific oligonucleotide binding represses transcription of the human c-myc gene in vitro. *Science* **241,** 456–459.
9. Cheng, A. J. and Van Dyke, M. W. (1993) Monovalent cation effects on intermolecular purine-purine-pyrimidine triple-helix formation. *Nucleic Acids Res.* **21,** 5630–5635.
10. Olivas, W. M. and Maher, L. J., 3rd. (1995) Overcoming potassium-mediated triplex inhibition. *Nucleic Acids Res.* **23,** 1936–1941.
11. Sen, D. and Gilbert, W. (1990) A sodium-potassium switch in the formation of four-stranded G4-DNA. *Nature* **344,** 410–414.
12. Williamson, J. R., Raghuraman, M. K., and Cech, T. R. (1989) Monovalent cation-induced structure of telomeric DNA: the G-quartet model. *Cell* **59,** 871–880.
13. Cheng, A. J. and Van Dyke, M. W. (1994) Oligodeoxyribonucleotide length and sequence effects on intermolecular purine-purine-pyrimidine triple-helix formation. *Nucleic Acids Res.* **22,** 4742–4747.
14. Povsic, T. J. and Dervan, P. B. (1989) Triple helix formation by oligonucleotides on DNA extended to the physiological pH range. *J. Am. Chem. Soc.* **111,** 3059–3061.
15. Ono, A., Ts'o, P. O. P., and Kan, L. S. (1991) Triplex formation of oligonucleotides containing 2'-O-methylpseudoisocytidine in substitution for 2'-deoxycytidine. *J. Am. Chem. Soc.* **113,** 4032–4033.
16. Miller, P. S., Bhan, P., Cushman, C. D., and Trapane, T. L. (1992) Recognition of a guanine-cytosine base pair by 8-oxoadenine. *Biochemistry* **31,** 6788–6793.
17. Jetter, M. C. and Hobbs, F. W. (1993) 7,8-Dihydro-8-oxoadenine as a replacement for cytosine in the third strand of triple helices. Triplex formation without hypochromicity. *Biochemistry* **32,** 3249–3254.
18. Berressem, R. and Engels, J. W. (1995) 6-Oxocytidine, a novel protonated C-base analogue for stable triple helix formation. *Nucleic Acids Res.* **23,** 3465–3472.
19. Ishibashi, T., Yamakawa, H., Wang, Q., et al. (1995) Triple helix formation with oligodeoxyribonucleotides containing 8-oxo-2'-deoxyadenosine and 2'-modified nucleoside derivatives. *Nucleic Acids Symp. Ser.* **34,** 127–128.
20. Krawczyk, S. H., Milligan, J. F., Wadwani, S., Moulds, C., Froehler, B. C., and Matteucci, M. D. (1992) Oligonucleotide-mediated triple helix formation using

an N3-protonated deoxycytidine analog exhibiting pH-independent binding within the physiological range. *Proc. Natl. Acad. Sci. USA* **89,** 3761–3764.
21. Soliva, R., Guimil Garcia, R., Blas, J. R., et al. (2000) DNA-triplex stabilizing properties of 8-aminoguanine. *Nucleic Acids Res.* **28,** 4531–4539.
22. Rao, T. S., Durland, R. H., Seth, D. M., Myrick, M. A., Bodepudi, V., and Revankar, G. R. (1995) Incorporation of 2'-deoxy-6-thioguanosine into G-rich oligodeoxyribonucleotides inhibits G-tetrad formation and facilitates triplex formation. *Biochemistry* **34,** 765–772.
23. Vasquez, K. M., Wensel, T. G., Hogan, M. E., and Wilson, J. H. (1995) High-affinity triple helix formation by synthetic oligonucleotides at a site within a selectable mammalian gene. *Biochemistry* **34,** 7243–7251.
24. Faruqi, A. F., Krawczyk, S. H., Matteucci, M. D., and Glazer, P. M. (1997) Potassium-resistant triple helix formation and improved intracellular gene targeting by oligodeoxyribonucleotides containing 7-deazaxanthine. *Nucleic Acids Res.* **25,** 633–640.
25. Milligan, J. F., Krawczyk, S. H., Wadwani, S., and Matteucci, M. D. (1993) An anti-parallel triple helix motif with oligodeoxynucleotides containing 2'-deoxyguanosine and 7-deaza-2'-deoxyxanthosine. *Nucleic Acids Res.* **21,** 327–333.
26. Sproat, B. S., Lamond, A. I., Beijer, B., Neuner, P., and Ryder, U. (1989) Highly efficient chemical synthesis of 2'-O-methyloligoribonucleotides and tetrabiotinylated derivatives; novel probes that are resistant to degradation by RNA or DNA specific nucleases. *Nucleic Acids Res.* **17,** 3373–3386.
27. Puri, N., Majumdar, A., Cuenoud, B., et al. (2001) Targeted gene knockout by 2'-O-aminoethyl modified triplex forming oligonucleotides. *J. Biol. Chem.* **276,** 28,991–28,998.
28. Dagle, J. M. and Weeks, D. L. (1996) Positively charged oligonucleotides overcome potassium-mediated inhibition of triplex DNA formation. *Nucleic Acids Res.* **24,** 2143–2149.
29. Vasquez, K. M., Dagle, J. M., Weeks, D. L., and Glazer, P. M. (2001) Chromosome targeting at short polypurine sites by cationic triplex-forming oligonucleotides. *J. Biol. Chem.* **276,** 38,536–38,541.
30. Good, L. and Nielsen, P. E. (1997) Progress in developing PNA as a gene-targeted drug. *Antisense Nucleic Acid Drug Dev.* **7,** 431–437.
31. Nielsen, P. E. (2001) Peptide nucleic acid: a versatile tool in genetic diagnostics and molecular biology. *Curr. Opin. Biotechnol.* **12,** 16–20.
32. Faruqi, A. F., Egholm, M., and Glazer, P. M. (1998) Peptide nucleic acid-targeted mutagenesis of a chromosomal gene in mouse cells. *Proc. Natl. Acad. Sci. USA* **95,** 1398–1403.
33. Nielsen, P. E., Egholm, M., Berg, R. H., and Buchardt, O. (1991) Sequence-selective recognition of DNA by strand displacement with a thymine-substituted polyamide. *Science* **254,** 1497–1500.
34. Peffer, N. J., Hanvey, J. C., Bisi, J. E., et al. (1993) Strand-invasion of duplex DNA by peptide nucleic acid oligomers. *Proc. Natl. Acad. Sci. USA* **90,** 10,648–10,652.

35. Wang, G., Xu, X., Pace, B., et al. (1999) Peptide nucleic acid (PNA) binding-mediated induction of human gamma-globin gene expression. *Nucleic Acids Res.* **27,** 2806–2813.
36. Maher, L. J. 3rd, Dervan, P. B., and Wold, B. J. (1990) Kinetic analysis of oligodeoxyribonucleotide-directed triple-helix formation on DNA. *Biochemistry* **29,** 8820–8826.
37. Maher, L. J. 3rd, Wold, B., and Dervan, P. B. (1989) Inhibition of DNA binding proteins by oligonucleotide-directed triple helix formation. *Science* **245,** 725–730.
38. Musso, M., Wang, J. C., and Van Dyke, M. W. (1996) In vivo persistence of DNA triple helices containing psoralen-conjugated oligodeoxyribonucleotides. *Nucleic Acids Res.* **24,** 4924–4932.
39. Rougee, M., Faucon, B., Mergny, J. L., et al. (1992) Kinetics and thermodynamics of triple-helix formation: effects of ionic strength and mismatches. *Biochemistry* **31,** 9269–9278.
40. Svinarchuk, F., Debin, A., Bertrand, J. R., and Malvy, C. (1996) Investigation of the intracellular stability and formation of a triple helix formed with a short purine oligonucleotide targeted to the murine c-pim-1 proto-oncogene promotor. *Nucleic Acids Res.* **24,** 295–302.
41. Postel, E. H., Flint, S. J., Kessler, D. J., and Hogan, M. E. (1991) Evidence that a triplex-forming oligodeoxyribonucleotide binds to the c-myc promoter in HeLa cells, thereby reducing c-myc mRNA levels. *Proc. Natl. Acad. Sci. USA* **88,** 8227–8231.
42. Bailey, C. P., Dagle, J. M., and Weeks, D. L. (1998) Cationic oligonucleotides can mediate specific inhibition of gene expression in *Xenopus* oocytes. *Nucleic Acids Res.* **26,** 4860–4867.
43. Duval-Valentin, G., Thuong, N. T., and Helene, C. (1992) Specific inhibition of transcription by triple helix-forming oligonucleotides. *Proc. Natl. Acad. Sci. USA* **89,** 504–508.
44. Ebbinghaus, S. W., Fortinberry, H., and Gamper, H. B., Jr. (1999) Inhibition of transcription elongation in the HER-2/neu coding sequence by triplex-directed covalent modification of the template strand. *Biochemistry* **38,** 619–628.
45. Krasilnikov, A. S., Panyutin, I. G., Samadashwily, G. M., Cox, R., Lazurkin, Y. S., and Mirkin, S. M. (1997) Mechanisms of triplex-caused polymerization arrest. *Nucleic Acids Res.* **25,** 1339–1346.
46. Macaulay, V. M., Bates, P. J., McLean, M. J., et al. (1995) Inhibition of aromatase expression by a psoralen-linked triplex-forming oligonucleotide targeted to a coding sequence. *FEBS Lett.* **372,** 222–228.
47. Wang, Z. and Rana, T. M. (1997) DNA damage-dependent transcriptional arrest and termination of RNA polymerase II elongation complexes in DNA template containing HIV-1 promoter. *Proc. Natl. Acad. Sci. USA* **94,** 6688–6693.
48. Young, S. L., Krawczyk, S. H., Matteucci, M. D., and Toole, J. J. (1991) Triple helix formation inhibits transcription elongation in vitro. *Proc. Natl. Acad. Sci. USA* **88,** 10,023–10,026.

49. Praseuth, D., Guieysse, A. L., and Helene, C. (1999) Triple helix formation and the antigene strategy for sequence-specific control of gene expression. *Biochim. Biophys. Acta* **1489,** 181–206.
50. Vasquez, K. M. and Wilson, J. H. (1998) Triplex-directed modification of genes and gene activity. *Trends Biochem. Sci.* **23,** 4–9.
51. Gasparro, F. P., Havre, P. A., Olack, G. A., Gunther, E. J., and Glazer, P. M. (1994) Site-specific targeting of psoralen photoadducts with a triple helix-forming oligonucleotide: characterization of psoralen monoadduct and crosslink formation. *Nucleic Acids Res.* **22,** 2845–2852.
52. Havre, P. A. and Glazer, P. M. (1993) Targeted mutagenesis of simian virus 40 DNA mediated by a triple helix-forming oligonucleotide. *J. Virol.* **67,** 7324–7331.
53. Havre, P. A., Gunther, E. J., Gasparro, F. P., and Glazer, P. M. (1993) Targeted mutagenesis of DNA using triple helix-forming oligonucleotides linked to psoralen. *Proc. Natl. Acad. Sci. USA* **90,** 7879–7883.
54. Sandor, Z. and Bredberg, A. (1994) Repair of triple helix directed psoralen adducts in human cells. *Nucleic Acids Res.* **22,** 2051–2056.
55. Takasugi, M., Guendouz, A., Chassignol, M., et al. (1991) Sequence-specific photo-induced cross-linking of the two strands of double-helical DNA by a psoralen covalently linked to a triple helix-forming oligonucleotide. *Proc. Natl. Acad. Sci. USA* **88,** 5602–5606.
56. Sancar, A. and Tang, M. S. (1993) Nucleotide excision repair. *Photochem. Photobiol.* **57,** 905–921.
57. Wang, G., Levy, D. D., Seidman, M. M., and Glazer, P. M. (1995) Targeted mutagenesis in mammalian cells mediated by intracellular triple helix formation. *Mol. Cell Biol.* **15,** 1759–1768.
58. Wood, R. D., Aboussekhra, A., Biggerstaff, M., et al. (1993) Nucleotide excision repair of DNA by mammalian cell extracts and purified proteins. *Cold Spring Harbor Symp. Quant. Biol.* **58,** 625–632.
59. Hanvey, J. C., Peffer, N. J., Bisi, J. E., et al. (1992) Antisense and antigene properties of peptide nucleic acids. *Science* **258,** 1481–1485.
60. Koppelhus, U., Zachar, V., Nielsen, P. E., Liu, X., Eugen-Olsen, J., and Ebbesen, P. (1997) Efficient in vitro inhibition of HIV-1 gag reverse transcription by peptide nucleic acid (PNA) at minimal ratios of PNA/RNA. *Nucleic Acids Res.* **25,** 2167–2173.
61. Praseuth, D., Grigoriev, M., Guieysse, A. L., et al. (1996) Peptide nucleic acids directed to the promoter of the alpha-chain of the interleukin-2 receptor. *Biochim. Biophys. Acta* **1309,** 226–238.
62. Nielsen, P. E., Egholm, M., Berg, R. H., and Buchardt, O. (1993) Peptide nucleic acids (PNAs): potential antisense and anti-gene agents. *Anticancer Drug Des.* **8,** 53–63.
63. Brown, P. M. and Fox, K. R. (1996) Nucleosome core particles inhibit DNA triple helix formation. *Biochem. J.* **319(Pt 2),** 607–611.
64. Espinas, M. L., Jimenez-Garcia, E., Martinez-Balbas, A., and Azorin, F. (1996) Formation of triple-stranded DNA at d(GA.TC)n sequences prevents nucleosome assembly and is hindered by nucleosomes. *J. Biol. Chem.* **271,** 31,807–31,812.

65. Westin, L., Blomquist, P., Milligan, J. F., and Wrange, O. (1995) Triple helix DNA alters nucleosomal histone-DNA interactions and acts as a nucleosome barrier. *Nucleic Acids Res.* **23,** 2184–2191.
66. Majumdar, A., Khorlin, A., Dyatkina, N., et al. (1998) Targeted gene knockout mediated by triple helix forming oligonucleotides. *Nat. Genet.* **20,** 212–214.
67. Vasquez, K. M., Wang, G., Havre, P. A., and Glazer, P. M. (1999) Chromosomal mutations induced by triplex-forming oligonucleotides in mammalian cells. *Nucleic Acids Res.* **27,** 1176–1181.
68. Wang, G., Seidman, M. M., and Glazer, P. M. (1996) Mutagenesis in mammalian cells induced by triple helix formation and transcription-coupled repair. *Science* **271,** 802–805.
69. Park, C. H. and Sancar, A. (1994) Formation of a ternary complex by human XPA, ERCC1, and ERCC4(XPF) excision repair proteins. *Proc. Natl. Acad. Sci. USA* **91,** 5017–5021.
70. Troelstra, C., van Gool, A., de Wit, J., Vermeulen, W., Bootsma, D., and Hoeijmakers, J. H. (1992) ERCC6, a member of a subfamily of putative helicases, is involved in Cockayne's syndrome and preferential repair of active genes. *Cell* **71,** 939–953.
71. Hanawalt, P. C. (1994) Transcription-coupled repair and human disease. *Science* **266,** 1957–1958.
72. Vasquez, K. M., Narayanan, L., and Glazer, P. M. (2000) Specific mutations induced by triplex-forming oligonucleotides in mice. *Science* **290,** 530–533.
73. Zendegui, J. G., Vasquez, K. M., Tinsley, J. H., Kessler, D. J., and Hogan, M. E. (1992) In vivo stability and kinetics of absorption and disposition of 3' phosphopropyl amine oligonucleotides. *Nucleic Acids Res.* **20,** 307–314.
74. Vasquez, K. M., Marburger, K., Intody, Z., and Wilson, J. H. (2001) Manipulating the mammalian genome by homologous recombination. *Proc. Natl. Acad. Sci. USA* **98,** 8403–8410.
75. Bhattacharyya, N. P., Maher, V. M., and McCormick, J. J. (1990) Intrachromosomal homologous recombination in human cells which differ in nucleotide excision-repair capacity. *Mutat. Res.* **234,** 31–41.
76. Tsujimura, T., Maher, V. M., Godwin, A. R., Liskay, R. M., and McCormick, J. J. (1990) Frequency of intrachromosomal homologous recombination induced by UV radiation in normally repairing and excision repair-deficient human cells. *Proc. Natl. Acad. Sci. USA* **87,** 1566–1570.
77. Vos, J. M. and Hanawalt, P. C. (1989) DNA interstrand cross-links promote chromosomal integration of a selected gene in human cells. *Mol. Cell. Biol.* **9,** 2897–2905.
78. Brenneman, M., Gimble, F. S., and Wilson, J. H. (1996) Stimulation of intrachromosomal homologous recombination in human cells by electroporation with site-specific endonucleases. *Proc. Natl. Acad. Sci. USA* **93,** 3608–3612.
79. Choulika, A., Perrin, A., Dujon, B., and Nicolas, J. F. (1995) Induction of homologous recombination in mammalian chromosomes by using the I-*Sce*I system of *Saccharomyces cerevisiae*. *Mol. Cell. Biol.* **15,** 1968–1973.

80. Lukacsovich, T., Yang, D., and Waldman, A. S. (1994) Repair of a specific double-strand break generated within a mammalian chromosome by yeast endonuclease I-*Sce*I. *Nucleic Acids Res.* **22,** 5649–5657.
81. Rouet, P., Smih, F., and Jasin, M. (1994) Introduction of double-strand breaks into the genome of mouse cells by expression of a rare-cutting endonuclease. *Mol. Cell. Biol.* **14,** 8096–8106.
82. Smih, F., Rouet, P., Romanienko, P. J., and Jasin, M. (1995) Double-strand breaks at the target locus stimulate gene targeting in embryonic stem cells. *Nucleic Acids Res.* **23,** 5012–5019.
83. Faruqi, A. F., Seidman, M. M., Segal, D. J., Carroll, D., and Glazer, P. M. (1996) Recombination induced by triple-helix-targeted DNA damage in mammalian cells. *Mol. Cell. Biol.* **16,** 6820–6828.
84. Luo, Z., Macris, M. A., Faruqi, A. F., and Glazer, P. M. (2000) High-frequency intrachromosomal gene conversion induced by triplex-forming oligonucleotides microinjected into mouse cells. *Proc. Natl. Acad. Sci. USA* **97,** 9003–9008.
85. Faruqi, A. F., Datta, H. J., Carroll, D., Seidman, M. M., and Glazer, P. M. (2000) Triple-helix formation induces recombination in mammalian cells via a nucleotide excision repair-dependent pathway. *Mol. Cell. Biol.* **20,** 990–1000.
86. Zhang, N., Lu, X., Zhang, X., Peterson, C. A., and Legerski, R. J. (2002) hMutSbeta is required for the recognition and uncoupling of psoralen interstrand cross-links in vitro. *Mol. Cell. Biol.* **22,** 2388–2397.
87. Colosimo, A., Goncz, K. K., Novelli, G., Dallapiccola, B., and Gruenert, D. C. (2001) Targeted correction of a defective selectable marker gene in human epithelial cells by small DNA fragments. *Mol. Ther.* **3,** 178–185.
88. Goncz, K. K., Prokopishyn, N. L., Chow, B. L., Davis, B. R., and Gruenert, D. C. (2002) Application of SFHR to gene therapy of monogenic disorders. *Gene Ther.* **9,** 691–694.
89. Sigurdsson, S., Van Komen, S., Petukhova, G., and Sung, P. (2002) Homologous DNA pairing by human recombination factors Rad51 and Rad54. *J. Biol. Chem.* **277,** 42,790–42,794.
90. Belousov, E. S., Afonina, I. A., Podyminogin, M. A., et al. (1997) Sequence-specific targeting and covalent modification of human genomic DNA. *Nucleic Acids Res.* **25,** 3440–3444.
91. Strobel, S. A., Moser, H. E., and Dervan, P. B. (1988) Double strand cleavage of genomic DNA at a single site by triple helix formation. *J. Am. Chem. Soc.* **110,** 7927–7929.
92. Chan, P. P., Lin, M., Faruqi, A. F., Powell, J., Seidman, M. M., and Glazer, P. M. (1999) Targeted correction of an episomal gene in mammalian cells by a short DNA fragment tethered to a triplex-forming oligonucleotide. *J. Biol. Chem.* **274,** 11,541–11,548.
93. Gamper, H. B., Hou, Y. M., Stamm, M. R., Podyminogin, M. A., and Meyer, R. B. (1998) Strand invasion of supercoiled DNA by oligonucleotides with a triplex guide sequence. *J. Am. Chem. Soc.* **120,** 2182–2183.

94. Biet, E., Maurisse, R., Dutreix, M., and Sun, J. (2001) Stimulation of RecA-mediated D-loop formation by oligonucleotide-directed triple-helix formation: guided homologous recombination (GOREC). *Biochemistry* **40,** 1779–1786.
95. Maurisse, R., Feugeas, J. P., Biet, E., et al. (2002) A new method (GOREC) for directed mutagenesis and gene repair by homologous recombination. *Gene Ther.* **9,** 703–707.
96. Datta, H. J., Chan, P. P., Vasquez, K. M., Gupta, R. C., and Glazer, P. M. (2001) Triplex-induced recombination in human cell-free extracts. Dependence on XPA and HsRad51. *J. Biol. Chem.* **276,** 18,018–18,023.
97. Rogers, F. A., Vasquez, K. M., Egholm, M., and Glazer, P. M. (2002) Site-directed recombination via bifunctional PNA-DNA conjugates. *Proc. Natl. Acad. Sci. USA* **99,** 16,695–16,700.
98. Egholm, M., Christensen, L., Dueholm, K. L., Buchardt, O., Coull, J., and Nielsen, P. E. (1995) Efficient pH-independent sequence-specific DNA binding by pseudoisocytosine-containing bis-PNA. *Nucleic Acids Res.* **23,** 217–222.
99. Nielsen, P. E., Egholm, M., and Buchardt, O. (1994) Evidence for (PNA)2/DNA triplex structure upon binding of PNA to dsDNA by strand displacement. *J. Mol. Recognit.* **7,** 165–170.
100. Wang, G., Chen, Z., Zhang, S., Wilson, G. L., and Jing, K. (2001) Detection and determination of oligonucleotide triplex formation-mediated transcription-coupled DNA repair in HeLa nuclear extracts. *Nucleic Acids Res.* **29,** 1801–1807.
101. Vasquez, K. M., Christensen, J., Li, L., Finch, R. A., and Glazer, P. M. (2002) Human XPA and RPA DNA repair proteins participate in specific recognition of triplex-induced helical distortions. *Proc. Natl. Acad. Sci. USA* **99,** 5848–5853.
102. Barre, F. X., Ait-Si-Ali, S., Giovannangeli, C., et al. (2000) Unambiguous demonstration of triple-helix-directed gene modification. *Proc. Natl. Acad. Sci. USA* **97,** 3084–3088.
103. Macris, M. A. and Glazer, P. M. (2003) Transcription dependence of chromosomal gene targeting by triplex-forming oligonucleotides. *J. Biol. Chem.* **278,** 3357–3362.
104. Bennett, C. F., Chiang, M. Y., Chan, H., Shoemaker, J. E., and Mirabelli, C. K. (1992) Cationic lipids enhance cellular uptake and activity of phosphorothioate antisense oligonucleotides. *Mol. Pharmacol.* **41,** 1023–1033.
105. Giovannangeli, C., Diviacco, S., Labrousse, V., Gryaznov, S., Charneau, P., and Helene, C. (1997) Accessibility of nuclear DNA to triplex-forming oligonucleotides: the integrated HIV-1 provirus as a target. *Proc. Natl. Acad. Sci. USA* **94,** 79–84.
106. Spiller, D. G. and Tidd, D. M. (1995) Nuclear delivery of antisense oligodeoxynucleotides through reversible permeabilization of human leukemia cells with streptolysin O. *Antisense Res. Dev.* **5,** 13–21.
107. Sazani, P., Gemignani, F., Kang, S. H., et al. (2002) Systemically delivered antisense oligomers upregulate gene expression in mouse tissues. *Nat. Biotechnol.* **20,** 1228–1233.

12

Using Nucleases to Stimulate Homologous Recombination

Dana Carroll

Summary

In essentially all organisms, double-strand breaks in chromosomal DNA stimulate repair by multiple mechanisms, including homologous recombination. It is possible to use site-specific reagents to produce a break or other recombinagenic damage at a unique site, which makes possible detailed analysis of the repair products. In addition, targeted mutagenesis and gene replacement are stimulated in the immediate vicinity of the break site. To utilize meganucleases with long recognition sequences, it is necessary to introduce the corresponding sequence prior to directed cleavage. The same is typically true of triplex-forming oligonucleotides that target polypurine-polypyrimidine tracts. Zinc-finger nucleases have the potential of being targetable to arbitrarily selected sites, owing to the flexibility of zinc finger recognition of DNA.

Key Words: genetic recombination, homologous recombination, nonhomologous end joining, I-*Sce*I, zinc fingers, zinc finger nucleases, double-strand break, DNA repair, gene targeting

1. Introduction

It has been known for many years that damage to DNA, particularly in the form of double-strand breaks (DSBs), increases the frequency of recombination. Although the underlying mechanism was not known at the time, H.J. Muller won a Nobel Prize for demonstrating in the 1920s that ionizing radiation causes mutations in *Drosophila* *(1)*. We now appreciate that DSBs produced by such treatment lead to deletions, inversions, and translocations (among other things), all formally products of recombination. In the last decade or so, it has become possible to examine the detailed molecular consequences of DSB formation and repair, through the use of targeted cleavage with highly specific enzymes *(2,3)*. This chapter focuses on the use of such procedures to investigate mechanisms of DSB repair and to stimulate targeted gene manipulations. Since experimental protocols will differ significantly among cell types and organisms, the discussion will remain conceptual.

From: *Methods in Molecular Biology, vol. 262, Genetic Recombination: Reviews and Protocols*
Edited by: A. S. Waldman © Humana Press Inc., Totowa, NJ

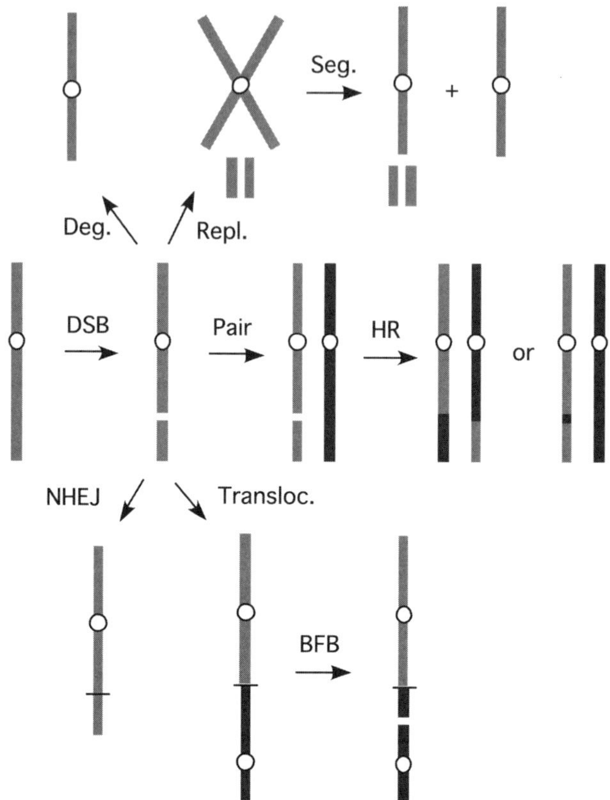

Fig. 1. In the center, a chromosomal DSB is illustrated. The broken chromosome may pair with its homolog or sister chromatid to effect repair by homologous recombination (HR), either with or without exchange of flanking segments. An alternative pathway is nonhomologous end-joining (NHEJ), shown in the lower tier; the fine horizontal line indicates an inexact joint. If a chromosomal fragment is joined, not to its original partner, but to an unrelated chromosome, a translocation results; if a dicentric product is formed, breakage will occur at the subsequent mitosis, and a cycle of breakage, fusion, and bridging (BFB) will result ***(63,64)***. In the top tier, two mechanisms are illustrated by which a noncentromeric fragment may be lost: degradation, or replication followed by mis-segregation.

Why do DSBs lead to recombination? In meiosis in most organisms, proper chromosome segregation is ensured by crossing over between homologous chromosomes. The underlying recombination events are mediated, at least in yeast and probably in other organisms ***(4)***, by programmed DSBs, followed by orderly exchange and resolution events. There is no need for such exchanges in somatic cells, so DSBs are treated as severe, potentially lethal, DNA damage.

If left unrepaired, chromosomal segments may be lost by exonucleolytic degradation, by incomplete replication, or by failure to segregate properly at mitosis; and repeated cycles of breakage and rejoining can propagate the damage to subsequent cell generations. Because of the severity of these consequences, cells have multiple pathways for repairing DSBs *(5–7)*. The outcomes of repair by these various mechanisms are illustrated in **Fig. 1**, along with the results of failure to repair.

The most conservative method of repair, which frequently restores the original information at the break site, is homologous recombination (HR) with a sister chromatid or homologous chromosome. HR with related sequences at ectopic sites would lead to translocations, but this reaction seems to be suppressed in organisms with high levels of reiterated sequences *(6)*. Nonhomologous end-joining (NHEJ) fuses DNA ends without regard for homology. When broken ends from a single event are rejoined, this is typically accompanied by small, localized deletions and/or insertions *(8)*. Although mutagenic if it occurs within coding sequences or other essential regions, this outcome is preferable to loss of larger chromosomal segments. NHEJ can also result in translocations, if a broken end is joined to an unrelated genomic locus.

Our understanding of DSB repair processes has been greatly advanced by the use of targeted cleavage reagents. When a single (or a limited number of) DSB(s) is introduced at a known site, it is possible to characterize in molecular detail the products of recombination *(2,3)*. The same target can be studied in various chromosomal contexts and in different genetic backgrounds to determine the effects of such manipulations. Two classes of endonucleases that have been used as targeted cleavage reagents are described here.

2. Methods

2.1. Targeted Cleavage With Meganucleases

Meganucleases have recognition sites that are large enough that they are typically not found to occur naturally in many genomes *(9)*. For example, I-*Sce*I, which is a yeast mitochondrial enzyme involved in intron homing, recognizes an 18-bp sequence *(10)*. Other members of this class include the yeast HO endonuclease *(2,11)* and the enzymes I-*Cre*I and PI-*Sce*I *(9)*. With the exception of the natural site for HO in the MAT locus of *Saccharomyces cerevisiae*, in order to use these agents to produce specific DSBs, their recognition sites must first be integrated into the genome of interest.

The basic experimental design of these studies is illustrated in **Fig. 2** for I-*Sce*I. The recognition site can be directed to a specific locus by gene targeting *(12)* or allowed to occur at random, followed by discovery of the integration site by physical mapping. Once the target site has been established,

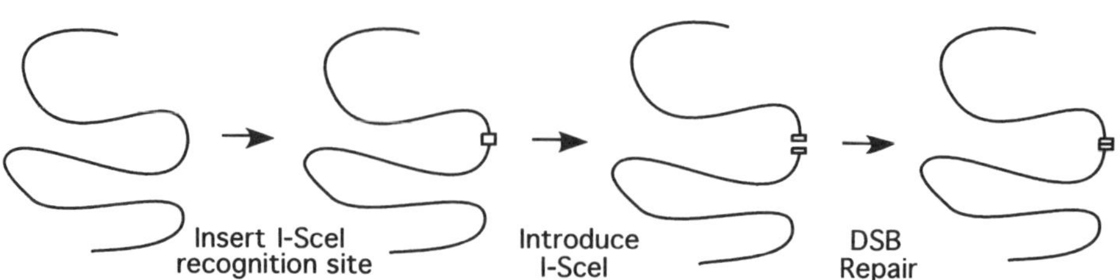

Fig. 2. A cellular chromosome is illustrated in the first diagram. Once a recognition site for I-*Sce*I is integrated into the chromosome, it can be cleaved specifically by that enzyme. In addition to rejoining of the broken ends as shown, other outcomes are possible, as illustrated in **Fig. 1**. DSB, double-strand break.

introduction of the corresponding endonuclease leads to a single, well-defined DSB *(2,3,13)*. In most experiments the enzyme is expressed transiently or in regulated fashion from a plasmid or transgene. Since natural I-*Sce*I is a mitochondrially encoded protein, its coding sequence had to be adjusted to match the universal genetic code, but a number of such I-*Sce*I genes now exist. Of course, sequences required for transcription, translation, and nuclear localization in the cells of interest must be provided. In some cases it may be possible to introduce the protein itself *(14)*, rather than its coding sequence.

The advantage to producing a single DSB at a known site is that the products of repair can be readily isolated and characterized *(15–18)*. As with any experiment, the answers you get depend on how you phrase the question. Specifically, the types of products recovered depend on the detailed design of the target, the availability of donors for homologous repair, and the selection (if any) applied to recover them. The examples shown in **Fig. 3** illustrate experiments based on the recovery in mammalian cells of a functional neo gene (by selection for resistance to G418) *(13)*, but many other designs are possible, including the reconstruction of an expressed GFP gene *(19)*, or the loss of a specific marker. In all cases NHEJ is possible and will compete with HR events, but this will only rarely yield a functional neo gene. The arrangement in **Fig. 3A** will score homology-dependent gene conversion *(20)*; that in **Fig. 3B** reflects single-strand annealing *(20)*; that in **Fig. 3C** models gene targeting of an introduced, extrachromosomal DNA *(16,21,22)*; and that in **Fig. 3D** reports on interhomolog conversion and/or exchange *(23)*. In all cases the frequency of recombination is greatly enhanced by inducing a DSB with I-*Sce*I. Following selection of cells carrying functional repair products, analysis is usually pursued by Southern blotting, polymerase chain reaction (PCR) amplification, and DNA sequencing.

The neo sequence carrying the I-*Sce*I site is the recipient of information by recombination in these experiments. Both it and the donor can be modified in

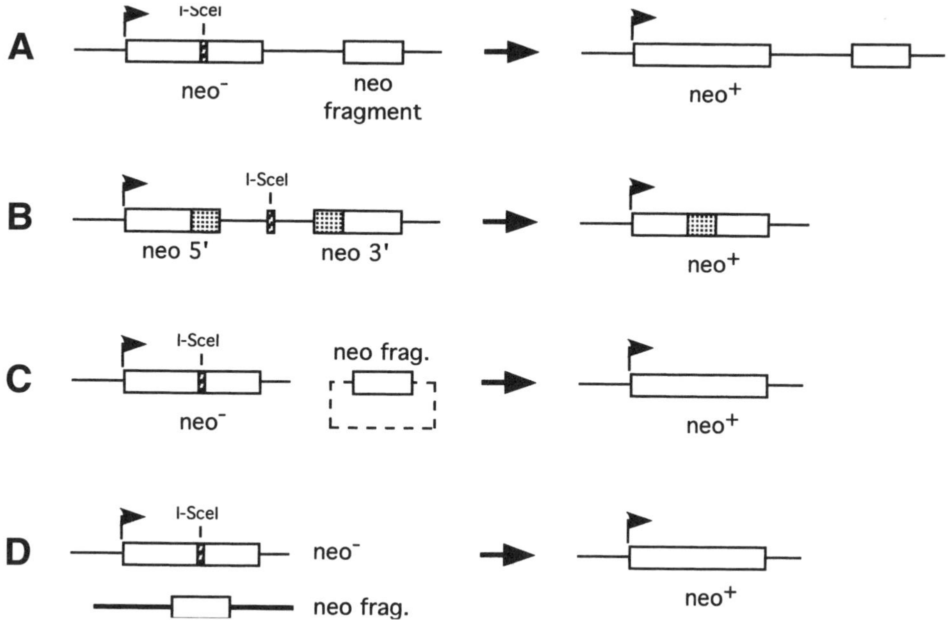

Fig. 3. Illustration of target and donor configurations for various types of repair of I-*Sce*I-induced DSBs ***(13)***. In each case the I-*Sce*I site is shown as a hatched rectangle; in **A**, **C**, and **D**, insertion of this site creates a nonfunctional neo gene. Neo frag is an incomplete internal fragment of the neo gene. In **B**, both the neo 5' and neo 3' fragments are incomplete but share homology indicated by the stippled box. The extrachromosomal donor in **C** may be part of a circular plasmid, as suggested by the dashed line, or a linear fragment. *See* text for additional descriptions.

various ways to provide additional insights into the repair process. For example, inclusion of multiple polymorphisms between donor and recipient across the region of homology allows one to map the extent of incorporation of donor sequences ***(24)***. In addition, introduction and cleavage of more than one I-*Sce*I site can report on more complex repair events ***(25)***.

The design shown in **Fig. 3C** is particularly relevant to attempts to improve the efficiency of gene targeting. Analogs of this experiment have been performed in many different organisms and cell types and essentially always show a substantial increase in targeting frequency upon cleavage of the chromosomal target, often by several orders of magnitude ***(16,21,22,26–31)***. Combined with the observation that manipulations of the donor DNA have modest effects on frequency, this indicates that efficiency is limited by the reactivity, or lack thereof, of the target. In fact, a target that lies within a continuous stretch of

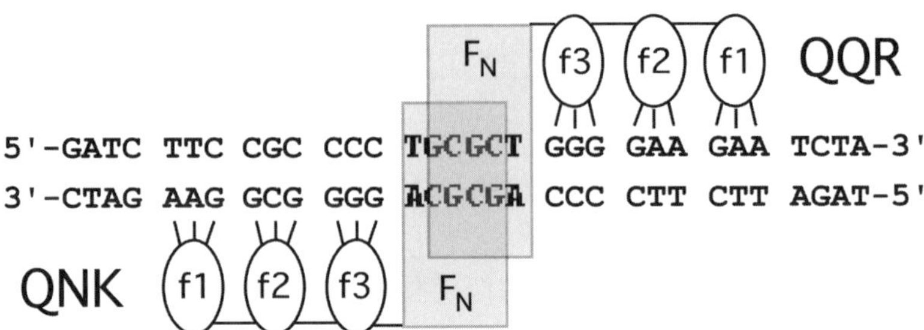

Fig. 4. Diagram of two ZFNs bound to their DNA target. The zinc fingers of each protein (f1, f2, f3) contact 3 bp apiece in the DNA, as indicated. Each set of three fingers is covalently linked to a cleavage domain (F_N) derived from FokI, which must dimerize to cleave DNA in the space between the two zinc finger binding sites. The ZFNs illustrated are QQR and QNK, which recognize the sequences shown *(37,38)*.

unbroken DNA would have no reason to engage in recombination, homologous or otherwise.

Although experiments with I-*Sce*I and similar meganucleases have been very informative in elucidating outcomes and mechanisms of DSB repair, they are limited in utility for gene targeting, since the recognition site must be introduced by a low-efficiency procedure before it can be used to enhance localized recombination. What is needed for this purpose is a class of reagents that can target a DSB, or other recombinagenic damage, to arbitrarily selected loci.

2.2. Zinc Finger Nucleases

The other targetable DSB-inducing reagents that have been developed recently are zinc finger nucleases (ZFNs) (**Fig. 4**). These are hybrid proteins comprised of a nonspecific DNA-cleavage domain linked to a DNA-binding module based on Cys_2His_2 zinc fingers *(32)*. Each finger contacts 3 bp of DNA in very modular fashion, and the ZFNs utilized to date contain a cluster of three fingers recognizing a 9-bp site. Because the binding specificity of zinc fingers can be manipulated *(33–35)*, the ZFNs can, in principle, be designed to bind and cleave a wide range of DNA sequences. The cleavage domain is derived from the type IIS restriction enzyme FokI. Because this domain must dimerize to cut DNA *(36)*, and because the dimer interface is quite weak, effective cleavage requires two ZFNs with fingers directed to a pair of 9-bp sites in close proximity and in opposite orientation *(37,38)*. As might be expected, there is a distinct relationship between the separation (in bp) of the binding sites in the target and the length of the peptide linker between the binding and cleavage domains in the protein *(38)*. With a very short linker, only paired sites separated by exactly 6 bp are efficiently cut in vivo.

The procedure for attacking a new genomic target with ZFNs is as follows ***(39,40)***:

1. Choose a target gene of interest.
2. Search it for plausible zinc finger binding sites. This consists of looking for pairs of inverted 9-bp sequences for which zinc fingers exist (in the literature) that bind each of the component DNA triplets. Fingers have been identified that recognize most of the GNN triplets ***(41,42)*** and some of the ANN triplets ***(43)*** with good specificity. Fingers for the TNN and CNN triplets are much rarer and less well characterized ***(35)***.
3. Construct coding sequences for the chosen sets of fingers, keeping in mind that the N-terminal finger binds the 3'-most triplet, and so on (**Fig. 4**).
4. Join the finger-coding sequences to the FokI cleavage domain. We usually express the ZFNs in bacteria and test their ability to cut the designed target sequence in plasmid DNA prior to any in vivo experiments.
5. Engineer the expression of the ZFNs in the desired host cell or organism. This is a critical step, which will differ for each experimental design. Expression from an appropriately chosen promoter has proved successful, and use of a conditional promoter may be advisable. One can also conceive of methods in which the proteins are delivered directly.
6. Include a donor DNA that carries a sequence alteration of interest. In addition to whatever mutation one may desire in the recombination product, it is important to modify the donor so that it is not cleaved by the ZFNs, i.e., the zinc finger recognition sites must be removed. Even in coding sequences this can be achieved with synonymous codon substitutions. As with the nucleases themselves, the method of delivery of the donor will be critical.

The possible outcomes of these experiments are the same as those illustrated in **Fig. 3** for cleavage by I-*Sce*I. If only the ZFNs are delivered to cells without a homologous donor DNA, accurate repair—by religation or by HR using a sister chromatid or allelic sequence as donor—and mutagenesis via NHEJ are the major outcomes ***(39)***. When the designed donor DNA is present, it must compete with those processes for repair of the DSB. We have had good success with ZFN-induced cleavage, mutagenesis, and gene targeting at the *yellow* locus of *Drosophila melanogaster* ***(39,40)***. Porteus and Baltimore ***(44)*** have reported ZFN-stimulated gene targeting in mammalian cells at levels comparable to those achieved with I-*Sce*I-induced target cleavage for cases in which the zinc finger recognition sites were placed in artificial constructs similar to those shown in **Fig. 3C** and then introduced into the genome.

A number of questions arise in conjunction with ZFN utilization:

1. How common are plausible ZFN sites in genomic DNA? Approximately half the RNN triplets (where R is either purine) have good fingers available ***(41,43)***. Therefore, sites of the form $(NNY)_3N_6(RNN)_3$ should be found about every 4 kb

in random sequence DNA. In my personal experience, it is rare that searching an arbitrarily chosen gene does not reveal at least one such site.

2. How often do novel combinations of existing fingers bind and cleave their target efficiently and uniquely? This important issue has been addressed empirically only once, with good success ***(39,40)***. Additional tests are clearly necessary to provide a satisfactory answer. Alternatives to simple design include selection of fingers for novel sites by phage display ***(33,34,41,43,45)***. In addition, it is possible to link more than three fingers to achieve greater affinity and specificity ***(46–48)***, and novel designs of multifinger proteins may help in discriminating against related DNA sequences ***(49–51)***.
3. Will excessive cleavage owing to inadequate specificity be a problem? One of the ZFNs we produced for the *Drosophila yellow* gene proved to be lethal if overexpressed ***(39)***. We were able to moderate its level of expression to a usable range, but the degree to which this will be a common problem has yet to be addressed.

2.3. Other Targeting Reagents

Common restriction enzymes can be introduced into cells to create DSBs that lead to the same classes of repair products discussed in **Subheadings 2.1.** and **2.2.** ***(52)***. The difficulty is that each restriction enzyme will have a very large number of recognition sites in a complex genome. For example, we would predict more than 400,000 *Bam*HI sites in the 3×10^9 bp of a mammalian genome. Cleavage at many sites in a single cell will be lethal, and finding the subset of sites that have been cut and repaired is very challenging. This can be done by selection for alterations in a particular gene ***(52)***, but the frequency of such events in the background of many other cleavage targets is quite low.

Triplex-forming oligonucleotides (TFOs) have also been used to induce recombinagenic damage at specific targets ***(53,54)***. (*See* Chapter 11.) A TFO linked to psoralen will direct UV-induced crosslinking to its preferred binding site, and the attempt to repair this lesion can lead to recombination ***(55,56)***. Surprisingly a TFO on its own, without psoralen, can also stimulate recombination ***(57)***. In neither case is the precise nature of the recombinagenic lesion known, although psoralen crosslinks have been shown to be converted to frank DSBs in yeast ***(58,59)***. The drawback to this approach is that the range of duplex sequences that can stably accommodate a third strand is limited to polypurine-polypyrimidine tracts of 20 bp or more, which are quite rare in genomes. In addition, TFOs with or without psoralen cause localized mutations, which are disadvantageous if the goal is gene targeting by homologous recombination. So far, only artificial substrates have been targeted for recombination with TFOs; no endogenous genes have been attacked successfully.

In principle, it is possible to link DNA-damaging moieties, or even complete endonucleases, to other sequence-specific agents. Such compounds include peptide nucleic acids ***(60)*** and minor groove-directed polyamides ***(61)***, but relatively little use has been made of these to date to induce recombination ***(62)***.

3. Concluding Remarks

If your goal is to stimulate homologous recombination at a particular chromosomal (or extrachromosomal) site by inducing a targeted DSB, you have several options available. You can introduce a site for an endonuclease of known specificity, and there are a number of these from which to choose. Various methods for expressing or introducing the corresponding enzyme lead to efficient cleavage, particularly in the case of I-*Sce*I. The newer method of designing zinc finger nucleases to target a pre-existing site is less thoroughly tested but holds great promise. Once a DSB is induced, the outcome will depend on the activities available for repair in the target cells and the DNA sequences available to participate in the reaction. To tip the balance strongly in favor of homologous recombination, it may ultimately be necessary to interfere with competing pathways, particularly NHEJ. DSB-stimulated gene targeting can be used to introduce designed mutations into genes of interest in experimental organisms. In the long run, it should be applicable to genetic manipulation of crop plants and to correction of human genetic diseases.

Acknowledgments

I am grateful to people who have worked in my lab on the zinc finger nucleases, particularly Marina Bibikova, and to Srinivasan Chandrasegaran, who developed these hybrid enzymes. Work in my lab is supported by USPHS research grants GM58504 and GM65173.

References

1. Muller, H. J. (1927) Artificial transmutation of the gene. *Science* **66,** 84–87.
2. Haber, J. E. (1995) In vivo biochemistry: physical monitoring of recombination induced by site-specific endonucleases. *Bioessays* **17,** 609–620.
3. Jasin, M. (1996) Genetic manipulation of genomes with rare-cutting endonucleases. *Trends Genet.* **12,** 224–228.
4. Paques, F. and Haber, J. E. (1999) Multiple pathways of recombination induced by double-strand breaks in *Saccharomyces cerevisiae. Microbiol. Mol. Biol. Rev.* **63,** 349–404.
5. Friedberg, E. C., Walker, G. C., and Siede, W. (1995) *DNA Repair and Mutagenesis.* ASM Press, Washington, DC.
6. Haber, J. E. (2000) Partners and pathways repairing a double-strand break. *Trends Genet.* **16,** 259–264.
7. van Gent, D. C., Hoeijmakers, J. H. J., and Kanaar, R. (2001) Chromosome stability and the double-strand break connection. *Nat. Rev. Genet.* **2,** 196–206.
8. Jeggo, P. A. (1998) DNA breakage and repair. *Adv. Genet.* **38,** 185–218.

9. Belfort, M. and Roberts, R. J. (1997) Homing endonucleases: keeping the house in order. *Nucleic Acids Res.* **25,** 3379–3388.
10. Colleaux, L., d'Auriol, L., Galibert, F., and Dujon, B. (1988) Recognition and cleavage site of the intron-encoded *omega* transposase. *Proc. Natl. Acad. Sci. USA* **85,** 6022–6026.
11. Rudin, N. and Haber, J. E. (1988) Efficient repair of HO-induced chromosomal breaks in *Saccharomyces cerevisiae* by recombination between flanking homologous sequences. *Mol. Cell. Biol.* **8,** 3918–3928.
12. Capecchi, M. R. (1989) Altering the genome by homologous recombination. *Science* **244,** 1288–1292.
13. Johnson, R. D. and Jasin, M. (2001) Double-strand-break-induced homologous recombination in mammalian cells. *Biochem. Soc. Trans.* **29,** 196–201.
14. Brenneman, M., Gimble, F. S., and Wilson, J. H. (1996) Stimulation of intrachromosomal homologous recombination in human cells by electroporation with site-specific endonucleases. *Proc. Natl. Acad. Sci. USA* **93,** 3608–3612.
15. Plessis, A., Perrin, A., Haber, J. E., and Dujon, B. (1992) Site-specific recombination determined by I-*SceI*, a mitochondrial group I intron-encoded endonuclease expressed in the yeast nucleus. *Genetics* **130,** 451–460.
16. Rouet, P., Smih, F., and Jasin, M. (1994) Introduction of double-strand breaks into the genome of mouse cells by expression of a rare-cutting endonuclease. *Mol. Cell. Biol.* **14,** 8096–8106.
17. Lukacsovich, T., Yang, D., and Waldman, A. S. (1994) Repair of a specific double-strand break generated within a mammalian chromosome by yeast endonuclease I-SceI. *Nucleic Acids Res.* **22,** 5649–5657.
18. Aylon, Y., Liefshitz, B., Bitan-Banin, G., and Kupiec, M. (2003) Molecular dissection of mitotic recombination in the yeast *Saccharomyces cerevisiae. Mol. Cell. Biol.* **23,** 1403–1417.
19. Pierce, A. J., Johnson, R. D., Thompson, L. H., and Jasin, M. (1999) XRCC3 promotes homology-directed repair of DNA damage in mammalian cells. *Genes Dev.* **13,** 2633–2638.
20. Liang, F., Romanienko, P. J., Weaver, D. T., Jeggo, P. A., and Jasin, M. (1996) Chromosomal double-strand break repair in Ku80-deficient cells. *Proc. Natl. Acad. Sci. USA* **93,** 8929–8933.
21. Smih, F., Rouet, P., Romanienko, P. J., and Jasin, M. (1995) Double-strand breaks at the target locus stimulate gene targeting in embryonic stem cells. *Nucleic Acids Res.* **23,** 5012–5019.
22. Donoho, G., Jasin, M., and Berg, P. (1998) Analysis of gene targeting and intrachromosomal homologous recombination stimulated by genomic double-strand breaks in mouse embryonic stem cell. *Mol. Cell. Biol.* **18,** 4070–4078.
23. Moynahan, M. E. and Jasin, M. (1997) Loss of heterozygosity induced by a chromosomal double-strand break. *Proc. Natl. Acad. Sci. USA* **94,** 8988–8993.
24. Elliott, B., Richardson, C., Winderbaum, J., Nickoloff, J. A., and Jasin, M. (1998) Gene conversion tracts from double-strand break repair in mammalian cells. *Mol. Cell. Biol.* **18,** 93–101.

25. Richardson, C. and Jasin, M. (2000) Frequent chromosomal translocations induced by DNA double-strand breaks. *Nature* **405,** 697–700.
26. Choulika, A., Perrin, A., Dujon, B., and Nicolas, J.-F. (1995) Induction of homologous recombination in mammalian chromosomes by using the I-*Sce*I system of *Saccharomyces cerevisiae*. *Mol. Cell. Biol.* **15,** 1968–1973.
27. Segal, D. J. and Carroll, D. (1995) Endonuclease-induced, targeted homologous extrachromosomal recombination in *Xenopus* oocytes. *Proc. Natl. Acad. Sci. USA* **92,** 806–810.
28. Puchta, H., Dujon, B., and Hohn, B. (1996) Two different but related mechanisms are used in plants for the repair of genomic double-strand breaks by homologous recombination. *Proc. Natl. Acad. Sci. USA* **93,** 5055–5060.
29. Cohen-Tannoudji, M., Robine, S., Choulika, A., et al. (1998) I-*Sce*I-induced gene replacement at a natural locus in embryonic stem cells. *Mol. Cell. Biol.* **18,** 1444–1448.
30. Lin, Y., Lukacsovich, T., and Waldman, A. S. (1999) Multiple pathways for repair of DNA double-strand breaks in mammalian chromosomes. *Mol. Cell. Biol.* **19,** 8353–8360.
31. Elliott, B. and Jasin, M. (2001) Repair of double-strand breaks by homologous recombination in mismatch repair-defective mammalian cells. *Mol. Cell. Biol.* **21,** 2671–2682.
32. Kim, Y.-G., Cha, J., and Chandrasegaran, S. (1996) Hybrid restriction enzymes: zinc finger fusions to *Fok*I cleavage domain. *Proc. Natl. Acad. Sci. USA* **93,** 1156–1160.
33. Isalan, M., Klug, A., and Choo, Y. (1998) Comprehensive DNA recognition through concerted interactions from adjacent zinc fingers. *Biochemistry* **37,** 12,026–12,033.
34. Pabo, C. O., Peisach, E. and Grant, R. A. (2001) Design and selection of novel Cys_2His_2 zinc finger proteins. *Annu. Rev. Biochem.* **70,** 313–340.
35. Segal, D. J. (2002) The use of zinc finger peptides to study the role of specific factor binding sites in the chromatin environment. *Methods* **26,** 76–83.
36. Bitinaite, J., Wah, D. A., Aggarwal, A. K., and Schildkraut, I. (1998) *Fok*I dimerization is required for DNA cleavage. *Proc. Natl. Acad. Sci. USA* **95,** 10,570–10,575.
37. Smith, J., Bibikova, M., Whitby, F. G., Reddy, A. R., Chandrasegaran, S., and Carroll, D. (2000) Requirements for double-strand cleavage by chimeric restriction enzymes with zinc finger DNA-recognition domains. *Nucleic Acids Res.* **28,** 3361–3369.
38. Bibikova, M., Carroll, D., Segal, D. J., et al. (2001) Stimulation of homologous recombination through targeted cleavage by chimeric nucleases. *Mol. Cell. Biol.* **21,** 289–297.
39. Bibikova, M., Golic, M., Golic, K. G., and Carroll, D. (2002) Targeted chromosomal cleavage and mutagenesis in *Drosophila* using zinc-finger nucleases. *Genetics* **161,** 1169–1175.
40. Bibikova, M., Beumer, K., Trautman, J. K., and Carroll, D. (2003) Enhancing gene targeting by target cleavage with designed zinc-finger nucleases. *Science* **300,** 764.

41. Segal, D. J., Dreier, B., Beerli, R. R., and Barbas III, C. F. (1999) Toward controlling gene expression at will: selection and design of zinc finger domains recognizing each of the 5'-GNN'-3' DNA target sequences. *Proc. Natl. Acad. Sci. USA* **96,** 2758–2763.
42. Liu, Q., Xia, Z. Q., Zhong, X., and Case, C. C. (2002) Validated zinc finger protein designs for all 16 GNN DNA triplet targets. *J. Biol. Chem.* **277,** 3850–3856.
43. Dreier, B., Beerli, R. R., Segal, D. J., Flippin, J. D., and Barbas III, C. F. (2001) Development of zinc finger domains for recognition of the 5'-ANN-3' family of DNA sequences and their use in the construction of artificial transcription factors. *J. Biol. Chem.* **276,** 29,466–29,478.
44. Porteus, M. and Baltimore, D. (2003) Chimeric nucleases stimulate gene targeting in human cells. *Science* **300,** 763.
45. Greisman, H. A. and Pabo, C. O. (1997) A general strategy for selecting high-affinity zinc finger proteins for diverse DNA target sites. *Science* **275,** 657–661.
46. Liu, Q., Segal, D. J., Ghiara, J. B., and Barbas III, C. F. (1997) Design of polydactyl zinc-finger proteins for unique addressing within complex genomes. *Proc. Natl. Acad. Sci. USA* **94,** 5525–5530.
47. Beerli, R. R., Segal, D. J., Dreier, B., and Barbas III, C. F. (1998) Toward controlling gene expression at will: specific regulation of the *erbB-2/HER-2* promoter by using polydactyl zinc finger proteins constructed from modular building blocks. *Proc. Natl. Acad. Sci. USA* **95,** 14,628–14,633.
48. Kim, J.-S. and Pabo, C. O. (1998) Getting a handhold on DNA: design of poly-zinc finger proteins with femtomolar dissociation constants. *Proc. Natl. Acad. Sci. USA* **95,** 2812–2817.
49. Isalan, M., Klug, A., and Choo, Y. (2001) A rapid, generally applicable method to engineer zinc fingers illustrated by targeting the HIV-1 promoter. *Nat. Biotech.* **19,** 656–660.
50. Moore, M., Choo, Y., and Klug, A. (2001) Design of polyzinc finger peptides with structured linkers. *Proc. Natl. Acad. Sci. USA* **98,** 1432–1436.
51. Moore, M., Klug, A., and Choo, Y. (2001) Improved DNA binding specificity from polyzinc finger peptides by using strings of two-finger units. *Proc. Natl. Acad. Sci. USA* **98,** 1437–1441.
52. Phillips, J. W. and Morgan, W. F. (1994) Illegitimate recombination induced by DNA double-strand breaks in a mammalian chromosome. *Mol. Cell. Biol.* **14,** 5794–5803.
53. Vasquez, K. M. and Wilson, J. H. (1998) Triplex-directed modifications of genes and gene activity. *Trends Biochem. Sci.* **23,** 4–9.
54. Casey, B. P. and Glazer, P. M. (2001) Gene targeting via triple-helix formation. *Prog. Nucleic Acid Res. Mol. Biol.* **67,** 163–192.
55. Sandor, Z. and Bredberg, A. (1995) Triple helix directed psoralen adducts induce a low frequency of recombination in an SV40 shuttle vector. *Biochim. Biophys. Acta* **1263,** 235–240.
56. Faruqi, A. F., Seidman, M. M., Segal, D. J., Carroll, D., and Glazer, P. M. (1996) Recombination induced by triple helix-targeted DNA damage in mammalian cells. *Mol. Cell. Biol.* **16,** 6820–6828.

57. Faruqi, A. F., Datta, H. J., Carroll, D., Seidman, M. M., and Glazer, P. M. (2000) Triple-helix formation induces recombination in mammalian cells via a nucleotide excision repair-dependent pathway. *Mol. Cell. Biol.* **20,** 990–1000.
58. Jachymczyk, W. J., von Borstel, R. C., Mowat, M. R. A., and Hastings, P. J. (1981) Repair of interstrand cross-links in DNA of *Saccharomyces cerevisiae* requires two systems for DNA repair: the *RAD3* system and the *RAD51* system. *Mol. Gen. Genet.* **182,** 196–205.
59. Dardalhon, M. and Averbeck, D. (1995) Pulsed-field gel electrophoresis analysis of the repair of psoralen plus UVA induced DNA photoadducts in *Saccharomyces cerevisiae. Mutat. Res.* **366,** 49–60.
60. Nielsen, P. E., Egholm, M., Berg, R. H., and Buchardt, O. (1993) Peptide nucleic acids (PNAs): potential antisense and anti-gene agents. *Anticancer Drug Des.* **8,** 53–63.
61. Dervan, P. B. (2001) Molecular recognition of DNA by small molecules. *Bioorg. Med. Chem.* **9,** 2215–2235.
62. Rogers, F. A., Vasquez, K. M., Egholm, M., and Glazer, P. M. (2002) Site-directed recombination via bifunctional PNA-DNA conjugates. *Proc. Natl. Acad. Sci. USA* **99,** 16,695–16,700.
63. McClintock, B. (1939) The behavior in successive nuclear divisions of a chromosome broken at meiosis. *Proc. Natl. Acad. Sci. USA* **25,** 405–416.
64. McClintock, B. (1942) The stability of broken ends of chromosomes in *Zea mays. Genetics* **26,** 234–282.

13

Enhancement of In Vivo Targeted Nucleotide Exchange by Nonspecific Carrier DNA

Katie K. Maguire and Eric B. Kmiec

Summary

Targeted nucleotide exchange (TNE) is a process in which an oligonucleotide bearing sequence complementarity aligns with the sequence of a target gene and directs the alteration of a single base. This technique can be used to repair a point mutation or mediate site-specific mutagenesis. A critical factor in the development of this approach centers around the elevation and stabilization of the frequencies with which these events occur. Here we describe a protocol for increasing the frequency of TNE in the true yeast, *Saccharomyces cerevisiae*, through the use of nonspecific, carrier oligonucleotides. These molecules, when added to the reaction, increase the TNE frequency up to 25-fold in some cases, perhaps by providing a molecular trap to bind factors, which may inactivate the specific targeting oligos.

Key Words: targeted nucleotide exchange, oligonucleotide, nonspecific DNA, eGFP, hygromycin resistance, yeast

1. Introduction

Oligonucleotide-directed DNA alteration is highly dependent on the efficiency with which the vector enters the cell and targets the DNA. In principle, the most useful vector would be one that: (1) is easily delivered into the nucleus; (2) is stable both within the cellular environment and at the target region within the DNA; and (3) gives the highest frequency of alteration with the least amount of nonspecific changes. The aforementioned challenges for vector development can be overcome by using a variety of methods that modulate either the transfection of the oligo or its conversion activity. For example, salmon sperm DNA can be added as a carrier to transfection mixtures that include the oligonucleotide, whereas cationic conjugates enhance the frequency of gene alteration, perhaps by escorting the DNA to the chromosomal target site ***(1,2)***.

From: *Methods in Molecular Biology, vol. 262, Genetic Recombination: Reviews and Protocols*
Edited by: A. S. Waldman © Humana Press Inc., Totowa, NJ

Modifications—such as 2'-*O*-methyl RNA, phosphorothioate linkages, and locked nucleic acids (LNAs)—protect against nuclease degradation and increase vector half-life within the cell, increasing the potential for conversion. Once transport barriers have been surpassed, the ability of the vector to convert is likely to be dependent, at least in part, on the sequence context surrounding the target site. It has been reported that the frequency of targeting with a single-stranded oligo will increase if a second, shorter, oligonucleotide modified with LNA bases *(3)* is added to the transformation protocol. The presence of this second oligo is thought to confer a higher level of stability onto the intermediate once the vector reaches its chromosomal target. The intermediate is then acted on by DNA repair proteins, which enable the alteration of the nucleotide. In this chapter, we describe a robust method for carrying out targeted nucleotide exchange (TNE) and specifically focus on the use of nonspecific oligos to enhance the frequency of repair of a point mutation in a hygromycin-eGFP fusion gene, directed by a *specific* oligonucleotide.

2. Materials

1. Yeast strain LSY678IntHyg-eGFP(rep).
2. Hyg3S/74NT 5'-G*T*A* GAA ACC ATC GGC GCA GCT ATT TAC CCG CAG GAC GTA TCC ACG CCC TCC TAC ATC GAA GCT GAA AGC AC*G*A*G-3' (* indicates phosphorothioate linkage).
3. Kan 3S/70 5'-C*A*T* CAG AGC AGC CAA TTG TCT GTT GTG CCC AGT CGT AGC CGA ATA GCC TCT CCA CCC AAA CGG CCG G*A*G*A-3' (* indicates phosphorothioate linkage).
4. Standard yeast complete medium (YPD): 20 g peptone, 20 g dextrose and 10 g yeast extract in 1 L H_2O. For plates, add 20 g agar/L.
5. 0.5 mg/mL Aureobasidin A.
6. 50 mg/mL Hygromycin B.
7. 1 *M* and 2 *M* sorbitol.
8. 1 *M* dithiothreitol (DTT).
9. Millipore dH_2O.
10. 2-mm Gap cuvet.
11. Primers to target gene Hyg-eGFP 5'-tctgcacaatatttcaagc-3' and pHyg 1560R 5'-aaatcagccatgtagtg-3'.
12. Polymerase chain reaction (PCR) kit (Invitrogen, Carlsbad, CA).
13. Sanger dideoxy sequencing kit.

3. Methods

The method of targeted nucleotide exchange is outlined below and categorized as follows: (1) description of the target gene and oligonucleotides used as vectors; (2) maintenance of the yeast strains; (3) growth conditions for yeast cells; (4) creation of competent yeast cells and electroporation conditions for introducing the oligonucleotides; (5) selection for the corrected cells; (6) visualization of corrected cells; and (7) colony PCR and SNaPSHOT (sequencing) analyses.

3.1. Description of the Target Plasmid and Oligonucleotides

A hygromycin-resistant eGFP fusion gene that carries a point mutation at nucleotide position 137 on the hyg gene is the target for TNE in this protocol. This point mutation is a TAG stop codon and will result in the translation of a truncated protein. A modified single-stranded oligonucleotide is used to direct the repair of the base, changing the codon from TA*G* to TA*C*. Wild-type hygromycin-resistant genes have a tyrosine encoded by TA*T* at this position. True conversion events are discerned by conversion to TAC, not TAT, avoiding issues regarding the possibility of contamination. There is no preference for either codon in yeast. The mutated fusion gene was cloned into the KpnI and *Sal*I sites of the integrative plasmid pAUR101 downstream of a nonfunctional aureobasidin gene (**Fig. 1A**). The plasmid was integrated into the yeast strain LSY678 at the nonfunctional chromosomal aureobasidin gene by homologous recombination, generating a functional gene *(4)*. The recombination event was scored by plating the transfected cells on aureobasidin plates (0.5 μg/mL), and Southern blot analysis confirmed that 1–2 copies of this plasmid were present in each yeast cell.

Previously, it has been shown that chimeric RNA/DNA oligonucleotides were efficient in directing the TNE reaction *(5)*. After a methodical analysis of the structure of this molecule, it was found that a single-stranded DNA oligonucleotide, modified on either end to protect it from nuclease degradation, enables the highest level of TNE. It has also been shown *(6)* that the frequency of correction is highest when the oligo is directed to the nontranscribed strand of the gene, as opposed to the transcribed strand, owing in part to the stability of the oligo-targeted DNA complex during transcription *(6)*. The length of the oligonucleotide is also an important aspect of this reaction, with optimal activity being 70–80 bases; however, shorter and longer oligos are active in the targeted nucleotide exchange reaction at a reduced level. In the experiment described below, a 74-nt oligonucleotide is used with a nonspecific oligonucleotide that has no homology to the target gene, and we now report that the presence of the nonspecific oligo increases the correction efficiency (**Fig. 1B**). Although the exact mechanism of stimulation engendered by the nonspecific oligo remains unknown, we suspect that the oligo may increase mass action within the cell, facilitating chromosomal localization. Alternatively, the nonspecific oligo could help bind cellular factors that would otherwise inhibit or suppress TNE by rendering the targeting oligo nonfunctional.

A

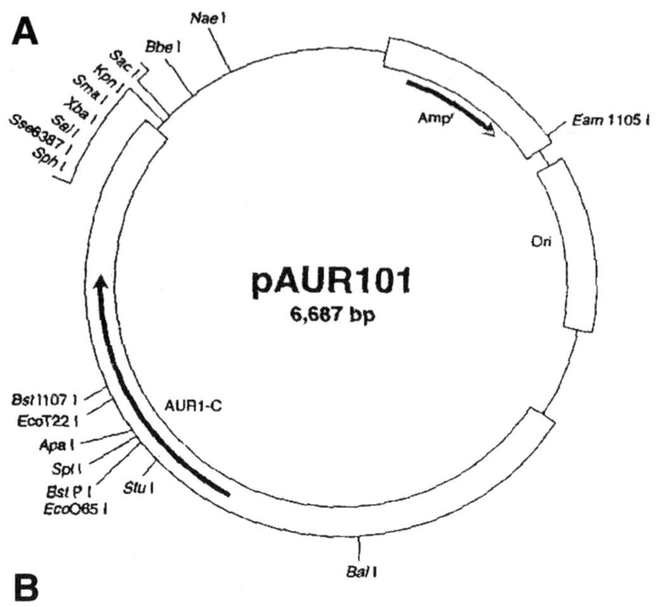

B

wild type

(NT) GATGTAGGAGGGCGTGGATATGTCCTGCGGGTAAATAGCTGC
(T) CTACATCC T CCCGCACCTATACAGGACGCCCATTTATCGACG

mutant
(NT) GATGTAGGAGGGCGTGGATAGGTCCTGCGGGTAAATAGCTGC
(T) CTACATCC T CCCGCACCTATCCAGGACGCCCATTTATCGACG

converted
(NT) GATGTAGGAGGGCGTGGATACGTCCTGCGGGTAAATAGCTGC
(T) CTACATCCT CCCG CACCTATGCAGGACGCCCATTTATCGACG

Hyg 3S 74NT
5'- G*T*A* GAA ACC ATC GGC GCA GCT ATT TAC CCG CAG GAC GTA TCC ACG CCC TCC TAC ATC GAA GCT GAA AGC AC*G*A*G -3'
Kan 3S 70G
5'- C*A*T* CAG AGC AGC CAA TTG TCT GTT GTG CCC AGT CGT AGC CGA ATA GCC TCT CCA CCC AAA CGG CCG G*A*G*A -3'

Fig. 1. (A) The yeast integrative plasmid pAUR101 carrying a mutant hygromycin eGFP fusion gene was integrated into the chromosomal aureobasidin gene of *LSY678* *(5)*. **(B)** The wild-type hygromycin has a TAT (tyrosine) codon at the target site, which has been mutated to a TAG stop codon. The oligo used in this reaction is the Hyg3S/74NT, directing the alteration of TAG to TAC. The nonspecific oligo Kan3S/70G is also shown.

3.2. Maintenance of Yeast Strain

The haploid yeast strain *LSY678* (mating type a) was a generous gift from Dr. Lorraine Symington (Columbia University). Plasmid pAUR101 Hyg eGFP was integrated at the aureobasidin locus, creating LSY678IntHyg-eGFP(rep), and serves as the target gene in this experiment; the method for integrating this plasmid has been described previously ***(4)***.

To maintain the yeast for TNE experiments, the following steps should be taken:

1. Streak a single colony on an Aureobasidin plate (0.5 µg/mL).
2. Incubate at 30°C for 3 d and store at 4°C.
3. Restreak the plate every week to maintain healthy cells (*see* **Note 1**).

3.3. Growth Conditions of the Yeast Cells

The targeting experiment uses 2×10^7 cells/mL per sample. To culture:

1. Inoculate four flasks containing 10 mL of YPD and 5 µL of 0.5 mg/mL stock of Aureobasidin A with one medium-sized colony of yeast.
2. Incubate these flasks for 16 h at 30°C and 300 rpm.
3. Centrifuge the cells at 3000*g* for 5 min, resuspend in 40 mL of room temperature YPD, and resuspend to an OD_{600} of 0.25.
4. Incubate the cultures shaking at 30°C for another 3 h.
5. At 3 h, take the OD_{600} and monitor this every hour until the cells reach an OD_{600} of 0.65–0.7 or 2×10^7 cells/mL (*see* **Note 2**).

3.4. Yeast Competent Cells and Electroporation

The following procedure describes the protocol for producing competent yeast cells prior to electroporation:

1. Yeast cells are grown as described and centrifuged at 3000*g* for 5 min at 4°C.
2. Cells are resuspended in 1 mL of YPD and 1 *M* DTT to a final concentration of 25 µ*M*.
3. Incubate at 30°C with shaking at 300 rpm for 20 min.
4. Wash the cells with 25 mL of cold Millipore purified dH_2O and centrifuge at 3000*g* for 5 min at 4°C.
5. Pour off the supernant and repeat step 4.
6. Pour off the supernant and resuspend in 1 mL of cold 1 *M* sorbitol.
7. Transfer the sample to a 1.7-mL Eppendorf and centrifuge at 5000*g* for 5 min.
8. Aspirate the supernant and gently resuspend the cells in 120 µL of cold 1 *M* sorbitol (*see* **Note 3**).
9. Aliquot 40 µL of the competent yeast cells into prechilled 0.7-mL microcentrifuge tubes.
10. Add the oligonucleotides in the volumes corresponding to the microgram amounts described in **Table 1**. Here, we demonstrate the effect of nonspecific or carrier DNA on TNE. Controls, which contain no nonspecific (Kan3S/70) oligonucleotide represent the standard reaction condition.

Table 1
Amounts of Oligonucleotides (in µg)[a]

Hyg3S/74NT	*Kan3S/70*
0	0
0	10
1	9
2	8
3	7
4	6
5	5
6	4
7	3
8	2
9	1
10	0
1	0
2	0
3	0
4	0
5	0
6	0
7	0
8	0
9	0

[a]The yeast cells are aliquoted in 40 µL of sample per tube. The oligos are added to the samples in the microgram amounts shown. The final volume of each sample should be the same.

11. Add water to the samples to maintain the same final volume.
12. Place the samples on ice for 5 min.
13. Transfer the samples to prechilled 2-mm gap cuvets and electroporate at 1.5 kV, 200 Ω, 25 µF, 1 pulse, 5 ms/pulse.
14. Transfer each sample to a separate culture tube containing 1.5 mL YPD, 1.5 mL 2 *M* sorbitol and 5 µL of the stock 0.5 mg/mL Aureobasidin A.
15. Incubate the tubes at 30°C shaking at 300 rpm for 16 h.

3.5. Selection of the Corrected Hygromycin Gene

To determine the efficiency of TNE of the hygromycin mutation, it is necessary to select by plating on YPD/hygromycin plates. The cells are also plated

Table 2
Mutated Hyg Targeted With a Specific Hyg3S/74NT in the Presence or Absence of the Nonspecific Oligo Kan3S/70G[a]

Sample	Avg. Hyg^r	Avg. Aureo^r/10^5	Avg. CE/10^5	Avg. fold	% Fold variance
0/0	0	108	0	0	0
0/10	25	120	0.208	0.08	0.0006
1/9	307	128	2.4	2.4	0.023
2/8	359	134	2.68	6.5	0.82
3/7	464	166	2.8	18.7	0.38
4/6	628	137	4.6	25.3	0.26
5/5	178	146	1.22	2.3	0.022
6/4	347	171	2.03	4	0.05
7/3	202	153	1.32	0.9	0.01
8/2	282	154	1.83	0.98	0.01
9/1	352	89	3.96	2.6	0.026
10/0	460	168	2.73	1	
1/0	0	136	0	1	
2/0	65	157	0.414	1	
3/0	26	179	0.15	1	
4/0	35	192	0.182	1	
5/0	56	104	0.54	1	
6/0	102	195	0.52	1	
7/0	251	172	1.46	1	
8/0	324	173	1.87	1	
9/0	314	207	1.52	1	

[a]The first column shows the oligo amounts (specific/nonspecific). The average fold was determined by dividing the correction efficiency (CE) of the nonspecific and specific oligos mixed by the corresponding specific oligo alone. The variance was determined by averaging the fold changes among all samples and presenting the percent difference. Samples 3/7 µg and 4/6 µg display the highest levels of stimulation.

on aureobasidin plates to determine how many cells survive electroporation (*see* **Note 4**). The ratio of Hyg^r colonies to aureo^r colonies generates the correction efficiency index (CE). This section describes how the cells are plated for selection of TNE activity within the cells.

1. Spin the cells down at 3000*g* for 5 min at 4°C.
2. Resuspend in 1 mL of YPD.
3. Plate 200 µL of cells on YPD-hygromycin plates (300 µg/mL).
4. Perform a 10^5 dilution and plate 200 µL on Aureobasidin plates (0.5 µg/mL).
5. Incubate the plates inverted for 3 d at 30°C.
6. Count the colonies and determine correction efficiency (*see* **Note 5** and **Table 2**).

3.6. Visualization of the Corrected Cells

The hygromycin gene is fused with an enhanced green fluorescent gene (eGFP). When the point mutation in the fusion gene is corrected, green fluorescent protein is produced and can be visualized using confocal microscopy. In this experiment the cells were viewed using a Zeiss inverted 100M Axioskop equipped with a Zeiss 510 LSM confocal microscope with a Coherent krypton argon laser and a helium neon laser. The laser excitation was 543 nm and the emission was a 560 nm longpass filter (**Fig. 2**). To prepare the cells, the following steps are taken:

1. Inoculate 1 mL of 1 *M* sorbitol with colonies isolated from a hygromycin plate.
2. Vortex this sample to resuspend the colony thoroughly.
3. Add 50 µL of this sample to 100 µL of 1 *M* sorbitol in a sterile Lab-Tek II chambered coverglass system.
4. Visualize cells (*see* **Note 6**).

3.7. Colony PCR and Sequencing

1. To ensure that the Hygromycin gene was corrected at the desired codon, colonies from hygromycin plates can be picked and added to a PCR reaction mix (1× PCR amplification buffer, 300 µ*M* dNTP, 0.2 pmol each of primers pAUR123F and pHyg1560R, and 3.5U *Taq* polymerase) to a final volume of 25 µL.
2. The samples are preheated for 2 min at 95°C and then 30 cycles of 95°C/20 s, 55°C/30 s, 72°C/1 min.
3. A final elongation at 72°C for 7 min and a final cool down to 4°C completes the reaction.
4. The samples are loaded onto a 0.7% agarose gel, and a band of 520 bp can be visualized with an Alpha Imager (data not shown).
5. The PCR products are then purified using a Qiagen PCR purification kit and sequenced in the region of the targeted base by the Sanger dideoxy sequencing method using an ABI Prism 3100 genetic analyzer (**Fig. 3**).

4. Notes

1. It is very important to maintain a fresh plate of the yeast cells. We suggest making glycerol stocks of the yeast once the target integration has been confirmed (store at –80°C). Two weeks is the maximal recommended time to keep the stock plate.
2. When culturing the cells, it is critical to monitor their growth for doubling time, optimally 90–120 min. It is also important to monitor the growth so that the cells can be targeted at the optimal OD_{600} of 0.65. If cells are targeted below or above this optical density, the correction efficiencies will be negatively impacted.
3. Aspirating the 1 *M* sorbitol before the final resuspension will help to ensure that an accurate density of cells is used for the electroporation process. When resuspending the cells, it is important to mix the cells in a circular fashion instead of using vertical pipeting. The cells are sensitive to disruption at this point, and if

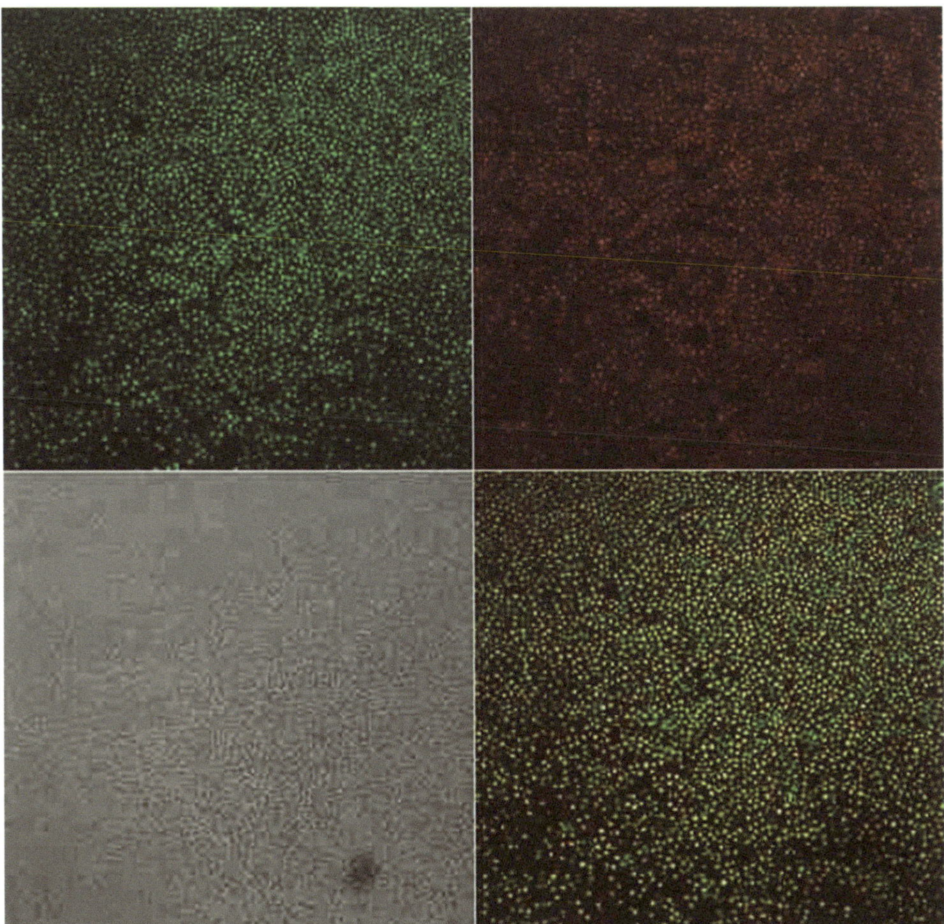

Fig. 2. Confocal image at a 200× magnification of yeast targeted with 3 µg Hyg3S/74NT and 7 µg of Kan3S/70NT. Top left: cells viewed with an FITC filter to determine the green fluorescence; top right: the same cells examined with a rhodamine filter to determine the amount of autofluorescence; bottom left: cells examined without filters; bottom right: superimposed images to minimize autofluorescence and visualize eGFP expression indicating corrected cells.

they are disturbed too much in this part of the experiment, cells may die and the results will be skewed.

4. It is important to cool the media before adding the antibiotic, and contamination can be common during this period. All solutions should be kept sterile. The plates used in the experiment should be fresh, no more than 2 wk old. These plates should be stored at 4°C and should be warmed before the plating process begins.

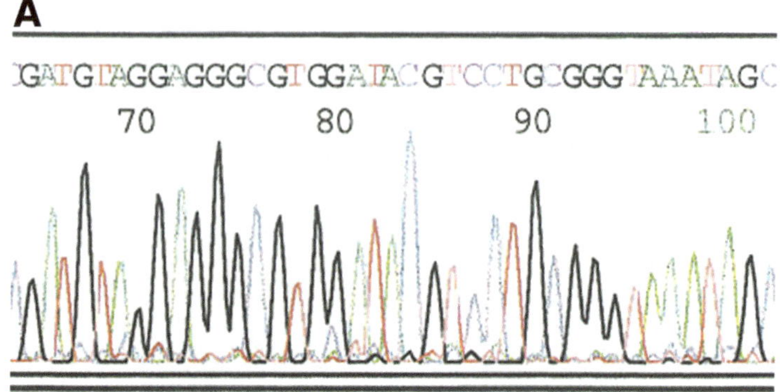

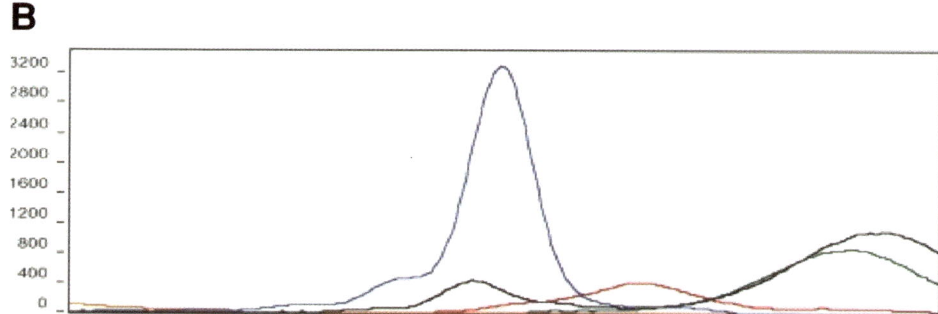

Fig. 3. The results from sequencing and SnAPSHOT sequencing, which will amplify a region of the target gene, show the single base change that was made by the oligonucleotide. **(A)** Sequence of the corrected Hyg target. The base change is located at position 83 in this figure. **(B)** The single base pair change at the nucleotide level; the base has been converted from a G to a C.

5. There will probably be some variation in the correction efficiencies from day to day, which is a result of minor differences encountered from experiment to experiment. However, the trends should remain the same and it is urged that one determine the fold change for each sample because *this should be similar* for all experiments. Statistical analyses using ANOVA by SPSS6.14 is strongly encouraged.
6. Yeast strains can often have a background of fluorescence (**Fig. 2**). For this reason it was necessary to visualize the cells under two filters, one to determine green fluorescence, which is what is screened for, and another to pick up another fluorescent tag, which is not present within the cell. In this experiment, a fluorescein isothiocyanate filter was used to determine green fluorescence, and a rhodamine filter was used to determine any red fluorescence, which would indicate the background level of fluorescence. The background level of red fluorescence

along with the green fluorescence will help determine what cells are truly corrected and those that are autofluorescing. Proteins can accumulate in lysosomes in cells and give this background. To distinguish true fluorescence from background, it is necessary to superimpose the filtered images. When examining the cells, the yellow color or a merger of the two should be localized at a small region within the cytoplasm. This is an indication of autofluorescence. Green fluorescence should be diffused throughout the cell, as well as at the periphery of the cell and should stand out without any yellow color.

References

1. Niidome, T. and Huang, L. (2002) Gene therapy progress and prospects: nonviral vectors. *Gene Ther.* **9,** 1647–1652.
2. Sin, F. Y., Walker, S. P., Symonds, J. E., Mukherjee, U. K., Khoo, J. G., and Sin, I. L. (2000) Electroporation of salmon sperm for gene transfer: efficiency, reliability, and fate of transgene. *Mol. Reprod. Dev.* **56,** 285–288.
3. Parekh-Olmedo, H., Drury, M., and Kmiec, E. B. (2002) Targeted nucleotide exchange in *Saccharomyces cerevisiae* directed by short oligonucleotides containing locked nucleic acids. *Chem. Biol.* **9,** 1073–1084.
4. Liu, L., Cheng, S., van Brabant, A. J., and Kmiec, E. B. (2002) Rad51p and Rad54p, but not Rad52p, elevate gene repair in *Saccharomyces cerevisiae* directed by modified single-stranded oligonucleotide vectors. *Nucleic Acids Res.* **31,** 2742–2750.
5. Kmiec, E. B. (1999) Gene therapy. *Am. Sci.* **87,** 240–247.
6. Liu, L., Rice, M. C., Drury, M., Cheng, S., Gamper, H., and Kmiec, E. B. (2002) Strand bias in targeted gene repair is influenced by transcriptional activity. *Mol. Cell Biol.* **22,** 3852–3863.

IV

Biochemistry of Recombination

14

Chromatin Immunoprecipitation to Investigate Protein–DNA Interactions During Genetic Recombination

Tamara Goldfarb and Eric Alani

Summary

Chromatin immunoprecipitation is a technique that allows one to examine the in vivo localization of proteins to DNA. This technique is well suited for studying genetic recombination since it can provide both a temporal and spatial assessment of the dynamic association of proteins with DNA in both wild-type and mutant backgrounds. To perform this procedure, cells undergoing a synchronous recombination event are treated with a crosslinking agent. Following cell lysis and shearing of the DNA, immunoprecipitation is used to isolate the protein of interest, along with any DNA that is crosslinked to the protein. Polymerase chain reaction (PCR) is then used to determine the relative amounts of DNA associated with the protein of interest throughout the recombination event. This in vivo chemical crosslinking technique can be used to localize proteins to both double-strand breaks and recombination intermediates.

Key Words: chromatin immunoprecipitation, recombination, mating-type switching, Msh2p, Msh3p, Rad1p, Rad10p, mismatch repair, nonhomologous end

1. Introduction

Genetic recombination plays an instrumental role in protecting the genome from DNA-damaging agents and is an integral component of the life cycle of both prokaryotes and eukaryotes. In the budding yeast *Saccharomyces cerevisiae*, the genetic recombination events that lead to mating-type switching have been very well characterized at the molecular level (reviewed in refs. ***1–5***; **Fig. 1**). This, along with the ability to synchronously induce double-strand break (DSB) formation through the use of a galactose-inducible HO endonuclease, make mating-type switching an ideal system in which to monitor the localization of proteins involved in genetic recombination by using a

From: *Methods in Molecular Biology, vol. 262, Genetic Recombination: Reviews and Protocols*
Edited by: A. S. Waldman © Humana Press Inc., Totowa, NJ

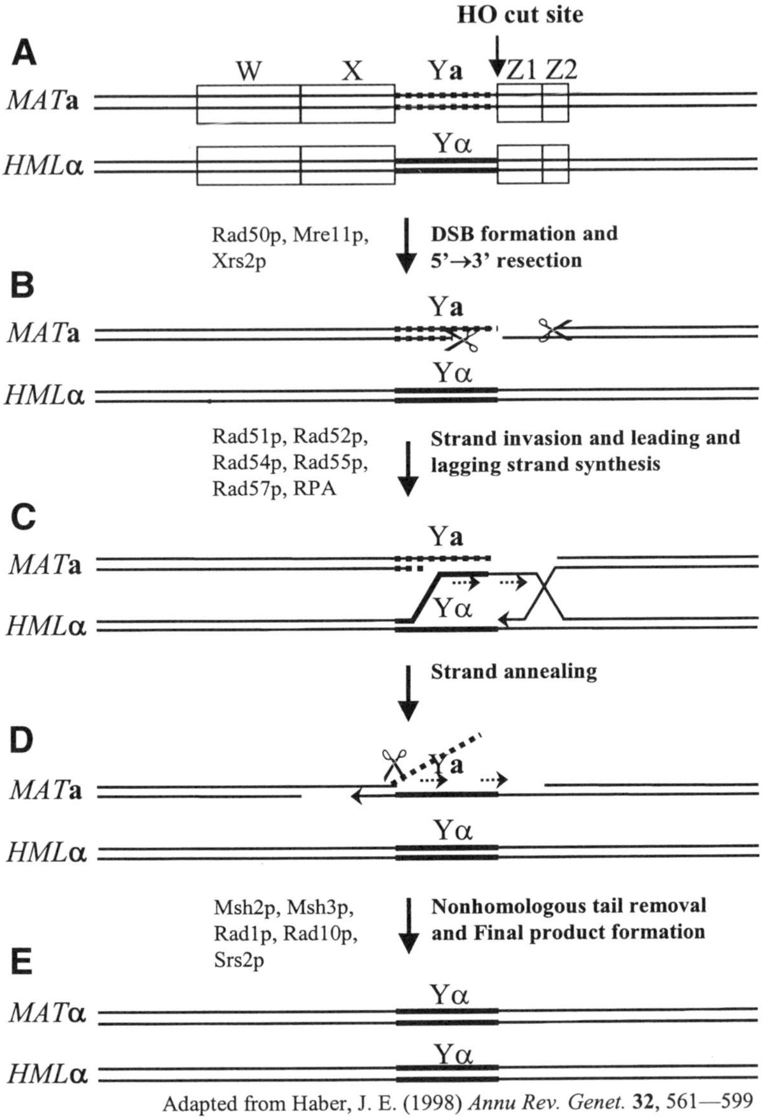

Fig. 1. Synthesis-dependent strand annealing model of mating-type switching. Shown is one model for mating type switching. The HO endonuclease creates a DSB in the Z1 region of *MAT* (**A**). The break is processed by 5' → 3' exonucleases to yield 3' single-stranded ends (**B**). The 3' end invades donor sequences, followed by both leading and lagging strand repair synthesis (**C**). The factors implicated in leading and lagging strand synthesis include PCNA, RFC, Polδ, Polε, Polα, Pri2p, and Rad27p *(1)*. Following strand annealing, the nonhomology is removed by Rad1p-Rad10p, with the help of Msh2p-Msh3p and Srs2p (**D**). Repair of the break can then be completed (**E**). Factors implicated in the individual steps of mating type switching are shown *(1)*.

chromatin immunoprecipitation technique. This method can be adapted to monitor the localization of a variety of proteins to synchronously induced recombination events.

In their haploid state, budding yeast possess one of two possible mating-types, **a** or α. The mating of two haploid cells of opposite mating type results in the formation of a diploid cell. The diploid state allows yeast to undergo meiosis and sporulation when they are placed in nutritionally limiting environments (reviewed in ref. ***2***). The mating type of a cell is determined by the allele present at the *MAT* locus on chromosome III. The Y**a** allele contains genetic information specific for **a** cells, whereas cells possessing the Yα allele at the *MAT* locus express the α mating type. These sequences, along with some of the flanking DNA sequences, are present at two transcriptionally silent donor loci on chromosome III, named HMR**a** and HMLα. The switching event involves a unidirectional transfer of information from either the HMR**a** or HMLα donors to the *MAT* locus. MAT**a** cells preferentially use *HML*α as a donor, whereas *MAT*α cells preferentially use *HMR***a** as their donor ***(6–9)***.

The mating-type switching event is initiated by an HO endonuclease-induced double-strand break (DSB) ***(10)*** within the Z1 region of *MAT* (**Fig. 1A**). Chromatin structure surrounding the donor loci prevents DSB formation in these regions. The break is processed by 5'→3' exonucleases, resulting in 3' single-stranded ends. Because the DSB is formed immediately adjacent to the Y sequences, only one of the two 3' ends shares homology with its donor sequences. The homologous end invades its donor to begin the gene conversion event. To complete repair of the break, the nonhomologous Y sequence must be removed. Removal of this 3' nonhomologous tail is thought to involve the mismatch repair proteins, Msh2p-Msh3p, the Srs2p helicase, and the Rad1p-Rad10p endonuclease ***(11–19)***. Sugawara et al. ***(12)*** have proposed a model in which Msh2p-Msh3p may act to stabilize and/or recruit Rad1p-Rad10p to substrates containing nonhomologous ends. More recent work by Evans et al. ***(20)*** has further suggested that Msh2p-Msh3p may also be acting at an early stage during recombination, before the formation of nonhomologous tails. Although Msh2p-Msh3p and Rad1p-Rad10p are involved in removal of 3' nonhomologous ends, no other components of either the mismatch repair or nucleotide excision repair pathways appear to be directly involved in this process ***(12,14,16)***.

The efficiency and kinetics of mating-type switching can be monitored through genetic and physical assays, respectively, that have been described in detail ***(19,21–24)***. Southern blot analysis can be used to monitor creation of the DSB at the *MAT* locus, its disappearance, and formation of the switched product (**Fig. 2C**). Polymerase chain reaction (PCR) can also be used to detect initial repair synthesis following strand invasion as well as formation of the final

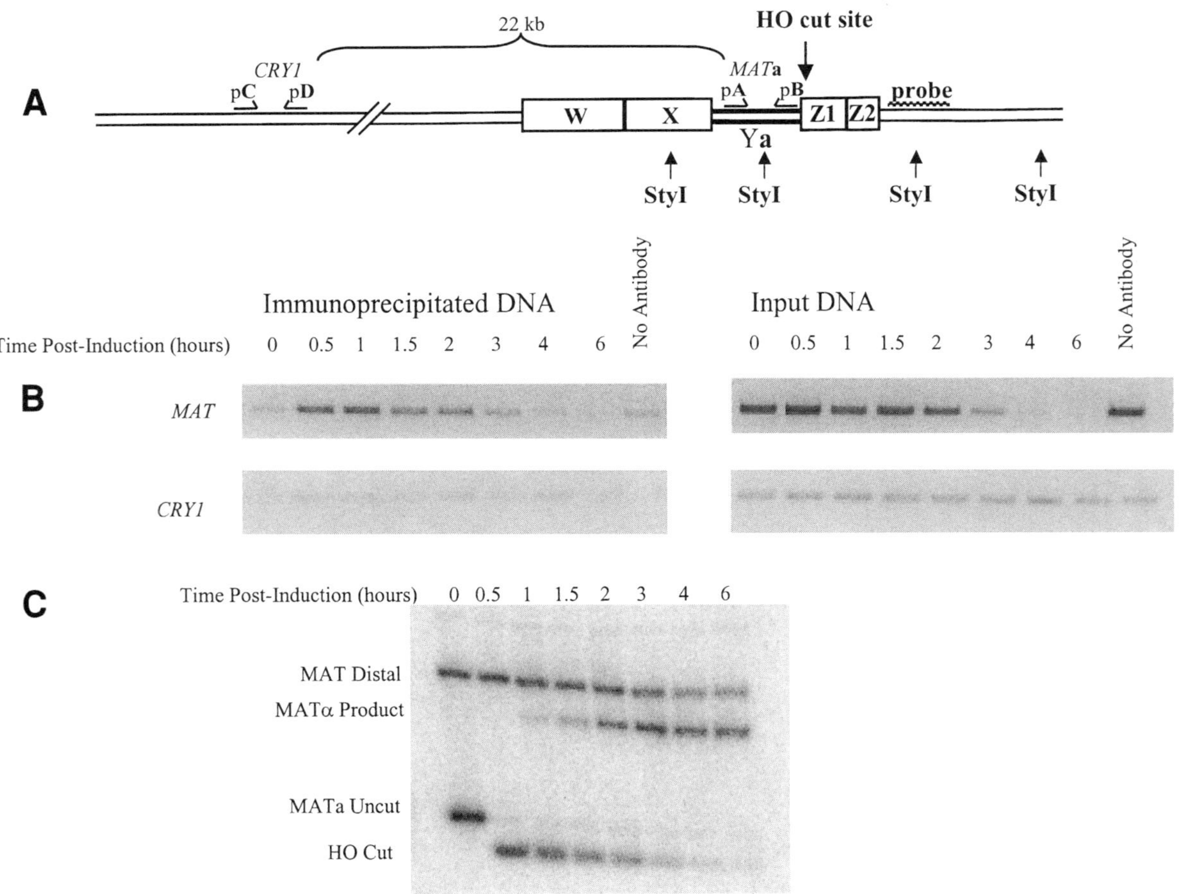
A
22 kb
HO cut site
CRY1
MATa
pC
pD
pA
pB
probe
W
X
Z1
Z2
Ya
StyI
StyI
StyI
StyI
B
Immunoprecipitated DNA
Input DNA
Time Post-Induction (hours)
0 0.5 1 1.5 2 3 4 6
No Antibody
MAT
CRY1
C
Time Post-Induction (hours)
0 0.5 1 1.5 2 3 4 6
MAT Distal
MATα Product
MATa Uncut
HO Cut

Fig. 2.

switched product *(22)*. Although much is known about the different DNA species formed during both mating-type switching *(19,21–24)* and meiotic recombination *(25–27)*, relatively little is known about the kinetics of association of the recombination proteins with these DNA structures.

Chromatin immunoprecipitation is a technique that allows one to monitor the association of proteins with specific DNA sequences. It has been used extensively in the last few years to study the association of transcription factors and replication proteins to their DNA substrates *(28–31)*. Studies are now beginning to emerge that focus on the dynamic association of proteins with DNA intermediates during recombination events *(20)*. This technique can be used to study the kinetics of association of proteins with DNA sequences. Yeast strains carrying specific mutations can be used to determine the genetic requirements of association as well as possible kinetic alterations of association. Mating-type switching is an ideal system in which to study the localization of proteins to DNA during recombination. It is a well-documented chromosomal recombination event that has the same genetic requirements as other mitotic recombination events *(2)* and shares many factors required in meiotic recombination. The kinetics of localization can be compared with those observed by physical assays, such as Southern blots (**Fig. 2**).

Fig. 2. *(previous page)* PCR from chromatin immunoprecipitation of Msh2-HA_4 associated with recombination intermediates during mating-type switching. (**A**) Schematic diagram of chromosome III indicating primer sets used for PCR amplification, relevant restriction sites, and the probe used for Southern blots. Primers p**A** and p**B** yield a 267-bp product at *MAT***a** and primers p**C** and p**D** amplify a 165-bp product at *CRY1*. (**B**) Timecourse experiment of EAY745 (Δ*ho*, *HMLalpha*, *MAT***a**, Δ*hmr::ADE1*, *ade1-100*, *leu2-3,112*, *lys5*, *trp1::hisG*, *ura3-52*, *ade3::GAL::HO*, *MSH2-HA*$_4$*::LEU2*) indicating association of Msh2-HA_4 near the DSB following HO endonuclease induction. Msh2-HA_4 associates near the break (*MAT*), but not at independent sequences (*CRY1*). Both immunoprecipitated DNA and non-immunoprecipitated DNA (Input) is shown. Input DNA with primers p**A** and p**B** is lost as the gene conversion event is completed and the nonhomologous **Ya** region is removed. The no antibody sample was collected at 1 h after HO induction. PCR reactions were loaded onto a 1.5% 1× TAE agarose gel stained with 0.5 µg/mL ethidium bromide. (**C**) Southern blot analysis reveals the kinetics of mating-type switching. Southern blots were performed on *Sty*I-digested DNA following HO induction as previously described *(19,21–24)*. The **Ya** sequence contains no *Sty*I restriction site so that the switched product can be differentiated from the MAT**a** original strain. The MAT distal band remains unchanged throughout the recombination event and acts as a loading control. The HO cut band appears at the first time point following HO induction, while the switched product begins to appear 1 h after HO induction.

The protocol presented here uses Msh2p as a model to study the in vivo localization of proteins to recombination intermediates using chromatin immunoprecipitation. Previous work using a plasmid-based recombination system has demonstrated that Msh2p localizes to recombination intermediates during DSB repair involving nonhomologous ends ***(20)***. This protocol can be easily adapted to study the localization of any recombination protein to DNA intermediates using an inducible DSB. Examples of other recombination events that could be studied in this way include I-*Sce*I-induced recombination, meiosis, single-strand annealing, HO-induced chromosomal recombination, and nonhomologous end-joining (reviewed in refs. ***2*** and ***3***). A prerequisite for studying the kinetics of protein association to recombination intermediates is that the DSB be synchronously induced in the culture.

In the mating-type switching example illustrated in this chapter, the HO endonuclease is expressed under the control of a galactose promoter that is integrated into the chromosome at the *ADE3* locus. The strain is initially *MAT***a**, and thus uses *HML*α as a donor to repair the DSB. The *HMR***a** locus is deleted to allow for unique PCR primer construction within the Y**a** region. The endogenous *HO* gene is deleted in these strains. For efficient DSB induction, cells are initially grown in rich medium containing lactate and are then induced with galactose. Glucose is added to the medium following a 30-min induction period to repress the galactose induction and to prevent a second round of mating-type switching. Samples are collected at times throughout the recombination event to monitor localization of proteins to DNA. Formaldehyde is added to the samples and acts as a general and nonspecific crosslinking agent, forming protein–protein and protein–DNA crosslinks. The crosslinking is then quenched by the addition of glycine. The cells are lysed under conditions that maintain the protein–DNA and protein–protein crosslinks. To study the localization of proteins to specific DNA sequences, the DNA must be sheared into smaller fragments. This is achieved by sonicating the lysate until the DNA reaches an average fragment size of about 500 bp. DNA fragmentation can be monitored by running a sample of DNA from the sonicated lysate on an agarose gel (**Fig. 3**). To isolate those DNA sequences that are associated with the protein of interest, the sonicated lysate is immunoprecipitated with an antibody directed toward the protein of interest. In this example, Msh2p is tagged with an HA epitope. The tagged protein is immunoprecipitated with a monoclonal antibody to the HA epitope. **Figure 4** demonstrates that the antibody efficiently clears Msh2p from the lysate. The immunoprecipitated sample is then heated to reverse the protein–DNA crosslinks. Following Proteinase K treatment, the free DNA is precipitated and analyzed by PCR. The PCR signal will be enhanced at times when the protein of interest is bound to the DNA contained within the primers used in the reaction. PCR is also performed on total input DNA that is

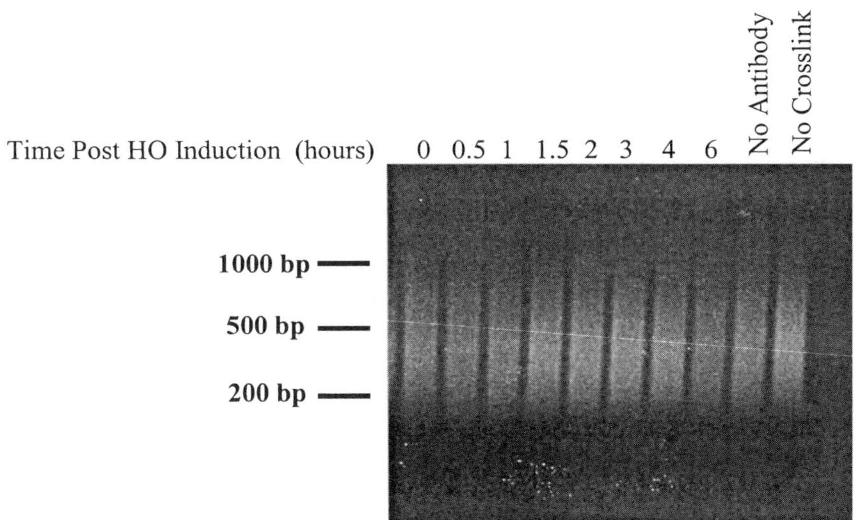

Fig. 3. Sonicated DNA. Sonicated DNA can be visualized to ensure that the fragment size is, on average, 500 bp. DNA samples from a mating-type switching timecourse were processed as described in the text. Twenty microliters of the non-immunoprecipitated DNA were treated with RNase and run on a 1X TAE agarose gel containing 0.5 μg/mL ethidium bromide. The no antibody and no crosslink control samples were collected 1 h after HO induction.

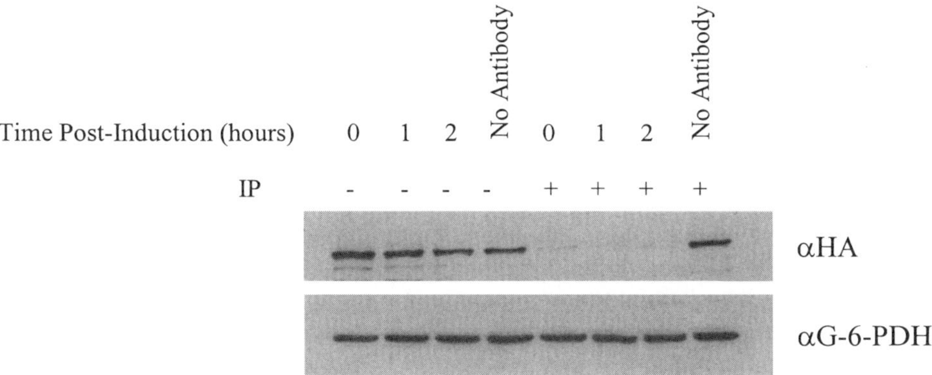

Fig. 4. Msh2-HA_4 is cleared from the lysate following immunoprecipitation. Equal volumes of total cell lysates collected either before or after HO induction were loaded onto an 8% SDS-PAGE gel both prior to (–) and following (+) removal of Msh2-HA_4 from the cell lysate by immunoprecipitation. Western blot analysis with an anti-HA antibody was performed to confirm clearing of Msh2-HA_4 from the lysates following immunoprecipitation *(20)*. The no antibody control sample was collected at 1 h after HO induction. Antibody to glucose-6-phosphate dehydrogenase was used as a loading control.

not subjected to immunoprecipitation to monitor the total amount of the DNA sequence of interest present in the sample throughout the timecourse.

Figure 2B shows that Msh2p is localized adjacent to the HO cut site following formation of the DSB. The PCR signal is lost over time, as this region of nonhomology is removed (**Fig. 1D**). Several controls should be performed to ensure that the PCR signal observed is specific to the recombination reaction and the antibody used and is dependent on crosslinked cells. To perform these controls, samples should be taken at times when the protein of interest is known to participate in recombination. To normalize for the amount of DNA present in each sample and to ensure that any increase in PCR signal is specific to the recombination reaction, PCR reactions can be performed with primers that amplify a region that is not associated with the recombination event. The kinetics of protein localization can also be compared with the formation of DNA intermediates and products. In **Fig. 2C**, Southern blots were performed on DNA samples taken from a timecourse similar to that performed for chromatin immunoprecipitation. When **Fig. 2B** and **Fig. 2C** are compared, it is clear that Msh2p localization corresponds to DSB formation (HO cut).

2. Materials

2.1. Double-Strand Break Induction and Sample Collection

1. YPD plates: 1% yeast extract (Difco), 2% peptone (Difco), 2% dextrose (US Biological), 2% agar (US Biological).
2. Minimal selective media with 2% dextrose *(32)*.
3. YP-lactate media: 1% yeast extract (Difco), 2% peptone (Difco), 2% lactate, pH 5.5. A stock of 40% lactate was made from 85% lactic acid (Sigma). Check the pH of the media before adding the lactate. After adding the lactate, bring the pH of the media back to its original pH using NaOH. Autoclave.
4. 20% Galactose (US Biological), filter-sterilized.
5. Formaldehyde, 37% stock.
6. 2.5 *M* Glycine, filter-sterilized.
7. TBS: 20 m*M* Tris-HCl (Sigma), pH 7.6, 150 m*M* NaCl.
8. 40% Glucose, autoclaved (US Biological).

2.2. Cell Lysis and Sonication of Samples

1. Acid-washed glass beads, 425–600 μm (Sigma).
2. Lysis buffer: 50 m*M* HEPES, pH 7.5, 1 m*M* ethylenediaminetetraacetic acid (EDTA), pH 8.0, 140 m*M* NaCl, 1% Triton X-100, 0.1% sodium deoxycholic acid, 1 mg/mL bacitracin, 1 m*M* benzamidine, 1 m*M* PMSF (phenylmethylsulfonyl fluoride; Sigma). **PMSF is toxic** and light sensitive. It is unstable in aqueous solutions. Store as a 0.1 *M* stock solution in isopropanol at –20°C in the dark.
3. Multivortexer (Eppendorf Mixer 5432).

4. 22G1 needle.
5. Branson Sonifier 250.

2.3. Immunoprecipitation of Cell Lysates

1. Antibody to protein of interest.
2. Protein A or Protein G agarose Beads (Roche).
3. Lysis buffer: as given in **Subheading 2.2.**
4. High salt lysis buffer: 50 m*M* HEPES, pH 7.5, 1 m*M* EDTA, pH 8.0, 500 m*M* NaCl, 1% Triton X-100, 0.1% sodium deoxycholic acid, 1 mg/mL Bacitracin, 1 m*M* benzamidine, 1 m*M* PMSF.
5. Wash buffer: 10 m*M* Tris-HCl (Sigma), pH 8.0, 1 m*M* EDTA, pH 8.0, 250 m*M* LiCl, 0.5% Nonidet-P 40 (Roche), 0.5% sodium deoxycholic acid (Sigma).
6. Elution buffer: 50 m*M* Tris-HCl pH 8.0, 10 m*M* EDTA pH 8.0, 1% SDS.
7. 1× TE: 10 m*M* Tris-HCl, pH 8.0, 1 m*M* EDTA pH 8.0.
8. 1× TE/1% sodium dodecyl sulfate (SDS): 10 m*M* Tris-HCl, pH 8.0, 1 m*M* EDTA, pH 8.0, 1% SDS.
9. 1× TE/0.67% SDS: 10 m*M* Tris-HCl, pH 8.0, 1 m*M* EDTA, pH 8.0, 0.67% SDS.
10. Nutator Mixer (Clay Adams).
11. 26G3/8 Needle.
12. Glycogen.
13. Proteinase K PCR Grade (Roche).

2.4. Extraction of DNA

1. 50% phenol/50% chloroform. **Phenol is toxic** and light sensitive. Store in the dark at 4°C.
2. Ethanol, 100 and 75%.

2.5. PCR

1. 0.5-mL thin-walled PCR tubes.
2. PCR Master Mix (per 50 µL reaction): 30.75 µL ddH_2O; 5 µL 10× thermophilic buffer, without $MgCl_2$ (Perkin Elmer); 5 µL 25 m*M* $MgCl_2$; 4 µL dNTPs 2.5 m*M* each (New England Biolabs); 2 µL primer I, 5 pmol/µL; 2 µL primer II, 5 pmol/µL; 1 µL DNA, 1 µL of immunoprecipitated DNA or 1 µL of a 1:100 dilution of input DNA; and 0.25 µL TaqGold polymerase (5 U/µL; Perkin Elmer).
3. PCR conditions will vary depending on the primers and templates that are used. A sample PCR reaction is as follows:
   Cycle 1: 95°C for 10 min, 55°C for 30 s, 72°C for 1 min.
   Cycles 2–28: 94°C for 1 min, 55°C for 30 s, 72°C for 1 min.
   Final extension: 72°C for 1 min.
4. Primer sequences:
   p**A**: tcaccccaagcacgggcatt (**Ya**)
   p**B**: ctgttgcggaaagctgaaac (**Ya**)
   p**C**: cgccagagttactggtggtatgaagg (*CRY1*)
   p**D**: ggagtcttggttctagtaccacgg (*CRY1*)

3. Methods

3.1. Double-Strand Break Induction and Sample Collection

3.1.1. Preparing Cells for the Timecourse Experiment

1. Patch a single colony taken from a frozen glycerol stock of the appropriate strain onto YPD plates (approx 2 × 2-cm patch) and incubate overnight at 30°C.
2. Resuspend the patch of cells into 20 mL of dextrose-containing selective synthetic minimal media to maintain any markers necessary for the experiment. Incubate overnight at 30°C with proper aeration until the culture is saturated.
3. Make a 1:100 dilution of the culture into 500 mL YP-lactate media and incubate overnight at 30°C with proper aeration until the culture reaches a cell density of approx 1×10^7 cells/mL (*see* **Note 1**).

3.1.2. Timecourse Experiment

4. Begin collecting samples by removing a 45-mL aliquot from the flask. This first sample will be used as the Time = 0 h (no induction) sample. Induce the remaining culture by adding galactose to a 2% final concentration.
5. Pour the 45 mL of culture into a tube containing 0.45 mL of formaldehyde (*see* **Note 2**).
6. Allow cells to crosslink, by incubating on Nutator rocker for 15 min at room temperature.
7. Quench the crosslinking for each 45-mL aliquot by adding 2.5 mL of 2.5 *M* glycine (final concentration of approx 125 m*M*).
8. Incubate on Nutator for 5 min at room temperature.
9. Centrifuge in a tabletop centrifuge at 2800*g* for 3 min and discard supernatant.
10. Wash the pellet twice with 20 mL of ice-cold TBS.
11. Resuspend the pellet in 1 mL of ice-cold TBS and transfer to a 1.5-mL centrifuge tube.
12. Centrifuge for 1 min at 16,046*g* (13,000 rpm) in a microcentrifuge.
13. Completely remove supernatant and freeze pellet at –80°C.
14. After 30 min of galactose induction, add glucose to a final concentration of 2% to prevent recutting of the DNA after repair of the break is complete.
15. Over the period of the timecourse, the cell density will increase. Over a 6-h timecourse, the cells will typically go through approx two doubling periods. To help maintain a more constant amount of DNA throughout the period of the timecourse, dilute each sample with YP-lactate so that it matches the cell density first measured at time = 0 h.

3.2. Cell Lysis and Sonication of Samples

3.2.1. Cell Lysis

1. Resuspend each cell pellet in 0.5 mL lysis buffer.
2. Transfer the cell lysate to a centrifuge tube containing 0.5 mL of 425–600-µm glass beads.

3. Lyse cells on multivortexer for 40 min at 4°C (*see* **Note 3**).
4. Before sonication, the lysate must be separated away from the glass beads. Invert tubes, and puncture the bottom of each tube with a flaming hot 22G1 needle. Place the punctured tube on top of a 15-mL conical collection tube on ice.
5. Centrifuge the tubes in a tabletop centrifuge at 385*g* for 1 min.
6. Resuspend the cell pellet in the supernatant (*see* **Note 4**).
7. Transfer the lysate to a 1.5-mL microcentrifuge tube for sonication.

3.2.2. Sonication

8. Shear chromatin to an average fragment size of 500 bp through sonication (*see* **Note 5**).
9. To clear lysate, centrifuge samples at 16,046*g* (13,000 rpm) in a microcentrifuge for 10 min at 4°C.
10. Transfer supernatant to a new tube.
11. Pellet cell debris once more by centrifugation of the supernatant at 16,046*g* (13,000 rpm) in a microcentrifuge for 15 min at 4°C.
12. Transfer supernatant to a new tube.
13. Quick-freeze lysates in a dry ice/ethanol bath, and store at –80°C until immunoprecipitation.

3.3. Immunoprecipitation of Cell Lysates

1. Allow cell lysates to thaw on ice.
2. For immunoprecipitation, remove an equal volume of lysate (approx 375 µL) from each tube and put lysate in chilled 1.5-mL microcentrifuge tubes containing the primary antibody (*see* **Note 6**), making sure that there is enough lysate left in each tube for the non-immunoprecipitated controls (*see* **step 3**).
3. For non-immunoprecipitated control, remove 50 µL of lysate to a new tube and add 200 µL of 1× TE/1% SDS.
4. Incubate the immunoprecipitation tubes at 4°C with slow rocking on a Vari-Mix (or equivalent machine) for 1 h.
5. Prepare either Protein G-agarose or Protein A-agarose beads (*see* **Note 7**) while the lysates are incubating with the antibody. Gently centrifuge the appropriate volume of agarose beads at 854*g* (3000 rpm) for 1 min in a microcentrifuge. Remove the storage buffer. Wash the beads with at least 2 vol of lysis buffer. Repeat once more. Make a 50% slurry by adding 1 vol of lysis buffer to the agarose beads.
6. Add 35 µL of the 50% agarose bead slurry to each sample (*see* **Notes 6** and **8**).
7. Incubate the mixture at 4°C with slow rocking on a Vari-Mix (or equivalent machine) for 1 h.
8. Gently pellet the agarose beads in a microcentrifuge at 854*g* (3000 rpm) for 1 min and remove supernatant. At this point, the immunoprecipitated lysate can be used to look for clearing of the protein of interest from the lysate (*see* **Note 6**). Remove remaining supernatant with a 26G3/8 needle.

9. Wash beads with 1 mL of lysis buffer by placing tubes on a Nutator with gentle rocking for 5 min. Pellet beads as instructed in **step 8**, and remove wash buffer with an aspirator attached to a 26G3/8 needle.
10. Wash beads three times more with lysis buffer.
11. Wash once with 1 mL high salt lysis buffer.
12. Wash once with 1 mL wash buffer.
13. Wash twice with 1 mL TE.
14. Elute precipitate from beads by adding 100 µL elution buffer and incubating at 65°C for 15 min.
15. Pellet beads by centrifugation at 16,046*g* (13,000 rpm) for 1 min and remove eluate to a fresh tube.
16. Add 150 µL TE/0.67% SDS to tubes containing agarose beads and incubate at 65°C for 10 min.
17. Pellet beads by centrifugation at 16,046*g* (13,000 rpm) for 1 min and remove eluate to the tubes containing the eluate with the elution buffer.
18. Incubate elution tubes at 65°C for at least 6 h to reverse crosslinks.
19. After the crosslinks are reversed, add 250 µL TE containing 0.08 µg/µL glycogen and 0.4 µg/µL Proteinase K to each tube.
20. Incubate for at least 2 h at 37°C.

3.4. Extraction of DNA

1. Add 55 µL of 4 *M* lithium chloride to each tube.
2. Add 500 µL of 50% phenol:50% chloroform to each tube, and mix well.
3. Centrifuge tubes at 16,046*g* (13,000 rpm) in a microcentrifuge for 10 min. Remove aqueous phase to new tubes
4. To precipitate DNA, add 1 mL ethanol to tubes. Mix well. Incubate for 15 min at room temperature.
5. Pellet DNA by centrifugation at 16,046*g* (13,000 rpm) in a microcentrifuge for 10 min. Remove supernatant (*see* **Note 9**).
6. Wash pellet with 1 mL of 75% ethanol. Centrifuge at 16,046*g* (13,000 rpm) for 10 min. Remove supernatant and allow pellet to dry completely.
7. Completely resuspend DNA in 50 µL TE.

3.5. PCR

1. PCR reactions of immunoprecipitated DNA are performed in 50 µL reaction volumes containing 1× thermophilic buffer, 2.5 m*M* $MgCl_2$, 0.2 m*M* of each dNTP, 10 pmol of each primer, 1.25 U TaqGold, and 1 µL of DNA (equivalent to 1/50th of the total DNA).
2. PCR reactions of the input DNA are performed similarly to those of the immunoprecipitated DNA, except that 1 µL of a 1:100 dilution of the input DNA is used. Primers should be designed to give approx 250-bp amplicons.
3. Typical PCR conditions are as follows. Cycle 1: 95°C for 10 min, 55°C for 30 s, 72°C for 1 min. Cycles 2–28: 94°C for 1 min, 55°C for 30 s, 72°C for 1 min. Final extension: 72°C for 1 min (*see* **Note 10**).

4. PCR reactions can be run on agarose gels and stained with ethidium bromide for visualization. Alternatively, RealTime PCR (A. K. Boehm, A. Saunders, J. Werner, and J. T. Lis, personal communication) can be used to obtain more quantitative results.

4. Notes

1. Growth times may vary from strain to strain. It may be necessary to make several different dilutions of the culture for overnight growth to reach the desired cell density. It is important that cells be in logarithmic growth phase for proper induction of the HO endonuclease.
2. It may be necessary to try a few different formaldehyde concentrations to find the appropriate concentration that yields the highest signal to background ratio. Typical concentrations used vary from 0.37% to 1.5% final concentration.
3. Conditions for complete cell lysis may need to be optimized. Cells can be visualized under the microscope to ensure that efficient lysis occurred. Cells that have been lysed appear as "ghosts." Most cells should appear as "ghosts" under the microscope.
4. Some proteins in the crosslinked material are found in the insoluble pellet at this stage. It may therefore be necessary to resuspend the pellet in the supernatant before sonication.
5. Sonication conditions required to yield an average fragment size of 500 bp will vary depending on the machine used and must be optimized. The conditions used in these experiments were as follows: Branson Sonifier 250, 100% Duty Cycle, Output Setting = 1.5. Three bursts were performed on each sample with a burst time of 12 s each. Samples were kept in a chilled block during the sonication bursts. Lysates must be incubated on ice for at least 2 min between bursts to prevent overheating of the sample. To determine the fragment size following sonication, 20 µL of DNA treated with RNase can be run on an agarose gel (**Fig. 3**).
6. One must determine how much antibody and protein agarose beads are required to clear the protein of interest from the lysate. This will depend on the concentration of protein in the lysate and the antibodies being used. Western blots can be performed on lysates both prior to and after incubation with the antibody and beads to determine the efficiency of clearing of the lysate (**Fig. 4**).
7. Using Protein A or Protein G in this step will depend on the relative affinity of the antibody for these reagents.
8. It can be difficult to pipet the Protein A and/or Protein G-agarose bead slurry using a small pipet, as the beads may not properly pass through the tip opening. One can more easily pipet the appropriate volume by cutting off the very tip of the pipet.
9. The DNA pellets from the immunoprecipitated samples often do not adhere well to the sides of the tubes. It is best to take extra caution when removing the supernatant, to ensure that the DNA pellet is not removed from the tube. Do not aspirate.
10. PCR conditions must be optimized for each primer set, and serial dilutions of DNA should be performed to ensure that each reaction is in the linear range.

Acknowledgments

We are grateful to Neal Sugawara and James Haber for providing the strains necessary for the mating-type switching experiments as well as for helpful discussions. We thank Jennifer Surtees and Juan Lucas Argueso for critical reading of the manuscript.

T.G. was supported by a Natural Sciences and Engineering Research Council of Canada PGSB Award. E.A. was supported by National Institutes of Health grant GM53085.

References

1. Haber, J. E. (1998) Mating-type gene switching in *Saccharomyces cerevisiae. Annu. Rev. Genet.* **32,** 561–599.
2. Pâques, F. and Haber, J. E (1999) Multiple pathways of recombination induced by double-strand breaks in *Saccharomyces cerevisiae. Microbiol. Mol. Biol. Rev.* **63,** 349–404.
3. Symington, L. S. (2002) Role of *RAD52* epistasis group genes in homologous recombination and double-strand break repair. *Microbiol. Mol. Biol. Rev.* **66,** 630–670.
4. Haber, J. E. (1992) Mating-type gene switching in *Saccharomyces cerevisiae. Trends Genet.* **8,** 446–452.
5. Herskowitz, I. (1988) Life cycle of the budding yeast *Saccharomyces cerevisiae. Microbiol. Rev.* **52,** 536–553.
6. Klar, A. J., Hicks, J. B., and Strathern, J. N. (1982) Directionality of yeast mating-type interconversion. *Cell* **28,** 551–561.
7. Wu, X. and Haber, J. E. (1995) *MAT*a donor preference in yeast mating-type switching: activation of a large chromosomal region for recombination. *Genes Dev.* **9,** 1922–1932.
8. Wu, X., Moore, K., and Haber J. E. (1996) Mechanism of *MAT*α donor preference during mating-type switching of *Saccharomyces cerevisiae. Mol. Cell. Biol.* **16,** 657–668.
9. Wu, X., Wu, C., and Haber J. E. (1997) Rules of donor preference in *Saccharomyces* mating-type gene switching revealed by competition assay involving two types of recombination. *Genetics* **147,** 399–407.
10. Strathern, J. N., Klar, A. J. S., Hicks, J. B., et al. (1982) Homothallic switching of yeast mating type cassettes is initiated by a double-stranded cut in the *MAT* locus. *Cell* **31,** 183–192.
11. Fishman-Lobell, J. and Haber, J. E. (1992) Removal of nonhomologous DNA ends in double-strand break recombination: the role of the yeast ultraviolet repair gene *RAD1*. *Science* **258,** 480–484.
12. Sugawara, N., Pâques F., Colaiacovo, M., and Haber, J. E. (1997) Role of the *Saccharomyces cerevisiae* Msh2 and Msh3 repair proteins in double-strand break-induced recombination. *Proc. Natl. Acad. Sci. USA* **94,** 9214–9219.
13. Pâques, F. and Haber, J. E. (1997) Two pathways for removal of nonhomologous DNA ends during double-strand break repair in *Saccharomyces cerevisiae. Mol. Cell. Biol.* **17,** 6765–6771.

14. Saperbaev, M., Prakash, L., and Prakash, S. (1996) Requirement of mismatch repair genes MSH2 and MSH3 in the RAD1-RAD10 pathway of mitotic recombination in *Saccharomyces cerevisiae*. *Genetics* **142,** 727–736.
15. Kirkpatrick, D. T. and Petes, T. D. (1997) Repair of DNA loops involves DNA-mismatch and nucleotide-excision repair proteins. *Nature* **387,** 929–931.
16. Ivanov, E. L. and Haber, J. E. (1995) *RAD1* and *RAD10*, but not other excision repair genes, are required for double-strand break-induced recombination in *Saccharomyces cerevisiae*. *Mol. Cell. Biol.* **15,** 2245–2251.
17. Evans, E. and Alani, E. (2000) Roles for mismatch repair factors in regulating genetic recombination. *Mol. Cell. Biol.* **20,** 7839–7844.
18. Prakash, S. and Prakash, L. (2000) Nucleotide excision repair in yeast. *Mutat. Res.* **451,** 13–24.
19. Holmes, A. M. and Haber, J. E. (1999) Double-strand break repair in yeast requires both leading and lagging strand DNA polymerases. *Cell* **96,** 415–424.
20. Evans, E., Sugawara, N., Haber, J. E., and Alani, E. (2000) The *Saccharomyces cerevisiae* Msh2 mismatch repair protein localizes to recombination intermediates in vivo. *Mol. Cell.* **5,** 789–799.
21. Connolly, B., White, C. I., and Haber, J. E. (1988) Physical monitoring of mating type switching in *Saccharomyces cerevisiae*. *Mol. Cell. Biol.* **8,** 2342–2349.
22. White, C. I. and Haber J. E. (1990) Intermediates of recombination during mating type switching in *Saccharomyces cerevisiae*. *EMBO J.* **9,** 663–673.
23. Ivanov, E. L., Sugawara, N., White, C. I., Fabre, F., and Haber, J. E. (1994) Mutations in *XRS2* and *RAD50* delay but do not prevent mating-type switching in *Saccharomyces cerevisiae*. *Mol. Cell. Biol.* **14,** 3414–3425.
24. Holmes, A. and Haber, J. E. (1999) Physical monitoring of HO-induced homologous recombination. *Methods Mol. Biol.* **113,** 403–415.
25. Allers, T. and Lichten, M. (2001) Intermediates of yeast meiotic recombination contain heteroduplex DNA. *Mol. Cell* **8,** 225–231.
26. Allers, T. and Lichten M. (2001) Differential timing and control of noncrossover and crossover recombination during meiosis. *Cell* **106,** 47–57.
27. Hunter, N. and Kleckner, N. (2001) The single-end invasion: an asymmetric intermediate at the double-strand break to double-Holliday junction transition of meiotic recombination. *Cell* **106,** 59–70.
28. Aparicio, O. M., Weinstein, D. M., and Bell, S. P. (1997) Components and dynamics of DNA replication complexes in *S. cerevisiae*: redistribution of MCM proteins and Cdc45p during S phase. *Cell* **91,** 56–69.
29. Strahl-Bolisnger, S., Hecht, A., Luo, K., and Grunstein, M. (1997) *SIR2* and *SIR4* interactions differ in core and extended telomeric heterochromatin in yeast. *Genes Dev.* **11,** 83–93.
30. Meluh, P. and Koshland, D. (1997) Budding yeast centromere composition and assembly as revealed by in vivo cross-linking. *Genes Dev.* **11,** 3401–3412.
31. Andrulis, E. D., Werner, J., Hazarian, A., Erdjument-Bromage, H., Tempest, P., and Lis, J. T. (2002) The RNA processing exosome is linked to elongating RNA polymerase II in *Drosophila*. *Nature* **420,** 837–841.
32. Rose, M. D., Winston, F., and Hieter, P. (1990) *Methods in Yeast Genetics*. Cold Spring Harbor Laboratory Press, Cold Spring Harbor, NY.

15

Holliday Junction Branch Migration and Resolution Assays

Angelos Constantinou and Stephen C. West

Summary

Holliday junctions are central intermediates in the process of genetic recombination; they form as a consequence of a reciprocal exchange of strands between paired DNA molecules. Enzymes that specifically recognize and process these junctions are necessary for the formation of recombinant products. In the methods described here, we detail the in vitro construction of two types of Holliday junction: (1) a small synthetic junction formed by the annealing of partially complementary oligonucleotides; and (2) a true recombination intermediate structure formed by RecA protein-mediated strand exchange. The use of these substrates in assays designed to detect Holliday junction branch migration and resolution activities is described.

Key Words: homologous recombination, double-strand break repair, recombination intermediates, RuvABC

1. Introduction

Recombination between homologous DNA molecules provides a mechanism for the repair of double-strand breaks in DNA in mitotic cells and for the exchange of genetic information at meiosis. A central intermediate structure in this process is a four-way junction that forms by the exchange of strands between homologous duplexes. Robin Holliday first proposed the idea that recombining molecules were linked by a crossover *(1)*, hence the recombination intermediate is often referred to as a Holliday junction. Recent studies have shown that Holliday junctions also arise when the progression of replication forks is impaired and that recombination plays an important role in replication restart *(2–5)*. As a consequence, DNA transactions involving Holliday junction intermediates are central to the maintenance of genome stability in mammalian cells. To facilitate the identification and biochemical analysis of Holliday junction processing activities in higher eukaryotes, we describe the

From: *Methods in Molecular Biology, vol. 262, Genetic Recombination: Reviews and Protocols*
Edited by: A. S. Waldman © Humana Press Inc., Totowa, NJ

preparation of two types of Holliday junction intermediates. These have been used extensively as substrates in studies of enzymes that process Holliday junctions in *Escherichia coli* ***(6–26)***, yeast ***(27–31)***, archaea ***(32–37)***, and mammalian cells ***(38–42)***.

2. Materials

2.1. Synthetic Holliday Junctions

1. Synthetic Holliday junction X26 is made up from four oligonucleotides, X26.1–X26.4. Residues in bold constitute the 26-bp homologous core through which the junction can move.

   X26.1: 5'-CCGCTACCAGTGATCAC**CAATGGATTGCTAGGACATCTTT GC**CCACCTGCAGGTTCACCC-3'

   X26.2: 5'-TGGGTGAACCTGCAGGTG**GGCAAAGATGTCCTAGCAATC CATTG**TCTATGACGTCAAGCT-3'

   X26.3: 5'-GAGCTTGACGTCATAGA**CAATGGATTGCTAGGACATCTTT GC**CGTCTTGTCAATATCGGC-3'

   X26.4: 5'-TGCCGATATTGACAAGAC**GGCAAAGATGTCCTAGCAATC CATTG**GTGATCACTGGTAGCGG-3'
2. T4 polynucleotide kinase (New England Biolabs) and [γ-^{32}P]ATP (10 mCi/mL, Amersham Pharmacia Biotech).
3. One-Phor-All buffer: 10 m*M* Tris-acetate, pH 7.5, 10 m*M* $Mg(OAc)_2$, 50 m*M* K(OAc).
4. TBE electrophoresis buffer: 89 m*M* Tris base, 89 m*M* boric acid, 2 m*M* EDTA.
5. Polyacrylamide gel equipment (Protean II xi cell, Bio-Rad).
6. TrackerTape™ (Amersham Pharmacia Biotech).
7. Biotrap BT100 electroelution chamber (Schleicher & Schuell).
8. Horizontal gel electrophoresis tank (BRL Horizon 20.25)
9. TMN dialysis buffer: 10 m*M* Tris-HCl, pH 8.0, 10 m*M* $MgCl_2$, 50 m*M* NaCl.
10. RuvABC reaction buffer: 20 m*M* Tris-acetate, pH 7.5, 15 m*M* $MgCl_2$, 2 m*M* ATP, 1 m*M* dithiothreitol (DTT), and 100 µg/mL bovine serum albumin (BSA).

2.2. α-Structures

Plasmid DNA. pDEA-7Z derives from the replacement of the *Sca*I-*Bsa*I fragment of pGEM-7Z f(+) (Promega) with the *Sca*I-*Bsa*I fragment of pBR322 (Promega) ***(22)***. pAKE-7Z derives from the insertion of the 1688-bp *Asp*700-*Hinc*II fragment of pACYC184 (Promega) into pDEA-7Z at the *Sca*I site ***(15)***.

1. *E. coli* strains DH5α and JM109.
2. Carbenicillin, chloramphenicol, and kanamycin.
3. 2× YT (double yeast tryptone) media. 20% glucose.

4. Helper phage VCS-M13 (Stratagene, La Jolla, CA).
5. PEG6000, NaCl; Na_2B_4; $CsCl_2$; phenol.
6. TNE buffer: 10 m*M* Tris-HCl, pH 8.0, 100 m*M* NaCl, 1 m*M* EDTA.
7. Restriction enzymes. *Bsa*I and *Pst*I (New England Biolabs).
8. Phenol/chloroform/iso-amyl alcohol (25:24:1); chloroform/iso-amyl alcohol (24:1); ethanol; 3 *M* Na(OAc) pH 5.2.
9. Agarose gel equipment.
10. Neutral sucrose solutions: 10 m*M* Tris-HCl, pH 7.5, 1 *M* NaCl, 10 m*M* EDTA, containing 5% or 20% (w/v) sucrose.
11. Annealing buffer: 10 m*M* Tris-HCl, pH 7.5, 10 m*M* $MgCl_2$, 50 m*M* NaCl.
12. TE buffer: 10 m*M* Tris-HCl, pH 8.0, 1 m*M* EDTA.
13. Slide-A-Lyzer cassette and dialysis tubing (Pierce).
14. Polystyrene igloo.
15. Ethidium bromide.
16. Mineral oil, parafilm.
17. Butan-2-ol, ether.
18. Agarose gel loading buffer: 50 m*M* Tris-HCl, pH 8.0, 50% (v/v) glycerol, and 0.2% (w/v) xylene cyanol.
19. TAE electrophoresis buffer: 40 m*M* Tris base, 1.1% (v/v) glacial acetic acid, and 1 m*M* EDTA.
20. Terminal transferase (New England Biolabs) and [α-^{32}P]ddATP (10 mCi/mL; Amersham Pharmacia Biotech).
21. MicroSpin™ S-400 HR columns (Amersham Pharmacia Biotech).
22. RecA protein (New England Biolabs).
23. Phosphocreatine, creatine phosphokinase, and proteinase K (Sigma).
24. Strand exchange buffer: 50 m*M* Tris-acetate, pH 8.0, 15 m*M* $Mg(OAc)_2$, 20 m*M* K(OAc), 2 m*M* adenosine triphosphate (ATP), 2 m*M* DTT, and 100 µg/mL BSA.
25. Sepharose CL-2B (Sigma), 10 mL Stripette (Costar), siliconized glasswool.
26. Sepharose column buffer: 20 m*M* Tris-HCl, pH 7.5, 1 m*M* DTT, 10 µg/mL BSA, 10 m*M* $MgCl_2$.

3. Methods

3.1. Synthetic Holliday Junctions: Preparation and Use in In Vitro Assays for Branch Migration and Resolution

The methods described in this subheading apply to the production and use of a variety of branched DNA substrates using oligonucleotides designed to meet specific experimental requirements (*see* **Note 1**). As an example, we describe four complementary oligonucleotides that constitute a synthetic Holliday junction with a central homologous core of 26 bp that is flanked by terminal regions of heterology (**Fig. 1**).

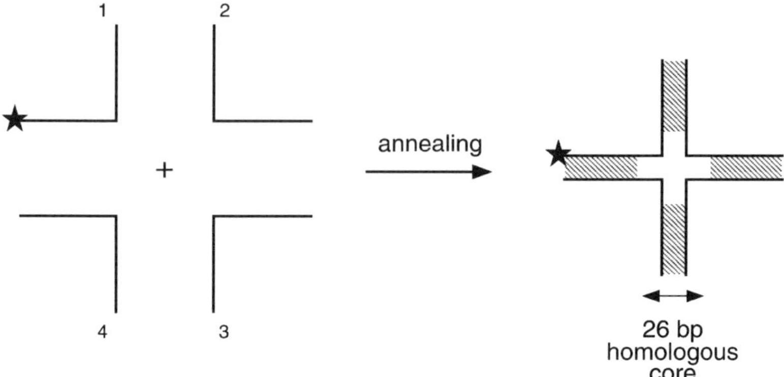

Fig. 1. Schematic representation of a synthetic Holliday junction produced by annealing four partially complementary oligonucleotides. This junction contains a central homologous core of 26 bp that is flanked by terminal regions of heterology (striated). One strand is 5'-^{32}P-end-labeled (★).

3.1.1. Substrate Preparation

1. One of the four oligonucleotides is 5'-^{32}P-end-labeled using T4 polynucleotide kinase and [γ-^{32}P]ATP (*see* **Note 2**). Typical labeling reactions (10 µL) contain 10 pmol oligonucleotide, 25 µCi [γ-^{32}P]ATP, and 10 U of T4 polynucleotide kinase. After 30 min of incubation at 37°C in One-Phor-All buffer, the reaction is stopped by addition of 25 m*M* EDTA, and the kinase is inactivated by incubation at 65°C for 15 min.
2. Add a fivefold excess (50 pmol) of the three partially complementary oligonucleotides and anneal by heating for 3 min at 95°C, followed by 10 min at 65°C, 10 min at 37°C, and 10 min at room temperature (*see* **Note 3**).
3. The ^{32}P-labeled Holliday junctions are then separated from incomplete products on a 10% neutral polyacrylamide gel. Electrophoresis is carried out for 1–2 h at 200 V.
4. The wet gel is covered with plastic wrap and the annealed products are visualized by autoradiography. Typical exposure times are for 1–2 min. With the help of phosphorescent tape (TrackerTape™), the autoradiograph is aligned precisely with the gel, allowing excision of the band corresponding to the Holliday junctions.
5. The junctions are electroeluted from the gel slice using a Biotrap BT1000 chamber. The gel slice is placed in 0.5 mL TBE buffer in a compartment between two semipermeable membranes, and the chamber is placed in a gel tank containing TBE buffer and electroelution proceeds for 1 h at 100 V at 4°C (*see* **Note 4**). The products are recovered in the next compartment containing 0.5 mL TBE buffer. Before collecting the DNA, the polarity of the current is reversed for 30 s to detach DNA molecules that are adsorbed onto the nonpermeable membrane.

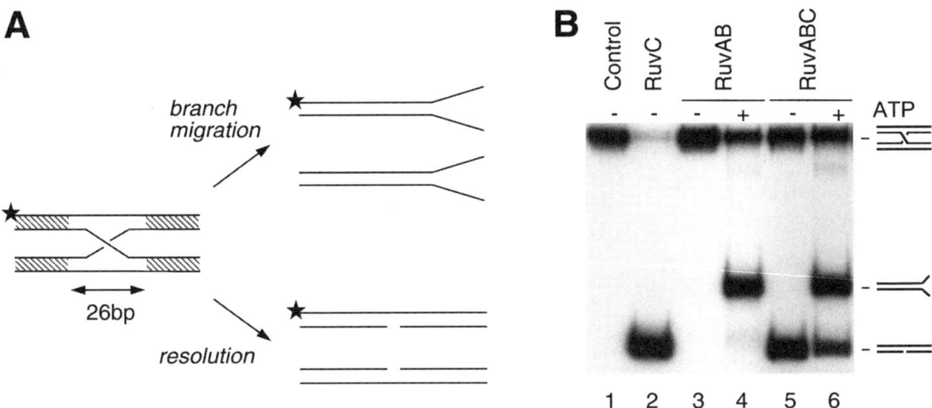

Fig. 2. (A) Schematic representation of the products of branch migration and Holliday junction resolution. Branch migration activities translocate the junction through the terminal regions of heterology and dissociate the substrate into splayed arm products. Holliday junction resolves introduce symmetrically related nicks in two strands of like polarity to yield nicked duplex products. **(B)** Branch migration and Holliday junction resolution catalyzed by *E. coli* RuvABC. AS indicated, ATP was present or omitted from the reaction buffer. RuvA (20 n*M*), RuvB (600 n*M*), and/or RuvC (10 n*M* in lanes 5 and 6, or 100 n*M* in lane 2) were present as indicated.

6. The DNA is dialyzed for 1 h against 2 L of TMN buffer. Substrate concentrations can be determined by spectrophotometry using 1 OD_{260} = 50 µg/mL. Aliquots are stored at –20°C. In general, each aliquot should be used only once (repeated freeze-thawing destabilizes the substrate) and can be used for about 2 wk.

3.1.2. In Vitro Assays

1. Branch migration and/or Holliday junction resolution activities are typically assayed in 20-µL reactions containing approx 1 n*M* ^{32}P-end-labeled synthetic Holliday junction DNA. Optimal protein concentrations, salt, buffer, and pH conditions have to be determined experimentally for each activity to be analyzed.
2. Incubate for 30 min at 37°C.
3. Stop the reactions and deproteinize DNA by addition of 0.8% (w/v) SDS and 1.6 mg/mL proteinase K for 15 min at 37°C.
4. The labeled Holliday junctions, splayed arm and/or nicked duplex products of branch migration and resolution, respectively, are separated by electrophoresis through a 10% neutral polyacrylamide gel (**Fig. 2**). If required, the precise sites of cleavage can be mapped, in comparison with sequencing ladders, by extracting the samples with phenol, followed by ethanol precipitation, before loading them onto an 8% denaturing gel containing 7 *M* urea.
5. Labeled products are revealed by autoradiography. **Figure 2** illustrates the processing of synthetic Holliday junction X26 by the *E. coli* RuvA, RuvB, and RuvC

proteins. RuvC resolves Holliday junctions (lanes 2, 5, and 6), whereas RuvAB promotes ATP-dependent branch migration (lanes 4 and 6).

3.2. Recombination Intermediates Containing Holliday Junctions

Recombination intermediates that resemble α-structures can be made by the exchange of strands between gapped circular and homologous linear duplex DNA molecules. Strand exchange is promoted by the *E. coli* RecA protein (New England Biolabs), so the Holliday junction present within the α-structure is a true recombination reaction intermediate.

Since RecA will not act upon fully duplex DNA molecules, it is necessary to use gapped circular duplex DNA that has a single-stranded region approx 100–200 nt in length. The linear duplex DNA molecule that we have chosen to illustrate the method includes a 1.7-kb region of heterology that prevents completion of strand exchange (**Fig. 3A**). The heterologous block leads to stabilization of the α-structure and facilitates studies of Holliday junction processing activities ***(15)***.

3.2.1. Preparation of Gapped Circular DNA

To generate gapped DNA, supercoiled plasmid DNA is cut such that a small fragment (minimum 150–200 bp) will be excised. The resulting linear DNA is then purified, denatured, and annealed with full-length single-stranded DNA plasmid (**Fig. 3A**). Gapped circular DNA molecules are purified from other DNA species by electrophoresis on agarose gels and recovered by electroelution.

3.2.1.1. Single-Stranded DNA

Single-stranded phagemid DNA is produced from *E. coli* JM109 cells transformed with plasmid pDEA-7Z using the helper phage VCS-M13.

1. Infect 3× 8 mL of an exponentially growing JM109 culture carrying pDEA-7Z (OD_{600} = 0.4, 2× YT media supplemented with 0.1% glucose and 100 µg/mL carbenicillin) with 80 µL VCS-M13 (10^{11} pfu/mL) and grow for 1 h at 37°C.
2. Use each 8-mL culture to inoculate 3× 800 mL of 2× YT supplemented with 0.1% glucose, 100 µg/mL carbenicillin, and 70 µg/mL kanamycin. Grow overnight at 37°C.
3. Remove cells by centrifugation at 8000 rpm for 15 min in a Sorvall GS3 rotor. The supernatant containing the phage is collected and respun. The volume is measured and the phage particles are precipitated by addition of 40g/L PEG6000 and 40 g/L NaCl for 2 h at 4°C. The precipitate is collected by centrifugation at 8000 rpm for 20 min in a GS3 rotor. The phage are resuspended in 25 mL TNE buffer per liter of culture.
4. The phage are then purified by cesium chloride equilibrium centrifugation. To do this, for the equivalent of a 1-L culture, add 17.5 g CsCl and make up to a final volume of 50 mL.

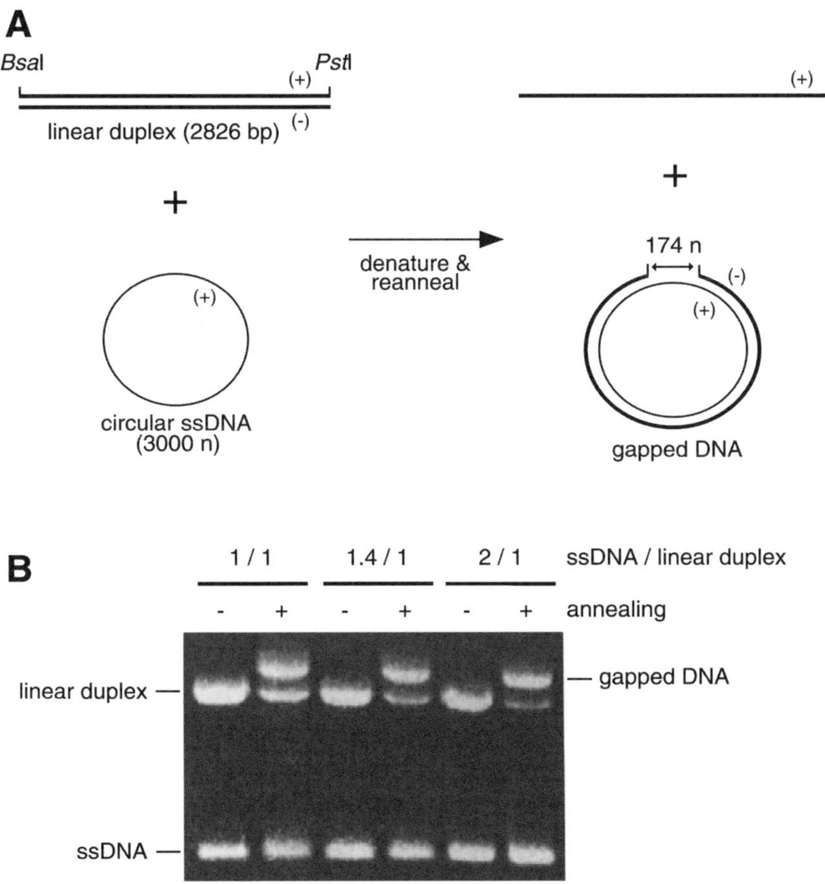

Fig. 3. **(A)** Denaturation of the *Bsa*I-*Pst*I 2826-bp fragment of pDEA-7Z and subsequent annealing with closed circular single-stranded pDEA-7Z yields gapped duplex DNA. **(B)** Trial annealing reactions using the indicated ratios of ssDNA:linear duplex DNA. Samples before (–) and after (+) annealing were separated on a 0.8% agarose gel. The positions of ssDNA, linear duplex DNA, and gapped DNA are indicated.

5. Divide among heat-sealable tubes and spin in a Beckman Ti70 rotor at 55,000 rpm for 18–20 h at 20°C.
6. Visualize the phage band against a black background and collect by withdrawal through the side of the tube using a syringe. The pDEA-7Z phagemid is the top band. Pool fractions and dialyze twice against 2 L of 50 m*M* Na_2B_4 for 1 h at room temperature (Na_2B_4 prevents nicking of the single-stranded DNA). Allow for slight expansion of dialysis tubing.
7. Add Na_2B_4 crystals to the phage solution until they are in excess. After saturation remove excess crystals. Prepare Na_2B_4-saturated phenol (*see* **Note 5**) and equili-

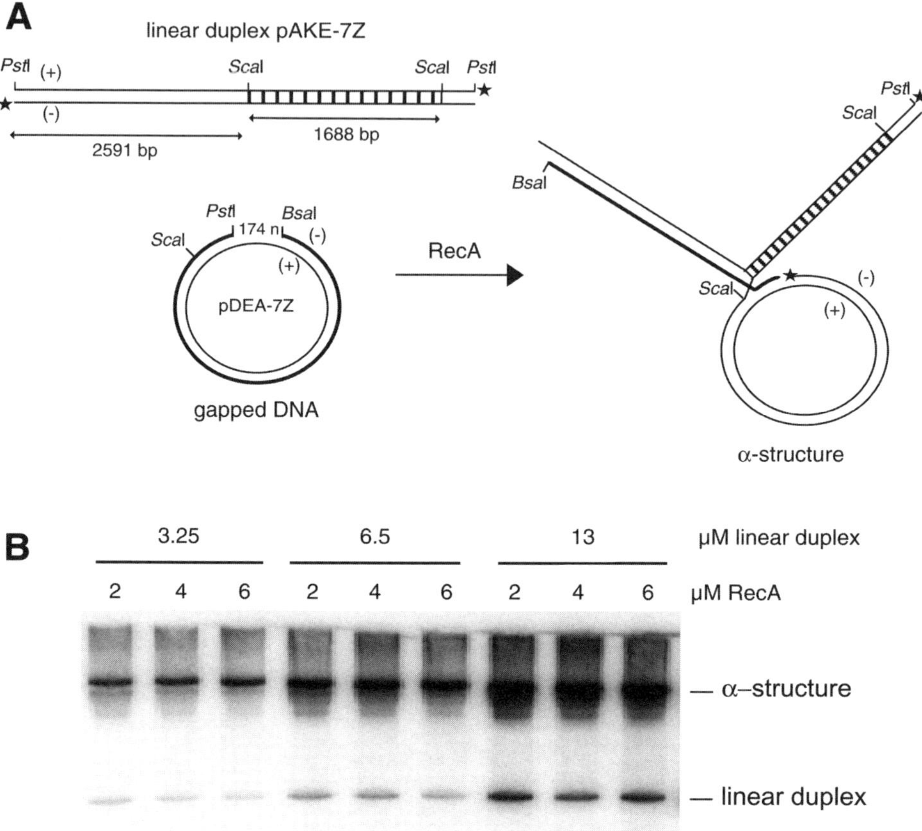

Fig. 4. (A) Schematic representation of the formation of recombination intermediates. RecA-mediated strand exchange between gapped circular pDEA-7Z and *Pst*I-linearized pAKE-7Z. The progression of strand exchange is blocked by the heterologous block (striated) such that recombination intermediates accumulate throughout the course of the reaction. ^{32}P end labels are indicated with asterisks. **(B)** Trial strand exchange reactions. Small-scale reactions were conducted using gapped circular DNA (10 µ*M*) and 3.25 µ*M*, 6.5 µ*M* and 13 µ*M* of 3'-^{32}P-end labeled linear duplex DNA. The concentrations of RecA are indicated. Reaction products were separated by electrophoresis through a 1.2% agarose gel in the presence of ethidium bromide.

brate to 70°C. Similarly, put the Na_2B_4-saturated phage solution separately at 70°C. Mix equal volumes of both solutions gently for 3 min and return to 70°C for a further 3 min.

8. Cool to room temperature and spin in a Sorvall SS34 rotor for 5 min at 6000 rpm. Re-extract the aqueous upper phase twice at room temperature with Na_2B_4-saturated phenol.

9. Dialyze the final aqueous phase twice against 2 L of TE buffer at 4°C.
10. Precipitate the DNA with ethanol and resuspend in 1 mL TE buffer. Determine the DNA concentration using 1 OD_{260} = 36 µg/mL.

3.2.1.2. The Complementary Strand

The strand that will be annealed to the ssDNA to make the gapped circular DNA is prepared by excising a 174-bp *Bsa*I-*Pst*I fragment from form I pDEA-7Z (prepared using a Qiagen Plasmid Maxi Kit as specified by the manufacturer), and the resulting 2826-bp linear DNA molecule is purified on a neutral sucrose gradient.

1. Digest 250 µg form I pDEA-7Z plasmid DNA in a 4-mL reaction using 750 U *Bsa*I for 2 h at 55°C. Check an aliquot of the reaction for complete digestion on a 0.8% agarose gel. Add 400 U *Pst*I, and continue the digestion for 2 h at 37°C.
2. Stop the reaction by addition of 10 m*M* EDTA and extract the DNA twice with phenol/chloroform/iso-amyl alcohol and once with chloroform/iso-amyl alcohol. Precipitate the DNA with ethanol. Resuspend in 200 µL of TE buffer.
3. Prepare a 5–20% (w/v) sucrose gradient in 40 mL polyallomer ultracentrifuge tubes (Beckman). Carefully apply the DNA sample (200 µL) to the surface and spin for 16 h at 26,000 rpm in a Beckman SW28 ultracentrifuge rotor) at 4°C. Pierce the bottom of each tube and collect 1-mL fractions. Analyze 10-µL aliquots on a 0.8% agarose gel. Pool fractions containing the 2826 bp DNA fragment and precipitate it with ethanol. Determine the DNA concentration using 1 OD_{260} = 50 µg/mL.

3.2.1.3. Preparation of Gapped DNA

The 2826-bp *Bsa*I-*Pst*I fragment of pDEA-7Z is denatured, and the (–) strand is annealed with full-length circular single-stranded pDEA-7Z (**Fig. 3A**).

1. Trial annealing reactions (40 µL) should be performed to determine the optimum ratio for generating gapped DNA (**Fig. 3B**) prior to carrying out the large-scale preparation. Typically, 1 µg of linear duplex fragment is mixed with 1, 1.4, or 2 µg of circular single-stranded pDEA-7Z in annealing buffer, and the reaction is covered with mineral oil. The tube is placed in a beaker of boiling water for 5 min and then cooled slowly to room temperature by placing the beaker in a polystyrene igloo. Analyze the DNA before and after annealing on a 0.8% agarose gel.
2. Based on this result, prepare a large-scale annealing reaction. A ratio of ssDNA:dsDNA of 1.4:1 is usually optimal. Mix 168 µg of ssDNA with 120 µg of the linear fragment and bring volume to 1 mL with ddH_2O. Dialyze in a Slide-A-Lyzer Cassette (Pierce) for 2 h against 2 L TE buffer at 4°C. Split into four 500-µL reactions (250 µL DNA mixture, 50 µL 10× annealing buffer, 200 µL ddH_2O). Cover each tube with 200 µL mineral oil, and anneal as described in **step 1**.
3. To each reaction mixture (*see* **Note 6**), add 1/10th volume of agarose gel loading buffer, and separate the products using two 300-mL agarose gels (1%) in a BRL

Horizon 20.25 gel apparatus (*see* **Note 7**). No ethidium bromide should be present in the gel or the buffer since any residual ethidium bromide in the gapped DNA will affect subsequent strand exchange reactions. Electrophoresis should be performed at 200 V for 3 h at 4°C in TAE buffer.

4. To localize the band corresponding to gapped DNA (the gDNA usually migrates just below the xylene cyanol dye), remove a 2-cm-wide slice from each side of the gel and stain with ethidium bromide (5 µg/mL). The position of the gapped DNA is identified using a UV transilluminator and used as a reference to excise the gDNA from the preparative gel.
5. The agarose gel slices are placed into dialysis tubing using as small a volume of TAE buffer as possible (approx 8 mL per tube; be careful to avoid trapping any air inside the tubing). The DNA is extracted by electroelution at 50 V overnight at 4°C using TAE buffer. Before collecting the sample into 50-mL Falcon tubes, reverse the polarity of the current for 2 min to detach any DNA that might be adsorbed onto the dialysis tubing.
6. The samples are then concentrated to a workable volume by two extractions with an equal volume of butan-2-ol (discard the upper phase). Extract once with diethyl ether and leave in a fume hood to evaporate any residual ether.
7. Precipitate the gapped DNA with ethanol and resuspend in 300 µL of TE buffer. Dialyze overnight against 2 L of TE buffer at 4°C.
8. Determine the DNA concentration using 1 OD_{260} = 50 µg/mL and dilute DNA to give a final working concentration of about 300 µ*M* (expressed in moles of nucleotides).

3.2.2. Preparation of 3'-^{32}P-End-Labeled Linear Duplex DNA

The plasmid pAKE-7Z is homologous to pDEA-7Z but contains a heterologous insertion of 1688 bp (**Fig. 4A**). It is linearized with *Pst*I.

1. Digest 250 µg of form I pAKE-7Z (prepared using a Qiagen Plasmid Maxi Kit as specified by the manufacturer) in a 4 mL reaction using 400 U *Pst*I for 2 h at 37°C. Stop the reaction with 10 m*M* EDTA.
2. Check a sample for complete digestion on a 0.8% agarose gel, and then extract once with phenol/chloroform/isoamyl-alcohol and once with chloroform/isoamyl-alcohol and precipitate the DNA with ethanol. Determine the DNA concentration.
3. The linearized pAKE-7Z (8 µg) is then 3'-^{32}P-end-labeled in a 50 µL reaction mixture containing 30 µCi [α-^{32}P]ddATP and 10 U terminal deoxynucleotidyl transferase in cacodylate buffer as described by the manufacturer (New England Biolabs). After 90 min at 37°C, the reaction is stopped by addition of 0.8% (w/v) SDS and 25 m*M* EDTA.
4. Extract with 1 vol of phenol/chloroform/isoamyl-alcohol and back extract with 50 µL TE buffer. Remove unincorporated nucleotides using an S-400 HR MicroSpin column as described by the manufacturer.
5. Add 1/2 vol of ethanol to the eluate and store the labeled DNA at 4°C until required. The ethanol reduces damage caused by radioactive decay. Determine the DNA concentration.

3.2.3. Preparation of α-Structures

Optimal conditions for strand exchange need to be established for each new batch of gapped circular and ^{32}P-labeled linear duplex DNA. Conduct small-scale reactions (10 µL) as described in **steps 1–3** by varying the relative concentration of each component of the reaction (**Fig. 4B**).

1. Add gapped circular DNA (10 µ*M*) to strand exchange buffer supplemented with 20 m*M* phosphocreatine and 5 U/mL creatine phosphokinase (an ATP regeneration system). Supplement with RecA protein (2–6 µ*M*), leave for 5 min at 4°C, and then move to 37°C for 5 min. Finally, add 3'-^{32}P-end-labeled linear DNA (3–13 µ*M*) and incubate for 90 min at 37°C.
2. Stop the reaction by addition of 0.8% (w/v) SDS and 1.6 mg/mL proteinase K. Continue incubation for 15 min at 37°C.
3. The reaction products are visualized by electrophoresis on 1.2% agarose gels (with buffer recircularization) containing 0.5 µg/mL ethidium bromide. The ethidium bromide blocks spontaneous branch migration and thereby maximizes the yield of recombination intermediates. Electrophoresis is carried out using a WIDE MINI-SUB™ CELL (Bio-Rad) for 3.5 h at 65 V in TAE buffer. ^{32}P-labeled products are detected by autoradiography.
4. Using the optimal amounts of DNA and RecA, scale-up the reaction to 200 µL and proceed as described in **steps 1–3**. Typical reactions contains 6 µ*M* RecA, 6.5 µ*M* 3'-end-labeled linear DNA, and 10 µ*M* gapped DNA.
5. Recombination intermediates are purified using a 3.5-mL Sepharose CL-2B column equilibrated in Sepharose column buffer (*see* **Note 8**). Two drop fractions are collected and monitored for Cerenkov counts using a hand-held monitor. Typically the DNA elutes in six to eight fractions that are collected in a total volume of approx 400 µL. Measure DNA concentration. Store at 4°C and use within 2–5 d.

3.2.4. In Vitro Assays

Branch migration and/or resolution reactions (20 µL) usually contain ^{32}P-end-labeled α-structures (0.1 n*M*) and buffers similar to those used for assays with synthetic Holliday junctions.

1. Incubate reactions for 90 min at 37°C. Deproteinize using SDS/proteinase K as described in **Subheading 3.2.3., step 2**.
2. DNA products are separated by electrophoresis on a 1% agarose gel in TAE buffer containing 0.5 µg/mL ethidium bromide as described, and ^{32}P-labeled products are detected by autoradiography. Complete branch migration results in the dissociation of the recombination intermediate, giving rise to ^{32}P-labeled linear and unlabeled gapped duplex DNA. Resolution can occur in two possible orientations to produce either ^{32}P-labeled linear dimers or gapped linear and nicked circle molecules, as illustrated (**Fig. 5**).

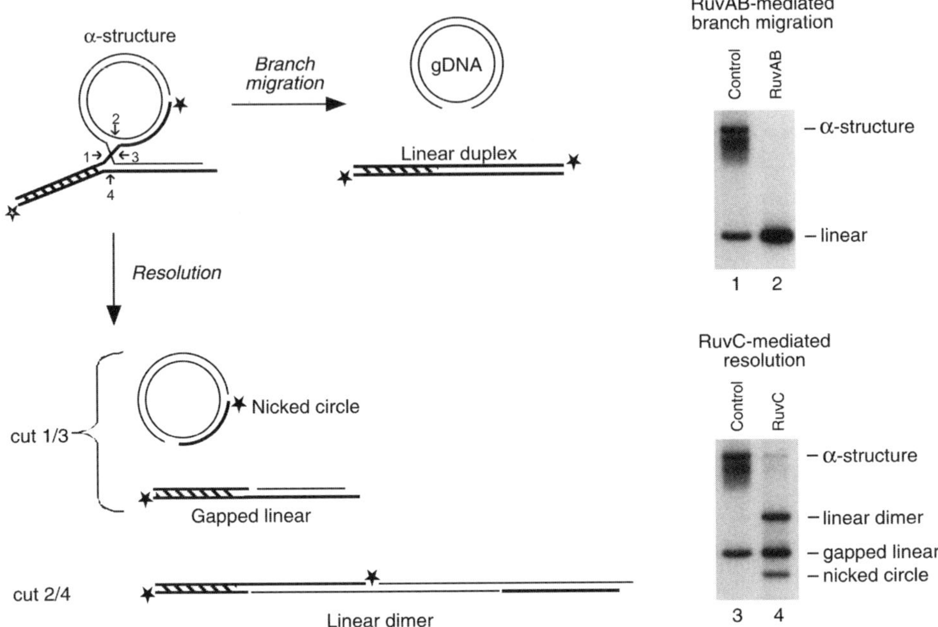

Fig. 5. Schematic representation of the products of branch migration and Holliday junction resolution. Branch migration activities translocate the junction back through 2.7 kb of homology to dissociate the recombination intermediate (α-structure) into ^{32}P-labeled linear duplex and unlabeled gapped DNA. A representative branch migration reaction catalyzed by RuvAB (20 n*M* RuvA and 600 n*M* RuvB) is shown in lane 2. Resolution in orientation (1/3) yields ^{32}P-labeled nicked circle and gapped linear DNA products, whereas resolution in orientation (2/4) produces ^{32}P-labeled linear dimer molecules. A representative resolution reaction catalyzed by RuvC (100 n*M*) is shown in lane 4).

4. Notes

1. Junctions should be designed such that the four oligonucleotides cannot self-anneal. This will maximize pairing with partner oligonucleotides.
2. Prior to labeling, full-length oligonucleotides should be purified from truncated products by denaturing polyacrylamide gel electrophoresis.
3. As an alternative, annealing can be performed by placing the reactions in a beaker containing water heated to 95°C. The beaker is then placed in a polystyrene igloo and left overnight to cool slowly to room temperature.
4. Alternatively, the product can be recovered by diffusion out of the gel slice. Place the slice in a 1.5-mL Eppendorf tube and cover it with 0.5 mL TMN buffer. Rotate slowly overnight at 4°C. Next day, remove the supernatant containing the DNA.

5. To prepare phenol saturated with Na_2B_4, saturate water with Na_2B_4 at room temperature. Mix phenol with an equal volume of Na_2B_4-saturated H_2O in a Falcon tube, shake, and allow the two phases to separate.
6. To get rid of residual mineral oil in the annealing reaction, pipet the contents of the tube onto parafilm and tilt to separate the aqueous layer from the oil.
7. The curviness of the metal electrodes in the gel box will influence how straight the DNA bands are. Take care to straighten them before use.
8. A plastic 10-mL Stripette (Costar) is convenient. Cut the Stripette at 4 mL, stand on a column holder, and pack a tiny amount of siliconized glasswool at the bottom of the tip. Pour in 3.5 mL of Sepharose CL-2B and equilibrate with 3 vol of Sepharose column buffer.

Acknowledgments

We thank our colleagues, past and present, who have developed substrates and methods for the analysis of Holliday junction processing enzymes. This work was supported by Cancer Research UK.

References

1. Holliday, R. (1964) A mechanism for gene conversion in fungi. *Genet. Res. Camb.* **5,** 282–304.
2. Seigneur, M., Bidnenko, V., Ehrlich, S. D., and Michel, B. (1998) RuvAB acts at arrested replication forks. *Cell* **95,** 419–430.
3. Cox, M. M. (2001) Recombinational DNA repair of damaged replication forks in *Escherichia coli*: questions. *Annu. Rev. Genet.* **35,** 53–82.
4. Sogo, J. M., Lopes, M., and Foiani, M. (2002) Fork reversal and ssDNA accumulation at stalled replication forks owing to checkpoint defects. *Science* **297,** 599–602.
5. Postow, L., Ullsperger, C., Keller, R. W., Bustamante, C., Vologodskii, A. V., and Cozzarelli, N. R. (2001) Positive torsional strain causes the formation of a four-way junction at replication forks. *J. Biol. Chem.* **276,** 2790–2796.
6. Bolt, E. L. and Lloyd, R. G. (2002) Substrate specificity of RusA resolvase reveals the DNA structures targeted by RuvAB and RecG *in vivo*. *Mol. Cell* **10,** 187–198.
7. Dunderdale, H. J., Benson, F. E., Parsons, C. A., Sharples, G. J., Lloyd, R. G., and West, S. C. (1991) Formation and resolution of recombination intermediates by *E. coli* RecA and RuvC proteins. *Nature* **354,** 506–510.
8. Connolly, B., Parsons, C. A., Benson, F. E., et al. (1991) Resolution of Holliday junctions *in vitro* requires the *Escherichia coli ruvC* gene product. *Proc. Natl. Acad. Sci. USA* **88,** 6063–6067.
9. Lloyd, R. G. and Sharples, G. J. (1993) Processing of recombination intermediates by the RecG and RuvAB proteins of *Escherichia coli*. *Nucleic Acids Res.* **21,** 1719–1725.
10. Whitby, M. C., Vincent, S. D., and Lloyd, R. G. (1994) Branch migration of Holliday junctions: identification of RecG protein as a junction specific DNA helicase. *EMBO J.* **13,** 5220–5228.

11. Bennett, R. J. and West, S. C. (1996) Resolution of Holliday junctions in genetic recombination: RuvC protein nicks DNA at the point of strand exchange. *Proc. Natl. Acad. Sci. USA* **93,** 12,217–12,222.
12. Bennett, R. J. and West, S. C. (1995) Structural analysis of the RuvC-Holliday junction complex reveals an unfolded junction. *J. Mol. Biol.* **252,** 213–226.
13. Bennett, R. J., Dunderdale, H. J., and West, S. C. (1993) Resolution of Holliday junctions by RuvC resolvase: cleavage specificity and DNA distortion. *Cell* **74,** 1021–1031.
14. Adams, D. E. and West, S. C. (1996) Bypass of DNA heterologies during RuvAB-mediated three- and four-strand branch migration. *J. Mol. Biol.* **263,** 582–596.
15. Eggleston, A. K., Mitchell, A. H., and West, S. C. (1997) *In vitro* reconstitution of the late steps of genetic recombination in *E. coli*. *Cell* **89,** 607–617.
16. Müller, B., Burdett, I., and West, S. C. (1992) Unusual stability of recombination intermediates made by *Escherichia coli* RecA protein. *EMBO J.* **11,** 2685–2693.
17. Müller, B., Tsaneva, I. R., and West, S. C. (1993) Branch migration of Holliday junctions promoted by the *Escherichia coli* RuvA and RuvB proteins: I. Comparison of the RuvAB- and RuvB-mediated reactions. *J. Biol. Chem.* **268,** 17,179–17,184.
18. Müller, B., Tsaneva, I. R., and West, S. C. (1993) Branch migration of Holliday junctions promoted by the *Escherichia coli* RuvA and RuvB proteins: II. Interaction of RuvB with DNA. *J. Biol. Chem.* **268,** 17,185–17,189.
19. Parsons, C. A., Tsaneva, I., Lloyd, R. G., and West, S. C. (1992) Interaction of *Escherichia coli* RuvA and RuvB proteins with synthetic Holliday junctions. *Proc. Natl. Acad. Sci. USA* **89,** 5452–5456.
20. Parsons, C. A. and West, S. C. (1993) Formation of a RuvAB-Holliday junction complex *in vitro*. *J. Mol. Biol.* **232,** 397–405.
21. Parsons, C. A., Stasiak, A., Bennett, R. J., and West, S. C. (1995) Structure of a multisubunit complex that promotes DNA branch migration. *Nature* **374,** 375–378.
22. Shah, R., Bennett, R. J., and West, S. C. (1994) Genetic recombination in *E. coli*: RuvC protein cleaves Holliday junctions at resolution hotspots in vitro. *Cell* **79,** 853–864.
23. Tsaneva, I. R., Müller, B., and West, S. C. (1992) ATP-dependent branch migration of Holliday junctions promoted by the RuvA and RuvB proteins of *E. coli*. *Cell* **69,** 1171–1180.
24. Van Gool, A. J., Hajibagheri, N. M. A., Stasiak, A., and West, S. C. (1999) Assembly of the *Escherichia coli* RuvABC resolvasome directs the orientation of Holliday junction resolution. *Genes Dev.* **13,** 1861–1870.
25. West, S. C. (1997) Processing of recombination intermediates by the RuvABC proteins. *Annu. Rev. Genet.* **31,** 213–244.
26. Iwasaki, H., Takahagi, M., Nakata, A., and Shinagawa, H. (1992) *Escherichia coli* RuvA and RuvB proteins specifically interact with Holliday junctions and promote branch migration. *Genes Dev.* **6,** 2214–2220.
27. White, M. F. and Lilley, D. M. J. (1997) The resolving enzyme Cce1 of yeast opens the structure of the 4-way DNA junction. *J. Mol. Biol.* **266,** 122–134.

28. White, M. F. and Lilley, D. M. J. (1997) Characterization of a Holliday junction-resolving enzyme from *Schizosaccharomyces pombe*. *Mol. Cell. Biol.* **17,** 6465–6471.
29. White, M. F. and Lilley, D. M. J. (1998) Interaction of the resolving enzyme Ydc2 with the four-way DNA junction. *Nucleic Acids Res.* **26,** 5609–5616.
30. Oram, M., Keeley, A., and Tsaneva, I. (1998) Holliday junction resolvase in *Schizosaccharomyces pombe* has identical endonuclease activity to the Cce1 homolog Ydc2. *Nucleic Acids Res.* **26,** 594-601.
31. Whitby, M. C. and Dixon, J. (1998) Substrate specificity of the Cce1 Holliday junction resolvase of *Schizosaccharomyces pombe*. *J. Biol. Chem.* **273,** 35,063–35,073.
32. Kvaratskhelia, M. and White, M. F. (2000) An archaeal Holliday junction resolving enzyme from *Sulfolobus solfataricus* exhibits unique properties. *J. Mol. Biol.* **295,** 193–202.
33. Kvaratskhelia, M. and White, M. F. (2000) Two Holliday junction resolving enzymes in *Sulfolobus solfataricus*. *J. Mol. Biol.* **297,** 923–932.
34. Bolt, E. L., Lloyd, R. G., and Sharples, G. J. (2001) Genetic analysis of an archaeal Holliday junction resolvase in *Escherichia coli*. *J. Mol. Biol.* **310,** 577–589.
35. Komori, K., Sakae, S., Shinagawa, H., Morikawa, K., and Ishino, Y. (1999) A Holliday junction resolvase from *Pyrococcus furiosus*: functional similarity to *Escherichia coli* RuvC provides evidence for conserved mechanism of homologous recombination in Bacteria, Eukarya, and Archaea. *Proc. Natl. Acad. Sci. USA* **96,** 8873–8878.
36. Komori, K., Sakae, S., Fujikane, R., Morikawa, K., Shinagawa, H., and Ishino, Y. (2000) Biochemical characterization of the Hjc Holliday junction resolvase of *Pyrococcus furiosus*. *Nucleic Acids Res.* **28,** 4544–4551.
37. Komori, K., Fujikane, R., Shinagawa, H., and Ishino, Y. (2002) Novel endonuclease in Archaea cleaving DNA with various branched structure. *Genes Genet. Syst.* **77,** 227–241.
38. Elborough, K. M. and West, S. C. (1990) Resolution of synthetic Holliday junctions in DNA by an endonuclease activity from calf thymus. *EMBO J.* **9,** 2931–2936.
39. Constantinou, A., Tarsounas, M., Karow, J. K., et al. (2000) Werner's syndrome protein (WRN) migrates Holliday junctions and co-localizes with RPA upon replication arrest. *EMBO R.* **1,** 80–84.
40. Constantinou, A., Davies, A. A., and West, S. C. (2001) Branch migration and Holliday junction resolution catalyzed by activities from mammalian cells. *Cell* **104,** 259–268.
41. Constantinou, A., Chen, X.-B., McGowan, C. H., and West, S. C. (2002) Holliday junction resolution in human cells: two junction endonucleases with distinct substrate specificities. *EMBO J.* **21,** 5577–5585.
42. Karow, J. K., Constantinou, A., Li, J.-L., West, S. C., and Hickson, I. D. (2000) The Bloom's syndrome gene product promotes branch migration of Holliday junctions. *Proc. Natl. Acad. Sci. USA* **97,** 6504–6508.

Index

G

H

Z